D1269490

Between Sputnik and the Shuttle
New Perspectives on American Astronautics

President John F. Kennedy making his historic address on the Space Program at Rice University, Houston, Texas, September 12, 1962:

"The exploration of space will go ahead, whether we join it or not, and it is one of the greatest adventures of all time, and no nation which expects to be the leader of other nations can expect to stay behind in this race for space...

"We set sail on this new sea because there is new knowledge to be gained and new rights to be won, and they must be won and used for all people....

"We choose to go to the moon. We choose to go to the moon in this decade and do the other things, not because they are easy, but because they are hard, because the goal will serve to organize and measure the best of our energies and skills ..."

Between Sputnik and the Shuttle

New Perspectives on American Astronautics

Frederick C. Durant, III, Editor

John H. Disher
Stephen E. Doyle
Frederick C. Durant, III
Eugene M. Emme
Eilene Galloway
Mary A. Holman
Frederick I. Ordway, III
Robert C. Seamans, Jr.
John L. Sloop
Theodore Suranyi-Unger, Jr.

AAS History Series, Volume 3

A Supplement to Advances in the Astronautical Sciences

AAS Publications Office
P. O. Box 28130
San Diego, California 92128

Affiliated with the American Association for the Advancement of Science
Member of the International Astronautical Federation

ISBN 0-87703-145-2 (Hard Cover)

ISBN 0-87703-149-5 (Soft Cover)

Published for the American Astronautical Society
by Univelt, Inc., P. O. Box 28130, San Diego, California 92128

Printed and Bound in the U.S.A.

PREFACE

From modest beginnings the exploration and exploitation of space science and technology have proven to be a dynamic and complex phenomenon broadly impacting the affairs of mankind on its homeland planet. Few have comprehended fully either its full scope at any particular moment, or foreseen the pace in its evolution. Major happenings, whether a technological advance, a strategic requirement, or a scientific insight, have seemed to provoke implications for the future with opportunities testing priority problems on Earth. Evident have been the presence of enthusiasts for space, as well as its severe critics, their warranted concerns, and nascent problems revolving around unprecedented unknowns born anew. Like most revolutions in the long view of man's history, their dynamics enforce some value on a broad viewpoint on major developments in evolution, past and future.

Each chapter in this volume may be viewed as a historical test boring offering perspectives. Each has been written by an avid student of a part of the larger story since the first earth satellites gained orbit. A so-called "space age" evolved. We are now on the eve of a new era of space transportation anticipated with the Space Shuttle system.

Collectively these essays perhaps offer a more comprehensive view of our space experience to date. This volume is the product of two symposia sponsored by the History Committee of the American Astronautical Society in March of 1979 and of 1980. They are now made available to a wider audience.

This volume would not have been possible without the unstinting labors of the authors represented herein. No royalties are involved, only the reward that shared knowledge is the pleasure of the genuine student in service of the advancement of collective knowledge. In turn, each author would appreciate critical comments by our readers.

Appreciation is here given to the American Astronautical Society for enabling non-engineers to gain a platform. Lee Saegaesser, NASA Archivist, Les Gaver, chief of NASA's Audio-Visuals, and many others have been of assistance to each author. Appreciation for reference to copyrighted materials is evident in citations. But we are particularly grateful for the use of copyrighted cartoons by Mr. Herb Block of *The Washington Post*, Francis Brennan of *Newsweek*, and Joaquin De Alba, formerly of *The Washington Daily News*.

The General Editor also expresses his appreciation for the efforts of the editors of Univelt, Inc., especially Robert H. Jacobs, and of Rosalie Brazee for typing much of the manuscript.

Eugene M. Emme
General Editor
AAS HISTORY SERIES

TABLE OF CONTENTS

CONTENTS (Contd.)

Front Cover Illustration

Composite picture of Saturn showing six of its moons, based on photographs taken by Voyager 2. (JPL-NASA Photo)

I

INTRODUCTION

Perspectives on American Astronautics

F. C. Durant, III *

The ancient art of rocketry led to the development of large liquid-fueled rockets as ballistic missiles and space launch vehicles in the 1950's. This was the technological prologue to the large-scale phenomenon associated with the swift evolution of astronautics within a few decades. To comprehend more fully this dramatic epoch in American history we need to gain valid perspectives on what has occurred and its impact.

The governmental space program of the United States began with the minimum earth satellite program in 1955 planned for the International Geophysical Year (1957-1958). But the shock of the launch of two Sputniks by the U.S.S.R. in October and November 1957, drove the United States to outline and undertake a full-scale space effort. Indeed, the so-called "Magna Carta" of the U.S. space program was the National Aeronautics and Space Act of 1958. It treated the National Aeronautics and Space Administration (NASA) to conduct exploration and utilization of space. In contrast to the Soviet space program, NASA was "open" and without security restrictions. Programs, plans, schedules and scientific findings were published as public information for use by all. It truly became a national enterprise.

The so-called "space race" spurred American astronautics and spawned the magnificent Apollo program which accomplished in eight years the landing of "man on the Moon and returning him safely to Earth." The U.S. clearly won this "race." But the Apollo program was just one of many space endeavors. Other projects resulted in tremendous accomplishments in the space sciences, in meteorology, communications, and observations of earth resources. It was found that from the vantage point of satellites many practical benefits obtained. Interplanetary probes to Mercury, Venus, Mars, Jupiter, and Saturn, as well as studies of the Sun and observations of other stars from space have expanded exponentially man's knowledge of the cosmos.

Beyond the instant chronicles and some worthy institutional histories by NASA and a few other scholars, there indeed seems much history yet to be made literate and assayed for the past three decades of the space venture. Perspectives, albeit only on yesterdays, permit useful insights.

** Author was Assistant Director for Astronautics of the Smithsonian Institution's National Air and Space Museum, 1965-1980. Biographical profiles of all authors appear elsewhere in this volume.*

Eight distinct, yet related, perspectives are offered in this volume. Each of the survey papers reviews chronologically a segment of the totality of the U.S. space program from its infancy. Treated as foci of the history are the White House, the Congress, the evolution of space law, the political-economic nature of astronautics, the evolution of space technology for space transportation, liquid-hydrogen propulsion development permitting the triumph of Apollo, managerial parameters on high technology demonstrated in Apollo, and astronautical art as a stimulus and reflection of space phenomena. A few words seem in order on each of them.

In the first survey, Eugene M. Emme, founder of the NASA History Program, examines the role of the "space age" Presidents from Dwight D. Eisenhower to Jimmy Carter. Most interestingly, his essay attempts to explain why the sputniks happened, why NASA was created, why it was decided to land Americans on the moon, and why the decision to develop the Space Shuttle became the major post-Apollo decision. His thesis is that the White House is the "mission control" of the American space program, and he offers much new documentation of interest to others.

Eilene Galloway's paper, drawing from her many years of involvement and experience as a legal consultant to space committees on Capitol Hill, illuminates the function of Congressional overseers of the space program's goals, and their budgetry control on the funding of authorized programs through the dedicated space committees of both Houses of the Congress. International cooperation was incorporated into the Space Act, and the Apollo priority sustained. Mrs. Galloway treats the post-Apollo shift to place space in a more subordinate position within the House and the Senate committees on science and technology, and trends resultant.

Stephen E. Doyle, who began his space career as a colleague of the late Andrew G. Haley, submits a distinguished and concise treatment of the evolution of space law, along with its implications. It is a perspective much too important to be left to the legal eagles. Juridical problems have attended the development of astronautics from the beginning, and emerge importantly for the future.

Political-economic interrelationships are examined by Professors Mary A. Holman and Theodore Suranyi-Unger of the department of economics of George Washington University. In the American society the cost and the goals of public funding for space efforts, particularly those other than for national defense, are the product of a complex of working processes, conflicts and consensuses.

John H. Disher, associated with NASA spacecraft since Project Mercury, including Apollo spacecraft development and test, and deputy director of Skylab, became director of advanced programs in the Office of Space Transportation in 1974. His presentation is a concise tracing of the evolution of manned space flight technology, a remarkable survey indeed. It is a contribution to the history of technology not otherwise so well focused.

Attention on liquid-hydrogen propulsion for the first U.S. satellite proposal in 1945, proved to be of seminal importance in the upper stages of the Apollo-Saturn V for the lunar-orbital mode. John L. Sloop, author of a distinguished history of this important innovation, submits a significant paper on this important technology in the space program.

And the lessons of the Apollo achievement as the largest non-military technological enterprise is outstandingly summarized in the paper of Dean Robert C. Seamans of M.I.T. and Frederick I. Ordway, III, the latter being a long-time associate of Wernher von Braun and is now with the Department of Energy.

And, lastly, the writer was given the opportunity to present a personal survey of "space art." This was an enjoyable exercise. All kinds of art works have developed in the past few decades, as illustrations of contemporary space projects, future concepts, artistic interpretations of events and so on. Such works have played a role in public education and awareness and focused attention and contemplation in ways which words alone cannot. Such art, moreover, is an inherent component of the cultural changes resulting from the space program. About 150 illustrations of such works were shown at the oral presentation of this paper.

Since the founding of the American Astronautical Society in 1954, it has become traditional to sponsor open meetings to spread appreciation and understanding of astronautics beyond involved specialists and interested laymen. The 17th Goddard Memorial Symposium in March 1979, and the 18th in March 1980, both held at the Washington Hilton Hotel, adhered to this tradition. The sessions sponsored by the AAS History Committee provoked as many questions as they answered by examining the breadth of American astronautics.

Study and reflection upon the events and experiences of the recent past, from the sputniks to the Shuttle, enable us to anticipate the dynamics of change as we view tomorrow.

II

PRESIDENTS AND SPACE

Eugene M. Emme *

This paper submits perspectives on the White House during the first twenty-five years of the space era of mankind. From the approval in 1955 of the minimum earth satellite for the International Geophysical Year, to the "DC-3 era" of space mobility with the Space Shuttle for the 1980's, the Chief Executive has had to determine the policies and resources for space-related activities in a strategic context of scientific, defense, economic, and diplomatic realities, and always with a view of their likely support on Capitol Hill and by the American public. From this review, perspectives are also gained on the evolution of the U.S. space program as well as its impact upon the Office of the President.

As "mission commander" of the American "spaceship of state," each of the six succeeding Presidents has perforce dealt with a cosmos of his own, one with critical responsibilities to serve the then current as well as the future interests of the American people on planet Earth. During a first term in office, re-election is the near future for accountability with the electorate. As the harsh dictates of history were to work, only Dwight D. Eisenhower was to complete two terms of office in the Space Age, though the sputniks appeared in his second, and James E. Carter has passed mid-term of his first.

Historically considered, the Presidency has a constitutional basis which in practice was a product of the political process animated by the rhetoric and the actions of each President. His roles are well recognized as a campaign winner, *de facto* party leader, chief of the Federal administrative apparatus, proposer of basic legislation which the Congress disposes and must fund, statesman on the summit, and the commander-in-chief of the armed and other defense services.

Great change has been both broad and swift because of technological advances. Among them, mobility in outer space became genuinely feasible because of development of rocket technology for TN-tipped missiles with

* *Dr. Emme was the NASA Historian, 1959-78; he is an AAS Director, chairman of the AAS History Committee, and co-chairman of the History Committee of the International Academy of Astronautics.*

intercontinental ranges. Before cosmonauts and astronauts gained weightlessness in space flight, the large jet transport had already revolutionized global transportation. Television also had brought the world into most American homes. Modestly begun with the Soviet sputnik in October 1957, a dynamic and novel space venture increasingly began to influence White House concerns. Not the least of these was the high visibility of space achievements in the news media. Space technology became a growing budget reality. As most will recall, space achievement assumed a historical trajectory of its own. It became a new tool for universal science above and beyond the Earth, a national instrumentality for power and prestige, a prod for genuine international cooperation as well as competition, and, a viable projection of man's historic quest to explore and utilize the furthest reaches of his accessible environment.

It is easy to state that President Eisenhower got the U.S. space program underway with the IGY satellite commitment and by the creation of the National Aeronautics and Space Administration, and that he was aided by Senate Majority Leader Lyndon B. Johnson. Or, to repeat again that President John F. Kennedy set the Apollo commitment in motion for the 1960's, while Richard M. Nixon granted the commitment for a Space Shuttle for the 1980's. But let us now examine the role of the White House in the American space enterprise, expanding our memories and testing our vintaged caricatures as we review what we now know today about the recent past. Historians never consider they have all of the evidence. And, indeed, this paper cannot be accepted as the final word on such recent events.

BACKGROUND

Vice President Harry S. Truman of Missouri became President on April 12, 1945. He explained to members of the press corps: "I don't know whether you have ever had a load of hay or a bull fall on you. But last night the moon, the stars, and all planets fell on me."[1] During his almost eight years as President, Harry S. Truman faced a series of unprecedented problems related to ending the war and for preserving a peace. He required no college degree to define his own job description as "the buck stops here."[2] As it turned out, serious proposals for an earth satellite development program were never transmitted to the Truman White House. They never got out of the Pentagon.

The technological momentum generated by World War II was only briefly demobilized during the early post-war years. The conflict in the Pacific concluded with the dawn of the Atomic Age, a shocking portent. The American armed forces were swiftly broken up in almost wrecking fashion. A new Department of Defense was soon established to integrate three military services with the creation of the U.S. Air Force. It was to make future arguments on service roles and missions and updating military technology a four-sided dialogue in the Pentagon before the President and the Congress got into policy and budgetary issues.[3]

For its part, the Soviet Union was propelled by ideological dogma growing out of its horrible losses during the Nazi war, and which dictated its view of the structure of peace in Europe. In July 1945, at the Potsdam summit, President Truman advised Josef Stalin and Winston Churchill that an atomic weapon would be used to hasten the end of the war in the Pacific. Stalin declared at Potsdam, as later political and Red Army actions enforced, that Western Europe, and particularly all of defeated Germany, was considered in the Kremlin's sphere of influence. An "iron curtain" clanged down around Eastern Europe, and included Czechoslovakia by 1948. George Kennan early advised Truman from Moscow that any hope of peace with Stalin looked like a "pipe dream."[4] In April 1946, the Joint Chiefs of Staff advised Truman: "The Red Army could conquer much, perhaps all, of Western Europe. The earliest response would be strategic air attack upon vital areas of the U.S.S.R."[5] It took two years before James Forrestal, first Secretary of Defense, bought reluctantly that view. But President Truman decided not to lose the peace. This resulted in a "Truman Doctrine" which drew a line in Iran, Turkey and Greece. The Marshall Plan provided economic aid for rebuilding war-devastated Western Europe. And, later came the genesis of the North Atlantic Treaty Organization.

In every salient area from the national budget to post-war organization of the government, President Truman's actions, with sometimes strained Congressional support, were geared to attaining a peacetime mode. Foreign problems intruded. The hard part was, by 1948, an election year, to meet minimal needs for securing peace and increasingly to avoid a World War III.

The Atomic Energy Commission was established in 1946, to retain civil control of weapons and development. International control of atomic energy with inspection safeguards was proposed by the U.S. in the new United Nations. It was rejected by the U.S.S.R. In 1949, it was the reality of Soviet atomic technology detected by American aircraft monitoring the atmosphere for radiation that enabled President Truman to announce the existence of the Soviet A-bomb. He immediately ordered the development of the hydrogen bomb. This reaction had followed on the heels of the threat of war during the Berlin crisis of 1948, and Mr. Truman's upset re-election. A few B-29's were deployed to England and an airlift of coal and potatoes to the political island of Berlin kept alive the four-power occupation of Germany. After communists gained control of China came the invasion of South Korea and the American "police action." The nightmare come alive in the Cold War provided the historical environment in which R&D funding was increased but space feasibilities lacking any early military utility were not to be financed or even to receive unanimous support by the military services.[6]

When Josef Stalin was briefed in the Kremlin in April 1947, on rocket potentials, he likewise was not interested in "sputniks" or obsolete German ballistic rockets. Stalin was attracted by the boost-glide long-range rocket vehicle design of Eugen Saenger. Stalin is quoted as saying" "Such a rocket [Saenger's] could change the fate of the war. Do you realize the tremendous strategic importance of machines of this sort? They could be an effective strait jacket for that noisy shopkeeper Harry Truman. We must go ahead with it, comrades."[7] At that time, of course, the Soviet Union had no long-range heavy bomber, prop or jet, nor an atomic bomb. But any rocket-propelled carrier capable of delivery of a heavy atomic bomb to intercontinental distances was indeed a far greater technological challenge than merely upgrading a German V-2.

As a technology bench mark the U.S. Army launched its first captured V-2 from White Sands in April 1946. It also launched the significant vertical sounding rocket scientific program that was to provide thrust for the creation of the International Geophysical Year a decade later.[8] The Soviet Union launched its first captured V-2 above Russian soil in October 1947, which was the same month that Charles Yeager made the first supersonic flight in the X-1 research airplane. On February 29, 1949, a Bumper-Wac (V-2 with JPL Corporal) reached a record altitude of 244 miles over White Sands, New Mexico. This led to U.S. Army success in gaining DOD approval for the development of the Redstone battlefield ballistic rocket at Huntsville, Alabama. A new Long Range Proving Ground under the Air Force for testing "pilotless aircraft" at Cape Canaveral, Florida, was also approved. Long-range and large missiles were gaining military favor.[9]

Prodded by the 10-year technological jump in rocketry demonstrated by the German V-2, theoretical feasibility of space flight was openly voiced before the end of World War II. In February 1945, Arthur C. Clarke of the British Interplanetary Society published his first prescient letter in *Wireless World*. It was headed, "Peacetime Uses for V-2." In October 1945, Clarke published his letter on communications satellites.[10] On May 5, 1945, just before V-E Day, Wernher von Braun was interrogated at Garmisch. He presented his paper, which he once said was written entirely by himself, entitled, "Survey of the Development of Liquid Rockets and their Future Prospects."[11] Von Braun said: "We are convinced that a complete mastery of the art of rockets will change conditions in the world in much the same way that mastery of aeronautics and that this change will apply to both the civilian and military aspects of their use." He proposed satellites, the building of orbital stations "to go to other planets, first of all to the moon." Intercontinental rockets and satellites were mentioned in American projections for the distant future. But it was Clarke and von Braun that sparked the first American satellite proposal.

Historians are indebted to R. Cargill Hall for ferreting out the sensitive story of the first official American satellite proposals.[12] First was the High Altitude Test Vehicle (HATV) of the U.S. Navy Bureau of Aeronautics. In July 1945, the von Braun paper was carried directly into BuAer by Abraham Hyatt when he returned from Europe. It was added to the intelligence on German wartime innovations and all other available space thoughts by Navy Lt. Robert P. Haviland. In August, Haviland prepared a paper recommending "Project Rex" be authorized for clustering or staging V-2-type rockets for a manned artificial earth satellite. It would perform scientific research, communication relays, mapping, and meteorological surveillance. It won the immediate support of his superior, Captain Lloyd V. Berkner, who favored the idea of "an artificial ionosphere" for long-range communications. On October 3, 1945, BuAer organized a Committee for Evaluating the Feasibility of Space Rocketry (CEFSR). It was determined that advanced rocketry was required for a single-stage satellite rocket. Aerojet Engineering Corp. was authorized to confirm feasibility of liquid hydrogen as a rocket fuel.[13] By late October 1945, CEFSR asked for BuAer support to develop a test satellite for flight testing by 1951. Profile studies were contracted for with the Jet Propulsion Laboratory, and later design studies with North American Aviation and the Glenn L. Martin Company. By June 1946, two HATV design studies were in hand.

In the meantime, projected funding appeared inadequate to complete the Navy satellite project. In March 1946, Comdr. Harvey Hall sought help from the Army Air Force for $5 to $8 million for a joint satellite effort. The Air Force declined and requested Project Rand at the Douglas Aircraft plant to prepare an Air Force satellite feasibility study. By June 1946, both the Navy BuAer HATV's and the Air Force's "Preliminary Design of an Experimental World-Circling Spaceship" were reviewed by the Aeronautical Board. Nothing happened. Money was needed to carry the Navy satellite effort beyond the paper stage. In January 1947, BuAer requested the satellite question be reviewed by the Joint Research and Development Board (JRDB), citing the great potential for cooperation with civilian scientists and military groups. As the new Department of Defense got going in July 1947 nothing happened.

In March 1948, the Technical Evaluation Group of the Research and Development Board (RDB) made a decision. It declared that neither the Navy nor the Air Force had offered military or scientific utility commensurate with the expected cost of an earth-satellite vehicle. The services were to continue feasibility studies after the BuAer and the Rand satellites were again reviewed by RDB's Guided Missiles Committee in September 1948. The Army with its Hermes Project would make "continuing analysis of the long-range rocket problem as an expansion of their [study] task on an earth satellite vehicle." Rand Corporation was to continue USAF studies. The Navy cancelled its project and transferred its remaining funds to the Office of Naval Research for the development of the Viking sounding rocket. It must be assumed that the Secretary of the Air Force, W. Stuart Symington, might have mentioned satellite items to his Missouri colleague in the White House. However, problems in the development of jet aircraft for the Air Force was the priority undertaking. It was no accident that the Oval Office of Mr. Truman contained many airplane models, and three Charles Hubbell paintings hung on each side of both doors. They were paintings of aircraft from the Thompson Products distinguished calendar.

In late December 1948, the first report of the Secretary of Defense was released. In an appendix, the report of the Research and Development Board was included, which said:[13]

> The Earth Satellite Vehicle Program, which is being carried out independently by each military service, was assigned to the Committee on Guided Missiles for coordination. To provide an integrated program with resultant elimination of duplication, the committee recommended that current efforts in this field be limited to studies and component designs; well-defined areas of such research have been allocated to each of the military services.

This publicity was widely noted and sparked two consequences: It made it known that an American satellite vehicle was being considered. To some consternation it indicated that a potential psychological weapon of launching the first earth satellite, which Rand studies had submitted, had now been compromised. It was an argument favoring the decision by the Eisenhower administration to launch a scientific satellite for the IGY in 1955.

The Kremlin propaganda mill did not neglect the opportunity, even after the tragic death of James Forrestal, to refer to that absurd satellite project sponsored by "that mad-man Forrestal." And, of course, official and unofficial space buffs were now more hopeful.

The appointment of K. T. Keller as Director of Guided Missiles in the Office of the Secretary of Defense in October 1950 has some relevance. Destruction of the American monopoly of atomic energy by the Soviet atomic bomb in 1949, and the beginning of the Korean War in June 1950, stimulated reactions not to again underestimate the competence of the U.S.S.R. In August 1950, Stewart Alsop sounded as if he had been talking with Wernher von Braun in a column on "'Man-Made Moon' Serious Project."[14] Alsop wrote: "a former Missouri farm boy will soon have to decide whether or not to create a new heavenly body." He referred to the research started by the "order" of James Forrestal, and "now that a great rearmament program is under way, a perfectly serious question has arisen whether or not to move ahead with the actual engineering phase." He concluded: "In the end, President Truman will order work started on this fantastic project if only for an old familiar reason--if we don't, the Russians will." But Truman did not, and it is not known if Alsop's effort to get his attention succeeded.

Not until February 1952, at the request of President Truman, did the White House request a survey of "the present state of the satellite program". It reached the Guided Missiles Section of the Department of Defense and Professor Ariste Grosse of Temple University, who had served on the Manhattan Project, made such a survey. He consulted offices in the Pentagon, the Naval Research Laboratory, Wright Field, and the Rand Corporation of the Air Force, Columbia University faculty, and "spent many hours" with Wernher von Braun at Army Ordnance in Huntsville, Alabama. Grosse's report was not completed before the election of President Eisenhower. It was submitted on August 25, 1953, to Deputy Secretary of Defense Donald Quarles. This was two weeks after the first thermonuclear reaction by the Soviet Union on August 12, 1953.[15]

That President Truman was not fully aware of space potentials seems not surprising with the continuance of the Korean War near the end of his first term when he decided not to run again. It is also confirmed by Truman's attitude toward the IGY satellite project after it had been announced by Eisenhower in July 1955. Interviewed while taking his brisk morning walk in New York City in February 1956, former President Truman was asked by an accompanying reporter what he thought about the American earth satellite program initiated by President Eisenhower. Truman replied that his opinion was exactly what Admiral William D. Leahy once thought about the atomic bomb--"a lot of hooey."[16]

Former President Harry S. Truman visits NASA Administrator James E. Webb, his former Director of the Budget and Under Secretary of State, November 3, 1961 (NASA Photo)

Dwight D. Eisenhower

General Dwight D. Eisenhower was the first Republican President in twenty years. It was a new experience for the Federal establishment, including the Congress. Like Harry Truman, Eisenhower was born West of the Mississippi River, and in Texas as a matter of fact. He had, truly, spent almost his entire career on active military service from West Point to NATO, only briefly interrupted by his 30-month presidency of Columbia University and his brief political campaign. "Ike" was the first President to learn to fly light aircraft. If President Eisenhower came to call his election as "a mandate for change," the impact of technical change was to greatly influence the policies, budgets, defenses, and world status of the Nation during his administrations. He approved the first space program to launch a scientific satellite during the International Geophysical Year, after he approved the priority Ballistic Missile Program made feasible by thermonuclear warheads. The first true submarine with nuclear power was also to be missile-equipped. And, after the Soviet sputniks in the fall of 1957, President Eisenhower was to recommend the creation of a civilian agency to conduct activities in outer space for their own sake, leaving space projects of military potential in the Department of Defense. The shock that the Soviet Union had first tested a rocket with intercontinental range in August, and the first earth satellite of the earth on October 4, 1957, proved a catalyst for a national decision shaping how the United States would organize itself for attaining well-considered objectives in outer space. The Congress, with Senate Majority Leader Lyndon B. Johnson, helped shape this decision in working on the outcome of the National Aeronautics and Space Act of 1958. Consequent creation of "NASA," and implementing the take-off of the U.S. Space Program, proved only prologue to almost two decades of sustained achievements in space exploration and exploitation not fully anticipated by the most avid space buff at the outset.

President Eisenhower's personal involvement with the evolution of space-related opportunities to tangible actions has been fairly treated in his public statements and his memoirs and those of his science advisers as well as the disparate thrusts of various viewpoints of those in each of the military services. But the percolation of official and informal tracings which historians would prefer seems not yet fully complete. Not until the sputniks and the subsequent creation of the National Aeronautics and Space Administration was the open record of science, technology, and policy deliberations sufficiently aired to give understanding and gain support for the interrelated and sometimes conflicting ways and means for space achievement.[17]

White House priorities involving a full spectrum of the national interests and space potentials have not often been on the top of the action list with regard to the Nation's survival and prosperity. The impetus for space flight had been clearly manifested at the time President Eisenhower began his eight-year administration. Arthur Clarke's *Exploration of Space* was a Book-of-the-Month Club selection. Willy Ley's *Rockets, Missiles, and Space Travel* was a perpetual seller. The Air Force School of Aviation Medicine and the Hayden Planetarium both held symposia on outer space which spread widely to the then popular weekly magazines such as *Collier's, Life, Look, Time* and *Newsweek*. Wernher von Braun lectured all over at every opportunity. As is well known, the American Rocket Society had an active committee on Satellites as the American Astronautical Society got going. When Fred Singer's "M.O.U.S.E." minimum satellite concept received world-wide attention, earth satellite proposals were beginning to gain toe-holds in official circles.

THE "NEW LOOK" AND "OPEN SKIES"

The take-off of the Eisenhower Administration was conditioned by priority problems on Earth, first of all to fulfill "Ike's" campaign promise to end the "no win" war in Korea. He flew to Korea before he was inaugurated. This chilled the personal take-over from Mr. Truman, though each held similar views on the cost and political affluence of the military services and their constituencies. President Eisenhower reorganized the White House staff, first the National Security Council. The NSC now included the Secretary of the Treasury, and all major proposals were costed. Ike appointed an entirely new Joint Chiefs of Staff, while a Planning Board worked all incoming NSC paper, and decisions and actions were monitored by the Operations Coordinating Board. Meetings of the full Cabinet and Legislative leaders were regular. A new Cold War strategic concept was early enunciated, not as a slogan but as the prevailing thrust of the Eisenhower administration. It became known as the "New Look." Formulation began upstairs in the White House, in the Solarium. The new American defense posture stemmed from avoiding future "Koreas" and it was severely tested when France was being thrown out of Indo-China in 1954. The New Look rested on the assumptions that (1) the United States would not start a war, (2) nuclear war was a "catastrophe beyond belief," (3) American armed forces should be developed and designed to deter nuclear war, and (4) military strength was only one factor in sustaining genuine national security for the "long haul" along with national will, intellectual genius and vigor, and fiscal solvency.[18]

Mr. Eisenhower declared that the United States would place greater reliance on nuclear weapons. By April 1953, after the death of Josef Stalin and before the armistice in Korea, President Eisenhower announced his reliance on nuclear weapons, particularly in the Strategic Air Command, in a public address.[19] Reliance on firm alliances among the Western democracies was elaborated by Secretary of State Dulles, while Secretary of Defense Charles Wilson instituted the reorganization of the Department of Defense. Functional unified commands were for NATO in Europe, the Far East, and Alaska, and Specified Commands in the Strategic Air Command and Continental Air Defense Command. Service budgets, particularly on R&D, increasingly were equated with the "New Look." Debates arose in the Pentagon and on Capitol Hill. Technological events intruded.

Already in August 1953, the U.S.S.R. exploded an H-bomb. The long-range Navaho two-stage cruise missile successfully achieved full-assembly thrust of 200,000 pounds for the first time. But the Navaho had been designed to carry a heavy atomic weapon. While it was a technological success, the Navaho was to be short-lived.[20] In December 1953, the proposal to create an International Atomic Energy Commission to develop peaceful uses was made at the United Nations by President Eisenhower.

A major decision not publicized by the Eisenhower administration was given "the highest possible priority in the use of talent, money, and material." It was the Intercontinental Ballistic Missile. In February 1954, the Teapot Committee under John von Neumann reported the possibility of a major breakthrough on nuclear war-head size and the technical problems associated with the development of intercontinental ballistic missiles which could be solved within a few years. The von Neumann Committee had initially been created by Trevor Gardner, assistant secretary of the Air Force for R&D, and moved up to the DOD level. In March, President Eisenhower met with the Science Advisory Committee of the White House Office of Defense Mobilization, a committee started in the Truman administration. Discussed were the implications of the ICBM and other recommendations made by Trevor Gardner. The President stressed the new danger of a surprise nuclear attack in the future, and the need "to reduce the ease with which a hostile nation with a closed society [could] plan an attack in secrecy," according to James R. Killian. Killian, President of M.I.T. was later asked to chair Technical Capabilities Panels, which he agreed to do with James B. Fisk as co-chairman. A series of interrelated studies was prepared by panels of leading scientists and engineers. Out of this process, which Eisenhower appreciated, came priority attention for ICBM development at the Air Force Western Development Division (later AFBMD), IRBM's for strategic use deployed overseas, early warning systems, study of anti-ballistic missiles, and improved intelligence capabilities. This is a significant and complex area yet to be assayed fully. However, no "Manhattan Project" was to be instituted. A new Air Force organization under Brig. General Bernard A. Schriever had the ICBM task; the Army Jupiter and the Air Force Thor and the Navy submarine-based Polaris were the IRBM's. Early warning networks were contracted beyond the early anti-bomber lines. And the U-2 photo-recce powered-glider became the responsibility of the CIA.[21]

Important for later developments, President Eisenhower came to have highest regard for the sound technical recommendations made to him by those not directly involved in on-going programs. In turn, they seemed to feel that they could communicate frankly and fully with the President.

As nuclear weapons testing continued, President Eisenhower took other initiatives to ameliorate the impact of thermonuclear capabilities 10,000 times more powerful than the A-bomb dropped on Hiroshima. After the death of Stalin, attempts to communicate with the new Kremlin were made. The President offered to go to Moscow to discuss disarmament or whatever might bridge the Iron Curtain, if some thawing of the cold war seemed feasible. International scientists were making progress toward an International Geophysical Year, and the National Science Foundation was modestly funded to support U.S. efforts. The Soviet Union suddenly agreed to neutralize Austria and withdrew its occupation forces. A Four Power summit meeting, including France, England, the U.S.S.R., and the United States was set up for Geneva in July 1955. At an informal dinner with Khrushchev, Bulganin, and Molotov, President Eisenhower said loudly, according to John Eisenhower who was present: "It is essential that we find some way of controlling the threat of the thermonuclear bomb. You know we both have enough weapons to wipe out the entire northern hemisphere from fall-out alone." But President Eisenhower had "a bombshell" for the formal session, as Nelson Rockefeller, special assistant for psychological warfare, told John Eisenhower.[22]

On July 21, 1955, President Eisenhower addressed the Geneva summit with few notes. He presented his proposal for "Open Skies." Each nation should give one another a complete blueprint of their respective military establishments, and provide "ample facilities for aerial reconnaissance, where you can make all the pictures you choose, and take them back to your own country for study...." By such a step, he submitted, "we would convince the world that provision against a surprise attack would be possible, and it would be but a beginning for effective inspection and disarmament."[23] Khrushchev told Eisenhower later: "That is a very bad idea, it is nothing more than a spy system."[24] As it turned out, Soviet-American cultural and scientific exchanges did increase. Eight days after Eisenhower's "Open Skies" proposal had been made in Geneva, Press Secretary James Hagerty in Washington, D.C., advised the American press that the United States would announce the next day that it would launch an earth satellite as a contribution to the IGY.

THE IGY SATELLITE

There are many here today, other than our Chairman Fred Durant, who appreciate that it could take several lectures to explain all aspects of how it was that President Eisenhower approved a minimum satellite as a contribution to the International Geophysical Year. It was passionately kicked about during the "sputnik crisis" after the Soviet Union launched the first satellite on an ICBM rocket. The United States, however, had decided to launch a scientific satellite with a non-missile rocket that had to be developed and tested. As it turned out, a U.S. Army Jupiter-C launched *Explorer I* almost four months after *Sputnik I*. The IGY *Vanguard* attained orbit on St. Patrick's Day in 1958. By then, James A. Van Allen's IGY experiment on the *Explorer* had made the most important discovery of the International Geophysical Year, and President Eisenhower had already decided to ask the Congress to create a civilian space agency. Fortunately, a distinguished history covering the White House involvement exists in Constance McL. Green's *Vanguard: A History*. She interviewed former President Eisenhower as well as most of the science leaders involved, which is in contrast to the rocket-oriented accounts.[25]

Each year, starting in 1952, the Space Flight Committee of the American Rocket Society, first chaired by Andrew Haley, had presented recommendations for the launching of a minimum-sized satellite to the Director of the National Science Foundation, Dr. Alan Waterman.[26] In June 1954, an informal group of rocket specialists was called together by Lt. Comdr. George Hoover of the Office of Naval Research, which included, among others, Wernher von Braun, and three "Freds"--Durant, Singer, and Whipple. This gave birth to the "Orbiter" concept for launching a small satellite using the Army Redstone with Loki upper stages. By August 1954, Orbiter was a joint Navy-Army study, von Braun versions of which, called "Project Slug," never got very far up the Army channels. But Orbiter proposals were widely circulated.[27] Encouragement appeared in October, when the International Council of Scientific Unions (ICSU), meeting in Rome, Italy, approved the recommendation of its Special Committee for the International Geophysical Year, that earth satellites should be launched. By November 24, 1954, the ARS Space Flight Committee, chaired by Milton W. Rosen, submitted a report to the National Science Foundation, entitled "On the Utility of an Artificial Unmanned Earth Satellite." It contained a series of signed appendices on the utility to science of experiments orbited above the earth's atmosphere.[28]

In the meantime, the U.S. National Committee for the IGY, chaired by Joseph Kaplan and functioning under the National Academy of Sciences, set up a Committee on Rockets in January 1955. Detlev Bronk, President of the National Academy of Science and NSF Director Waterman determined that the cost and the international implications of an IGY satellite required the approval of President Eisenhower. This would have to be preceded by review of the NSF National Science Board and the Department of State. The White House staff would circulate the matter for comment by the ODM Scientific Advisory Committee, the director of the Budget, and all the other representatives on the National Security Council. Only in this manner would approval by the President be gained, and with it the necessary funding and technical support required.[29] Bronk and Waterman contacted the Department of State, and, with Lloyd Berkner, made a preliminary presentation to President Eisenhower on March 22, 1955. When they visited the Pentagon, Secretary of Defense Wilson referred them to Assistant Secretary of Defense for Research and Development Donald Quarles. Quarles already had copies of the Orbiter proposal with an endorsement of the Assistant Secretary of the Navy for Air, and two memoranda from the Naval Research Laboratory. One of the NRL memos was by Milton Rosen on the M-10 Viking as a satellite launcher, the other on the Minitrack ground system by John Mengel and Roger Easton. Waterman, who impressed Eisenhower, also convinced Quarles that the satellite project was worth doing. By May, both Waterman and Quarles had received the estimate of Kaplan's committee that a little less than $10-million would fund ten satellites and five tracking stations, including staffing, and instrumentation.

By mid-May, the staff paper had been worked up in the White House for review by the President. ODM's Scientific Advisory Committee recommended approval of the scientific merit of the satellite, and also because it would test "freedom of space" as a principle of international law. Nelson Rockefeller's memorandum gave enthusiastic endorsement, stating that the satellite was "a race we cannot afford to lose," and a scientific project with results for all nations would deny any attempt by the U.S.S.R. to use a satellite as a threat to peace. Secretary Quarles just doubled the estimated cost to $15- to $20-million. On May 26, 1955, the National Security Council approved the proposal to launch an IGY satellite during the IGY in 1957-58. Two specific conditions were specified: (1) The peaceful purposes of the satellite must be stressed; and (2) it must not in the slightest interfere with work on IRBM and ICBM programs. It was also clear that overall responsibilities for the launching would rest with the Department of Defense, and some IGY money might be available in this regard.[30]

Secretary Quarles, already hard-pressed with the priority ballistic missile program as well as the crisis subsequent to the flypast on May Day in Moscow of heavy jet bombers for the first time, looked to the Ad Hoc Advisory Committee, chaired by Homer J. Stewart, to come up with the recommendation on how the U.S. IGY satellite would be launched. No recommendation was made by the Stewart Committee, when, on July 27, 1955, President Eisenhower agreed to announce the American satellite program on the 29th. The word was hastened to CSAGI to make certain that the U.S. announcement would precede any that might be made by the Soviet Union.[31]

White House Press Secretary invited White House correspondents in for an advanced announcement. It ran two hours. Explanations were elaborated by Alan Waterman and others. The next day, Hagerty released the White House statement, which read in part:

> "On behalf of the President, I am now announcing that the President has approved plans by this country for going ahead with launching small earth-circling satellites as part of the United States contribution to the International Geophysical Year....
>
> This program will for the first time in history enable scientists throughout the world to make sustained observations in the regions beyond the earth's atmosphere. The President expressed personal gratification that the American program will provide scientists of all nations this important and unique opportunity for the advancement of science."[32]

Four days later, Moscow announced that the U.S.S.R. would also launch a satellite during the IGY. Yet on August 3, 1955, the Stewart Committee had not yet selected the launcher for the American satellites.

Assistant Secretary of Defense Donald Quarles awaited the final recommendations of the Stewart Committee to institute the DOD obligation to launch the IGY satellite. The Stewart Committee was unaware of the context of the President's decision, other than the scientific portions of the IGY operations would be open and that the launcher should not conflict with ballistic missile programs. Briefings were heard in July. The revised ONR-Army Ordnance "Orbiter" was presented by von Braun, the NRL Viking-10 and Viking-15 designs by Martin were given by Milton Rosen, which included the Minitrack stations and NRL sounding rocket experience. The Air Force, pleading possible interference, essentially proposed that a complete Atlas could be orbited in late 1958 (which they did on Project Score). It was not a question to the Stewart Committee which system could launch a satellite the earliest, only that it serve the IGY period from mid-1957 through 1958. On August 4, 1955, the Stewart Committee brought forth a split recommendation favoring the Naval Research Laboratory proposed launcher and operation for the IGY satellite. Repeated Army requests for re-review of Orbiter were to no avail. Quarles made the decision on September 9, 1955.[33] Several days later, President Eisenhower suffered a heart attack in Denver, an unrelated event,but it meant that no problem of national scope such as the bomber gap or the missile race would have his attention for a while.

A year later, the Army Jupiter-C made its spectacular reentry test down the Atlantic Missile Range, a demonstration, in effect, of an in-being satellite launching capability. The IGY satellite question was reviewed again. The National Science Foundation remained confident that the NRL-managed Vanguard would achieve its IGY objectives. DOD made it known to the Army Ballistic Missile Agency in Huntsville, Alabama, that no live fourth stage would be placed into orbit. The American satellite was to be orbited by a non-military booster. References to Soviet missile progress and to satellites cropped up in the lengthy and open "Air Power" hearings conducted by Senator Stuart Symington. President Eisenhower was easily reelected in November 1956.

The Pentagon and the White House were alerted when the Soviet Union conducted a series of long-range rocket tests in August 1957. Moscow announced that it completed ICBM tests. The "missile gap" was now added to the "bomber gap" to challenge the budget estimates as well as Congressional intent to cut them. Then came the "beep-beep" of Sputnik on October 4, 1957. It was the end of the prelude of the American space program.

CREATION OF THE NATIONAL AERONAUTICS AND SPACE ADMINISTRATION

The day of *Sputnik I*, a Friday, President Eisenhower had gone to his Gettysburg farm in mid-morning, after having had meetings concerning the federalization of the Arkansas National Guard for the crisis in Little Rock. He played golf in the afternoon. He received advisory calls on *Sputnik* from the White House at 6:30 p.m. As Americans learned of the Soviet feat, Press Secretary Hagerty advised the news correspondents that "the Soviet satellite, of course, is of great scientific interest," and that it was clear that the Russian announcement "did not come as any surprise; we have never thought of our program as in a race with the Soviets." But it was a surprise to most Americans and the subsequent news media orgy consistently had a "space race" as its beat.[34]

The Soviet test satellite lofted by an ICBM rocket erected a double-barreled crisis to those not surprised. Much confusion in the post-Sputnik environment derived from the twin concerns of the "missile race" as a strategic threat and the dawn of the "Space Age" as a historic beginning. For the first time in American History, the sputnik confirmed that a potentially hostile nation on another continent possessed the technical means using an ICBM to strike directly and suddenly the North American continent with thermonuclear firepower. Army General James Gavin called it "a technological Pearl Harbor." But it was the sheer novelty, the intellectual and romantic attraction of the idea of space mobility, if only for science at first, that spawned the sustained concerns which did not go away. Space advocates, scientists and cadets alike, could not have possibly compounded any more dramatic or widely effective catalyst for a turning point in history.

The "sputnik crisis"--and we are interested mainly in the White House role here--had three distinct phases. These phases were to to be punctuated by the launching of *Sputnik II* with a little Russian dog on November 4, 1957, which prodded actions on Capitol Hill and in the White House, and the spectacular explosion of the Vanguard TV-3 test that blew up on the pad at Cape Canaveral on December 6, which produced a consensus that something basic for the American Space Program must be done. And, with the successful orbiting of *Explorer I*, launched on January 31, 1958, which sufficiently got the United States into space, the time came for organizing a space program for the future. The National Aeronautics and Space Act of 1958 was the result of a national decision, fully made in the open, which President Eisenhower signed into law on July 29, 1958.

As the Space Age got underway, President Eisenhower spent a long October 8th with military, science, and space advisers, reviewing policies and possible actions, and preparing for the regular news conference the next day. Validity of the separation of the IGY satellite from a military mode was sustained, as he wrote in his memoirs.[35] Secretary Quarles pointed out that the Soviet satellite had unintentionally established "the concept of freedom of international space" by its overflight. He talked with John Hagen about Vanguard. It was indicated that, "as a bonus," if everything worked in the TV-3 test in late November or December a small test satellite would gain orbit. Later, Academy President Bronk reasserted the importance of serving international science. James Hagerty was advised to prepare a release containing the stance of the White House on space policy for the press conference. President Eisenhower also wrote in his memoirs: "I met with Secretary Wilson. I directed him to have the Army prepare its Redstone at once as a backup for the Navy Vanguard." Yet it was to be a month later, after the second Soviet satellite had been launched, for reasons not yet clear, that the President's order was received at the Army Ballistic Missile Agency in Huntsville from Army headquarters.[36] In any case, they had been dusting off the Jupiter-C in Huntsville for some weeks.

The press received President Eisenhower's explanations fairly the next day, October 9, despite pointed questions on satellites and the school integration crisis in Little Rock. Beyond Hagerty's handout on the lack of surprise, and scientific interest guiding the American effort, not a race, the President stated the reasons for separating the IGY effort in space from a military project. He congratulated the Soviet scientists. He said the Soviet Union had a "very powerful thrust in their rocketry," hinting but not saying that the Russians had had to build a larger rocket for their ballistic missile. As far as national security was concerned the sputnik did not "raise my apprehensions." Khrushchev's claim that now bombers were museum pieces was not a sound viewpoint. But he made a blooper, which escaped most of the press, when he referred to the long interest of the Soviet Union in rocketry and space resulting from its capture of the V-2 assets, material and personnel, in 1945. This was based on an advisory from Secretary of State Dulles. The aerospace community knew that the key German engineers had been working for the U.S. Army, and were now in Huntsville.[37] The Redstone advocates never forgot this.

President Eisenhower closed his news conference with reference to the position of the Western democracies taken in August, to which the Soviet Union would be asked to participate, to make a multinational study so that "outer space shall be used only for peaceful, not military, purposes." This basic juridical principle also came from the Secretary of State, and proved significant later.[38]

Beyond the quip that grew out of Eisenhower's reference to V-2 engineers--"The Russians got all the good Germans"--there were other statements

HERBLOCK'S EDITORIAL CARTOON

HERBLOCK
©1957 THE WASHINGTON POST CO.

----from Herblock's Special For Today (Simon & Schuster, 1958)

HERBLOCK'S EDITORIAL CARTOON

Moonglow

----from Herblock's Special For Today (Simon & Schuster, 1958)

Used with permission of Mr. Herb Block, *Washington Post*

by Administration spokesmen that did not help the President. Secretary Wilson's reference to the sputnik as "a neat scientific trick," and Admiral Rawson Bennett's "hunk of iron," and Sherman Adam's disinterest "in the basketball score in outer space"--these were short-lived perks as the space crisis continued.

President Eisenhower met with the Scientific Advisory Committee, chaired by I. Rabi of Columbia University, on October 15, 1957. It was a long session, out of which the recommendation was made that the President should have a Science Adviser resident in the White House. After *Sputnik II*, James R. Killian was named. He formed panels on the outstanding military and space areas on which the President would need to meet the Soviet challenge in science and technology now evident.[39]

The second sputnik, heavier and carrying the dog, *Laika*, intensified the crisis as well as provoking a rash of speculation concerning the intent of the Soviet Union with regard to manned space flight. President Eisenhower made the first of nation-wide television talks on science and defense on November 7. Besides naming James R. Killian as his Science Adviser and upgrading the Scientific Advisory Committee, Eisenhower traced the augmentation of the New Look with missiles including submarine-based missiles with nuclear warheads, continental warning systems, and the like. He said: "although they [the Soviets] are ahead of us in satellite development, as of today the overall strength of the Free World is distinctly greater than that of the Communist countries." He debated whether to disclose at this time the U-2 capable of photographing the Soviet Union, but decided against it.[40]

On November 13, in Oklahoma City, the President's TV address was on scientific education and greater concentration on research, and their role in deterrence and defense. In the meantime, his Science Adviser Killian was spending much time in the Pentagon sorting out roles and missions. Both the Army and the Air Force had vocal space champions, who claimed total R&D jurisdiction because of competence in rocketry and related technology. The Air Force project 117L got moving. Creation of an Advanced Research and Development Agency (ARPA) by the Department of Defense was announced as a means of managing dollar control of the services' R&D as well as an interim focus for selecting satellite and space probes from among the hundreds proposed after the sputniks.

Sputnik II turned the possibility of a full-scale investigation on missiles and satellites by the Senate Investigating Subcommittee into a reality. Since the first sputnik, Senator Lyndon B. Johnson, who seemed to believe that space was the greatest challenge in human history, had conducted interviews and staffed up. The new Secretary of Defense, Neil McElroy, and his associates spent hours daily serving the Senate preparations. One of McElroy's aides quipped: "No sooner had Sputnik's first beep been heard--via the press--than the nation's legislators leaped forward like heavy drinkers hearing a cork pop. Never has a department [DOD] had so much help in facing up to a problem."[41] Eilene Galloway will cover this well. The Senate hearings were set for November 25th, and television coverage was arranged. This was the day that President Eisenhower suffered a "minor stroke," once again placing responsibility on his staff.

Early in December, Vanguard was ready for the first test launch with three live stages and a grapefruit-sized test satellite. The explosion of the first stage and the downward crash of the test payload was carried on live television. The 60-day sputnik crisis continued. PSAC panels debated Christmas week 1957 in the Executive Office Building on the components and the organization which a well-conceived American space program must possess. Proposals of the American Rocket Society and the Rocket and Satellite Research Panel (RSRP) both separately and then jointly proposed a civilian space organization. Difficulties would arise if the space mission was assigned to either the Army or the Air Force. Minimal disruption by giving all space responsibilities to ARPA only solved the problem of interservice rivalry, not the civilian non-military elements. Major laboratories like those of the Atomic Energy Commission might not be suitable for the numerous disciplines involved in space science and technology. The thought arose early that perhaps the National Advisory Committee for Aeronautics organization, with a tradition of demonstrated dealing with the military services, industry, and academe, might offer a means to get underway. PSAC had no clear-cut recommendations on overall space trends yet, at least until a Vanguard or a Jupiter-C team, both under severe pressure, succeeded in launching the first U.S. satellite. The Congress was most impatient. Meanwhile, the Atomic Energy Commission and the National Advisory Committee for Aeronautics (NACA) were stepping up their planning to take a place in a space program.[42] Then came January 31, 1958. Everyone heaved a sign of relief when the Army Jupiter-C put *Explorer I* into orbit. The White House even assigned its name, rejecting those preferred by the Army. "Top Kick" sounded too military. The President presented Wernher von Braun with the Distinguished Federal Civilian Award. Reorganization of military R&D remained a top problem.

But the way was opened with the decline of the hysteria to get the Space Program organized. President Eisenhower announced that Dr. Killian would come up with a recommended organization. On March 5, a month later, he approved the proposal that a "reconstituted NACA" (National Advisory Committee for Aeronautics) now would be the nucleus of a new civilian space agency, leaving military concerns in the Department of Defense. Coupled with preparation of the proposed legislation to send to the Congress for prompt action, was the writing of a rationale for the National Space Program by the President's Science Advisory Committee. It was called "Introduction to Outer Space," and was published and released on March 26, 1958, with a preface by President Eisenhower. To members of PSAC it was a "space primer," in a sense to help dispel some self-serving oral nonsense aired during the sputnik crisis. Most importantly it sought to explain what space exploration meant to the long-term interests of the nation and for mankind. Four factors gave "urgency and inevitability to advancement of space technology: (1) the compelling urge of man to explore the unknown; (2) the need to assure that full advantage is taken of the military potential of space; (3) the effect of national prestige of accomplishment of space science and exploration; and (4) the opportunities for scientific observation and experimentation which will add to our knowledge of the earth, the solar system, and the universe." This "white paper" was as well received in the major newspapers as the President himself had enjoyed the briefings on space potentials leading to his decision. Right or wrong, President Eisenhower was confident his science advisers had no axe of their own to grind.[43]

On April 2, 1958, President Eisenhower submitted a special message to the Congress on "Space Science and Exploration." It declared his intent as contained in the administration bill to create a "National Aeronautics and Space Agency." His message was a preamble, which said:

> I recommend that aeronautical and space science activities sponsored by the United States be conducted under the direction of a civilian agency, except for those projects primarily associated with military requirements. I have reached this conclusion because space exploration holds promise of adding importantly to our knowledge of the earth, the solar system, and the universe, and because it is of great importance to have the fullest cooperation of the scientific community at home and abroad in moving forward in the fields of space science and technology. Moreover, a civilian setting for the administration of space functions will emphasize the concern of our Nation that outer space be devoted to peaceful and scientific purpose.[44]

Both special committees of the Congress were to act promptly on the space legislation, retaining the President's intent but adding stronger authority to the new "administration." In particular, Senator Lyndon B. Johnson, chairman of the Senate Special Committee, insisted on the creation of a National Aeronautics and Space Council in the White House, chaired by the President to insure full concern of space policy and NASA-DOD cooperation. Eisenhower objected to this intrusion upon his White House staffing. There were discussions between the President and Senator Johnson, fellow "Texans." Johnson refused to move the legislation without a Space Council. Eisenhower reluctantly surrendered to LBJ as he wanted the space legislation passed, and Senator Johnson expedited its passage.[45]

President Eisenhower moved swiftly in the Executive Branch. Before the National Aeronautics and Space Act of 1958 became law, signed on July 29th, joint NACA-ARPA panels sorted out the non-military satellite and space probe projects to be transferred to the new NASA, to be covered by subsequent authorizations provided in the legislation. The manned satellite program was made the responsibility of NASA, and the President personally directed that the space test pilots, later called "astronauts," would be selected military officers. The Air Force was unhappy about the loss of its MISS program funding. The Army trenchantly resisted NASA's attempt to negotiate the transfer of the von Braun development team of the Army Ballistic Missile Agency into NASA. But the support of the President was to prove indispensable for the take-off of the new space agency.

T. Keith Glennan, former AEC commissioner and president of Case Institute of Technology, suggested by Killian, became the first administrator of the National Aeronautics and Space Administration in August 1958. Glennan carried a copy of the Space Act in his pocket. He needed it in his lengthy efforts in the Pentagon and even in the halls of NASA to organize a coherent structure and program for an agency dedicated to space development for its own sake. Beyond the basic legislation and the support of the White House, the NASA administrator was successful in recruiting space-dedicated specialists from many agencies to add to the NACA nucleus. But the space committees of both houses of the Congress, now controlled by the Democrats, exercised its overseer role with persistent zeal as NASA

FIRST NASA ADMINISTRATOR T. KEITH GLENNAN AND DEPUTY HUGH L. DRYDEN SWORN-IN BY PRESIDENT EISENHOWER, AUGUST 19,1958.

President Eisenhower hands commissions to the first leaders of the new National Aeronautics and Space Administration following swearing in ceremonies in the White House (Left to Right: Dryden, Eisenhower, and Glennan). Created atop the nucleus of the NACA, NASA immediately began to organize itself and assume direction of non-military space projects transferred to it, which were authorized on October 1, 1958. Its headquarters was the Dolley Madison building, a block from the White House; it inherited the NACA laboratories, and was authorized a new space research center in Beltsville, Maryland. The Jet Propulsion Laboratory of Cal Tech in Pasadena, California, was transferred to NASA from ABMA in December 1958. (NASA Photo No. 58-Adm.-1) gathered itself and formulated its program and its intent. Administrator Glennan was hard pressed on Capitol Hill, and particularly as the non-military space program under NASA was not, per Eisenhower's dictum, modeled according to a "space race" with the Soviet Union.

Glennan was effective at the White House. Eisenhower made the NASA Administrator his space officer for preparing the agenda for the initial meetings of the National Aeronautics and Space Council, later folded into the established White House staffing. When controversies arose with the Pentagon, Glennan contacted directly presidential assistants--and frequently the President--across Lafayette Square from the Dolley Madison House, which was NASA Headquarters. Army and Air Force officers had to route their their space arguments through the Secretary of Defense or the Joint Chiefs of Staff, although they had their friends on Capitol Hill.

When Administrator Glennan settled with the Army for only the transfer of the Jet Propulsion Laboratory in late 1958, President Eisenhower told him it was a mistake and indicated that NASA should also get the von Braun development team at ABMA, "lock, stock, and barrel." Glennan and Herbert York, chief scientist of DOD's Advanced Research Project Agency, however, wisely felt that this transfer should be postponed a year for more effective results. ABMA was made responsive to NASA planning requirements but the delay of the transfer of the Army rocket technologists for a year slowed the coalescence of a fully-viable agency. NASA's first major development contract was for the single-chamber one-million pound thrust rocket engine, which, with its NASA-sponsored liquid-hydrogen upper stages, were ultimately to power the giant Apollo-Saturn V. In the public eye, orbiting a man on a converted Atlas ICBM rocket in what became Project Mercury, was NASA's priority program, much to the dismay of scientists. But more money from the Congress could not solve the technical developments for the safety of manned space flight, a mission fatality holding the likelihood that all of NASA would be abolished. President Eisenhower strictly viewed the orbiting of an astronaut by Mercury was necessary before subsequent manned flight programs could be considered. Mercury was thus a dead end program when the Eisenhower administration ended.

There are adequate histories covering the space story during the Eisenhower and the Kennedy administrations.[46] But the most significant catalyst in the evolution of NASA was the development of its long-range plan. It meshed current programs and development with interim projects leading to the attainment, when feasible, of long-range goals in all major classes of space missions. Running out inherited DOD satellites and probes and maintaining the momentum of the IGY science came first. But to select the next phase of space missions made little sense unless they were consistent with long-range goals of what could be accomplished for good purpose. It took all of 1959 to develop NASA's Ten Year Plan. It was used as the basis for its Congressional presentations in 1960. And it came to be regarded as a good plan on Capitol Hill but it did not go far enough.[47]

President Eisenhower repeatedly asked how much should be spent on space, and at what rate and scale it should develop. He was merely the first President not to obtain a clear answer. Though Mercury had a long way to go before initial success, NASA selected a manned circumlunar mission as the goal of manned space flight in 1959. Administrator Glennan approved circumlunar Apollo as a planning concept in January 1960, clearly stating that White House approval would be required before it had status as a funded project. In 1959, Glennan scheduled a briefing of a direct ascent-to-the-moon concept for President Eisenhower, mainly to indicate that long-range planning was not being neglected. Eisenhower's immediate reaction was that such a program would surely cost a lot of money.[48] In December 1960, the circumlunar Apollo concept was briefed to Eisenhower, by way of supporting a modest request for money to support industry preliminary design studies. The President retorted that he "was not going to hock his jewels like Isabella."[49] Though he was just weeks away from living full-time on his Gettysburg farm, Eisenhower had to be persuaded to remove language in his budget message that Project Mercury was all the manned space flight that the United States required. On January 1, 1961, he enthusiastically approved the release calling for the early development of communication satellites preferably by industry participating and funding, as a national objective.

President Eisenhower was pleased with NASA before he left office. At Glennan's recommendation NASA launch vehicles had "UNITED STATES" on them, although they were Air Force or Army boosters. He also endorsed Glennan's recommendation to name NASA's space flight center in Huntsville, Alabama, after George C. Marshall, the first military man to receive a Nobel Peace Prize. He was pleased with *Tiros,* the weather satellite inherited from the Army; *Echo,* the balloon satellite optically viewed around the world; *Pioneer V,* an Air Force contractor interplanetary track probe launched for NASA. He also appreciated *Project Score,* a broadcast satellite, and the capsule of *Discoverer XIII* (first object recovered from orbit) launched by the U.S. Air Force. But the National Aeronautics and Space Administration had started from scratch as a new agency and had established a trajectory and momentum, blessed with its open information charter, and supported by Congressional committees dedicated to its success.

PERSPECTIVES ON EISENHOWER

As mid-wife President at the birth of the Space Age, Dwight D. Eisenhower was neither converted as an all-out space enthusiast nor was he inattentative to technological imperatives prodded by innovations. Elected as a war hero, Eisenhower's White House was organized, staffed, and managed thoroughly, and with regard to fiscal solvency. He sought to serve all Americans and often exhibited distaste for partisan politics. He did not hesitate to send in the Federal troops into Little Rock at the time of the sputnik crisis, or to place the Strategic Air Command on alert status. At the same time he "waged peace" during the rising hazards of a Cold War buttressed by global-legged thermonuclear weapons. He would not commit a military intervention in Vietnam but landed Marines at several warm places. He equated the "New Look" with advocating "open skies," including the launching a satellite for the International Geophysical Year and his "atoms for peace" program. He later owned up to the U-2 overflights of the U.S.S.R. before the world. There was no bomber gap, "missile gap," or "reconnaisance gap" when he left the White House.

President Eisenhower being briefed on the Saturn C-1 by Wernher von Braun, during the dedication of the George C. Marshall Space Flight Center, Huntsville, Alabama, September 8, 1960. Saturn was the first large rocket not developed as a military missile. (NASA Photo)

President Eisenhower survived three threats to his health while in office, and he was often angered by those, both military and civilian, who sought specious gain, including several of the loudest space advocates. "What was more important than keeping space peaceful?--he thrice stated when he was interviewed after he left the White House. The surprise shock of the sputniks was what he had sought to avoid when he approved the IGY satellite program in 1955. And he was proud of having started Project Mercury, perhaps as much as the Interstate Highways which laced the nation and contained as much concrete as would pave six sidewalks to the moon. Reaction to Soviet initiatives was not the way to design a space program to race in space. No one expected Dwight D. Eisenhower to advocate a massive space program after he left the White House for Gettysburg. He thought that a mere billion dollars was enough for NASA, and DOD's space needs were not neglected. And, upon reflection, he later stated: "How much easier it would have been if I had let the Redstone satellite go from the beginning." But then, of course if he had, there might not have been the crunch of Sputnik, no "birth of the Space Age" or no NASA. Premier Khrushchev's career in the Soviet Union was directly tied to missile and rocket success but he unleashed new political forces that were not subsequently undone. That's another story.

Right after the sputniks, President Eisenhower said on television, "What the world needs today even more than a giant leap into outer space, is a giant step toward peace," Within a dozen years American astronauts on the moon made "a giant leap for mankind." Eisenhower's contribution was the foundation for its success. Eisenhower's successor decided to race into space.

When President Eisenhower dedicated the NASA Marshall Space Flight Center in September 1960, he seemed to express his personal perspectives on the space challenge:

> All that we have already accomplished, and all in the future that we shall achieve, is the outgrowth not of a soulless, barren technology, nor of a grasping state imperialism. Rather it is the product of an unrestrained human talent and energy restlessly groping for the betterment of humanity....
>
> I find that the leaders of the new space science feel as if Venus and Mars are more accessible to them than a regimental headquarters was to me as a platoon commander forty years ago. To move conceptually, in one generation from the hundreds of yards that once bounded my tactical world to the unending millions of miles that beckon these men onwards, is a startling transformation.[50]

By the end of his administration, Dwight D. Eisenhower considered that the U.S. space program, "a comprehensive and costly venture," had turned out to be a reasonable effort. He had requested that the United Nations work to preserve outer space "for peaceful use and development for the benefit of all mankind," as accomplished for Antarctica. He had listened to his science advisers while the DOD-NASA Aeronautics and Coordinating Board (AACB) moderated development of the ways and means to wage peace, and he had requested the Congress to abolish the Aeronautics and Space Council. Also apparent was that space issues were not really partisan arguments.

Most remembered, perhaps, was his Farewell Address to Americans on January 17, 1961, on the pervading influence of the Cold War and its threat to the cause of liberty. He warned "against the acquisition of unwarranted influence, whether sought or unsought, by the military-industrial complex." Ike also warned that "public policy could itself become the captive of a scientific-technological elite." These words were not those coined by a speechsmith, for they appear to have reflected thoughts expressed by President Eisenhower to his intimate advisers. They were also reflected in actions to provide for national defense and for space exploration and exploitation. They derived from what James R. Killian called the "sputnik panic" in 1957. This included champions of space flight to the neglect of national defense and prestige in science, some in the military services. They derived also in the unfounded "missile gap" issue during the presidential political campaign during 1960. There was a danger that the arms race could be extended into outer space. Eisenhower apparently felt that the sputnik crisis and the threat of thermonuclear war sustained a drive for absolute national security exploiting the psychological vulnerability of the American people at the price of their liberty and prosperity. Historians and savants alike will appreciate forthcoming biographies and studies of the administration of Dwight D. Eisenhower as new unavailable documentation might become available.[51]

As it turned out, President Eisenhower was also the first Space Age resident of the White House whose successor inherited a major news media happening when the initial Mercury manned flight of Astronaut Alan Shepard took place on May 5, 1961. No President was ever to be in office when major space missions happened as a result of his initiatives or decision.

BACKGROUND AND DWIGHT D. EISENHOWER

REFERENCE NOTES

1. *Time Capsule/1950* (Time-Life Books, 1967), pp. 11-12.

2. This was a desk plaque in his White House office, borrowed from the Oval Office restoration at the Truman Library by President Jimmy Carter.

3. *The Forrestal Diaries* remain a valuable source for the early Truman administration and the Pentagon where the power of the Secretary of Defense proved difficult to sublimate henceforward. Walter Millis (ed.), *The Forrestal Diaries* (NY: 1951); and, of course, the *Memoirs by Harry S. Truman,* I, *Year of Decisions,* II, *Years of Trial and Hope,* 1946-1952 (Garden City, NY: 1955). Best political thrust of Truman is related in Robert J. Donovan, *Conflict and Crisis: The Presidency of Harry S. Truman* (NY: 1977).

4. Donovan, p. 187.

5. Walter S. Poole, "From Conciliation to Containment: The Joint Chiefs of Staff and the Coming of the Cold War, 1945-1946," *Military Affairs* (February 1978), pp. 12-16.

6. Cf. Harland B. Moulton, *From Superiority to Parity: The U.S. and the Strategic Arms Race* (Westport, Conn.: 1973), pp. 3-14; Eric F. Goldman, *The Crucial Decade - And After, 1945-1960* (NY: 1960).

7. G. A. Tokaty, "Soviet Rocket Technology," in *History of Rocket Technology,* edited by E. Emme (Detroit: 1964), pp. 271-83; Nicholas Daniloff, *The Kremlin and the Cosmos* (New York: 1972), pp. 46-49, 219-223.

8. See forthcoming volume, Homer E. Newell, "Beyond the Atmosphere: The Early Years of Space Science," NASA SP-4211, in press.

9. E. Emme, *Aeronautics and Astronautics, 1915-1960* (NASA: 1961), pp. 63-73.

10. Letter to author from A. C. Clarke, 8/8/68, in which he explains his "great surprise" in being reminded of his February 1945 article, in which he also cited concept of geocentric orbit comsats. His "Voices from the Sky," *Spaceflight* (March 1968), he pointed out, did not have it cited in a chronology of comsat development. Clarke's writings were well-known to space buffs in the Pentagon in 1945 and on.

11. Interview of Wernher von Braun by author, hand-written answers to questions, February 18, 1974, p. 2.

12. R. Cargill Hall, "Early U.S. Satellite Proposals," NSC's first Goddard Prize Essay, in *History of Rocket Technology*, pp. 67-93; and, "Earth Satellites: A First Look by the U.S. Navy," in *Essays on the History of Rocketry and Astronautics*, also edited by R. C. Hall (NASA Conference Publication 2014: 1977), II, pp. 253-77.

13. *First Report of the Secretary of Defense* (1948), p. 129; and, *New York Times*, 12/29/48, p. 11.

14. Stewart Alsop, "Man-Made 'Moon' Serious Project," *Washington Post*, 8/13/50.

15. A. V. Grosse, Letter to author, 1/12/73, 4 pp., with enclosure, "Report on the Present Status of the Satellite Problem," 8/25/53, 7 pp.

16. Milton Bracken, "Truman Varies -- Airy to Mundane," *New York Times*, 2/5/56, p. 56. Mr. Truman's hostility to President Eisenhower was total, and became active after sputnik to the extent he began a newspaper column. Recent historical scholarship enhances the importance of heretofore unavailable documentation on the Truman period. For example, note David A. Rosenberg, "American Atomic Strategy and the Hydrogen Bomb Decision, *Journal of American History*. 66 (June 1979), pp. 62-87; Daniel J. Kevles, *The Physicists* (NY: 1978); David MacIsaac, "The Air Force and Strategic Thought, 1945-1951," Wilson Center International Security Studies Program (Working Paper No. 8, June 21, 1979).

17. Dwight D. Eisenhower's memoirs, like Mr. Truman's, are fairly comprehensive on political explanations but not fully detailed on space-related phenomena not yet made available: *Mandate for Change, 1953-1956*, and II, *Waging Peace, 1957-1961* (NY: 1963). These volumes and his *Public Papers* volumes are abstracted on space subjects in E. Emme (ed.), *Statements by Presidents of the U.S. on International Cooperation in Space, October 1957-August 1971*, published by the Senate Committee on Aeronautical and Space Sciences, Document 92-40, 9/24/61, pp. 5-17. Significant documentation unavailable to historians is over twenty years old. Best collection of open documentation is found in the NASA Historical Archives, Washington, D.C.

18. Robert Cutler was responsible for the first strategic study conducted in the Solarium of the White House by the National Security Council Staff. It is not elaborated in his memoir, *No Time for Rest* (Boston: 1965), or "The Development of the National Security Council," *Foreign Affairs* (April 1960). On the U.S. Air Force enthusiastic view of the genesis of the New Look, see Herman Wolk, "The New Look in Retrospect," *Air Force* (March 1974), pp. 48-51; D. O. Smith, *U.S. Military Doctrine* (NY: 1955).

19. Policies and their existence were meticulously staffed and processed in the Eisenhower White House. Reliance on nuclear weapons was announced by Eisenhower in his classic "new look" speech to the Society of Newspaper Editors in April.

20. *Aeronautics and Astronautics, 1915-1960* (NASA: 1961), pp. 72-73; Lewis Straus, *Men and Decisions* (NY: 1963), pp. 358-75.

21. Eisenhower, *Memoirs*, I, pp. 436f.; James R. Killian, *Sputnik, Scientists, and Eisenhower* (Boston: 1977), pp. 67-88. On the U-2 reconnaissance role, see John Sloop, *Liquid Hydrogen as a Propulsion Fuel* (NASA SP-4404, 1979), pp. 113-17; Herbert F. York and G. A. Greb, "Strategic Reconnaissance," *Bulletin of the Atomic Scientists*, 33 (April 1977), pp. 33-42.

22. John Eisenhower, *Strictly Personal* (NY: 1975), pp. 176-78. Authorship for the Open Skies proposal was claimed by the President, Harold Stassen, special assistant for disarmament, and by Nelson Rockefeller, assistant for psychological warfare. But it was unacceptable to the U.S.S.R. just as had been the proposal to internationalize nuclear energy by the United States in United Nations.

23. Cf. Eisenhower, *Memoirs*, I, pp. 509-30; text in *Statements by Presidents on International Cooperation in Space*, p. 5.

24. John Eisenhower, p. 178.

25. Eisenhower, *Memoirs*, II, pp. 208-9; interview of D.D. Eisenhower by the author and the late Constance McL. Green, in Gettysburg, November 8, 1966 (no transcript of tape permitted). Cf. C. McL. Green, *Vanguard: A History* (NASA SP-4202, 1970).

26. Andrew G. Haley, "International Cooperation in Rocketry and Astronautics," *Jet Propulsion* (March 1955), pp. 631f.

27. Interview with Werhner von Braun by the author, NSI meeting, Titusville, Florida, July 14, 1975. "Project Slug" was never forwarded by Huntsville to Washington, but later "Orbiter" proposal was.

28. Milton W. Rosen, letter to author, 11/5/71; "Proceedings, First Meeting of the ARS Ad Hoc Committee on Space Flight, 5/17/52, 121 pp.; M. W. Rosen, "Down-to-Earth View of Space Flight," Jet Propulsion (December 1952), reprinted in ONR, *Research Reviews* (February 1957), pp. 8-13.

29. It is also important to note that Waterman and Bronk nominally attended meetings of the National Security Council. Cf. Green, *Vanguard*, pp. 29-31.

30. Eisenhower, *Memoirs*, II, pp. 208-209. It should be noted that his perspectives on satellites were consistently unrelated to priority consideration of ballistic missile development. Interservice rivalry on competing missile systems, particularly the Army Jupiter and the Air Force Thor IRBM's, was perpetuated, contrary to White House initial intent to eliminate the duplication, because of the urgency of gaining earliest operational capabilities. The Atlas ICBM had even greater technical development problems. Thus, interservice rivalry on development of an earth satellite not using a missile, and thus hampering least military priorities, compounded the space decision. Most appealing was the very low estimate of the cost of the IGY satellite program, which actually did not include development of the booster in real terms.

31. Green, *Vanguard,* pp. 31-37.

32. Green, pp. 37-38.

33. Green, pp. 41-56. Best post-sputnik review of Stewart Committee member of the minority regarding the recommendation favorable for the NRL proposal is Clifford C. Furnas, "Birthpangs of the First Satellite," *Research Trends* (ONR) (Spring 1970), pp. 15-18. It was the CIA, Furnas revealed, that estimated that the U.S.S.R. in May 1955 was well along in satellite development, and recommended that international prestige required consideration that the U.S. initiate a serious satellite program. Quarles favored the Army Redstone proposal, according to Furnas, but called for a vote of the members of his advisory staff. Furnas stated: "For the first time I can recall, the Air Force and the Navy outvoted the Army." Quarles went along with the majority opinion. Unfortunately for historians documentation of the White House and the Pentagon handling of the first satellite program decision remain to be revealed. This remains in severe contrast to the major proportion of U.S. space activities since 1957.

34. Richard Witkin (ed.), *The Challenge of the Sputniks* (NY: 1958); Cf. William Manchester, *The Glory and the Dream, 1932-72* (Boston: 1974), pp. 797-847; E. M. Emme (ed.), *Two Hundred Years of Flight in America* (San Diego, CA: AAS HS-1, 1977), pp. 23-39.

35. Eisenhower, *Memoirs,* II, pp. 210-11.

36. John B. Medaris, *Countdown for Decision* (NY: 1960), pp. 159-67.

37. John Foster Dulles, Memorandum to James C. Hagerty, 10/8/57, 3 pp. (DDE Library, Office of the President, Education File). Memo explained that the Red Army had captured Peenemuende in May 1945, which was true but the V-2 development personnel and all production had moved elsewhere many months before. Cover note said: "Here is a draft which I wrote...."

38. Eisenhower, *Public Papers,* 1957, pp. 719-35.

39. Eisenhower, *Memoirs,* II, pp. 210-12, 224; Killian, pp. 12-16

40. Eisenhower, *Memoirs,* II, pp. 223-25.

41. Oliver M. Gale, "Post-Sputnik Washington from an Inside Office," *Cincinnati Historical Society Bulletin,* vol. 31 (Winter 1973), pp. 225-52.

42. First space organization decision in the White House was rejection of the proposal to place it under the Atomic Energy Commission, which was the first bill received in the Senate. Cf. Eisenhower, *Memoirs,* II, pp. 225-26; Killian, pp. 119f.

43. Excepted interviews by author; cf. Killian, pp. 122-32, George Kistiakowsky, *A Scientist at the White House,* with introduction by C. S. Maier (Cambridge, MA: 1970), pp. v-vi, xxxi, lv-lvii.

44. Full text in *Statements of Presidents,* p. 12.

45. Lyndon B. Johnson, *Vantage Point* (NY: 1971), p. 277; Eisenhower, *Memoirs*, II, p. 257; cf. Killian, pp. 136-38.

46. Policy considerations are generally not ignored in NASA histories and documented references. Only available listing is Appendix C of "A Guide to Research in NASA History," compiled by A. Roland (NASA HHR-50, February 1979), pp. 15-17. It does not include NASA historical works published in academic journals or in educational and other NASA outlets. Cf. Joseph Shortal, *New Dimensions: Wallops Flight Test Range* (NASA Reference Publication 1028, 1978). There are likewise numerous unpublished histories in the NASA History Office archives.

47. NASA's long-range planning, intimate with program evaluation by a select staff of specialists, was successively headed by Homer J. Stewart, John P. Hagen, and Abraham Hyatt until 1962. They interfaced across the spectrum of systems development, program goals and feasibilities, and budgetary and manning realities. It had a ten-year plan for the new president, whomever might be elected in November 1960. Cf. R. L. Rosholt, *An Administrative History of NASA, 1958-1963* (NASA SP-4101, 1966), pp. 49, 89, 105-7, 130-31; E. Emme, "Historical Perspectives on Apollo," *Journal of Spacecraft and Rockets* (AIAA Paer 67-839), 4 (April 1968), which documents NASA's drive to institute Apollo, pp. 371-75.

48. Manned lunar concept briefed to Eisenhower was M. W. Rosen and F. C. Schwenk, "A Rocket for Manned Lunar Exploration," *Proceedings of the Xth International Astronautical Congress* (Vienna: 1960), pp. 311-26.

49. Excepted interviews; cf. Emme, "Introduction," *Two Hundred Years of Flight in America* (San Diego: AAS HS-1, 1977), pp. 22-29.

50. Text in Appendix F, D. S. Akens, *Historical Origins of the Marshall Space Flight Center* (NASA MSFC HM-1, December 1960), 4 pp., and not otherwise published.

51. Eisenhower, *Memoirs*, II, pp. 515-16; *Killian, Sputnik, Scientists, and Eisenhower* (Cambridge: 1978), pp. 237-38; Herbert F. York, *Race to Oblivion* (NY: 1970), p. 12; and excepted interviews. It must be noted that Eisenhower's was not an anti-technology stance, only the tilting of the recommended priorities for national decision by those with special interests. Emmet John Hughes opined after Watergate that Eisenhower's "antiseptic White House atmosphere sustained its own special disciplines and values. If this atmosphere was not creative, it was also not divisive." "Eisenhower's White House: Confidence--With Open Doors," *Washington Post* (June 3, 1974), p. A-22.

The appearance of additional studies since the completion of this paper may be of interest to readers.

E. M. Emme, "Six Space Age Presidents," AIAA Paper No. 80-931, Fifth AIAA History Lecture, Baltimore, MD, May 6, 1980 (Pre-print only), 48 pp.

Fred I. Greenstein, "Eisenhower as an Activist President: A New Look at the Evidence," *Political Science Quarterly* (Winter 1979-80), reviewed in the *Wilson Quarterly*, IV (Winter 1980), pp. 14-15, and by George F. Will, "Still Liking Ike," *Washington Post*, 5/11/80, p. C-7.

Brian R. Page, "The Creation of NASA," *Journal of the British Interplanetary Society*, 32 (December 1979), pp. 449-51.

John F. Kennedy

John F. Kennedy was the first President born in the Twentieth Century. Among his attributes, he read much, wrote some history. But he contributed to the history of astronautics despite his tragically-terminated presence. Perhaps his inner sense of history as well as the greatness of his contribution to man's initial voyages away from planet Earth are not even fully appreciated today. The horror of Dallas yet seems to block some memories of what happened thereafter of those most close to him. The aerospace community remembers his history. And, a hundred years from now, what else will be remembered? Because of President Kennedy's choice and commitment to "go", two Americans made rendezvous with the surface of our moon sixty-eight months after he was dead.

He was a Cape Cod sailor and drove *PT-109*. But it was an unknown Russian Major of the Soviet Air Force who first circled the Earth in space on April 12, 1961, that moved Kennedy to a space decision long before the first American orbited the Earth. And it was he who moved the Congress and the American people and NASA with his recommendation on May 25, 1961, "to take longer strides" and "achieving the goal, before this decade is out, of landing a man on the moon, and returning him safely to earth." That very same day, President Kennedy was seated in his rocking chair in the White House, with members of his family. He told them, "I firmly expect this commitment [of going to the moon] to be kept. And if I die before it is, all of you here now just remember when it happens, I will be sitting up there in heaven in a rocking chair just like this one, and I'll have a better view of it than anybody."[1]

Jack Kennedy's sense of history was enlivened in 1940 when his father was Ambassador in England planning to go for the nomination against Franklin D. Roosevelt while Hitler's *Luftwaffe* threatened and then took air war to England. His senior thesis with some editorial help was published as *While England Slept*. His own combat experience and "Brother Joe" who was lost before bail-out of a guided-bomber did not make him philosophical about the risks test-pilot astronauts were to freely take--"A man must do what he must," he said in *Profiles of Courage*.

Senator John F. Kennedy had met Wernher von Braun in Boston at the making of a "Person to Person" with Edward R. Murrow in October 1953. Von Braun remembered but there seems to be no record that the young-looking

Senator recalled the meeting. On "Person to Person," Senator Kennedy did refer to his reading a great deal, and that he was enjoying Alan Seeger's, "I have a Rendezvous with Death," a poem by one cut down in his youth.[2] Right after the humiliation of the sputniks in October 1957, Senator Kennedy was undoubtedly aided in shaping his later campaign thrust for getting the country moving again by his Cambridge friends, such as James M. Gavin, Jerome C. Wiesner, and others. But he was not an early space advocate as is indicated by an argument he had with C. Stark Draper in Lock Ober's famous stag eatery right after Sputnik. The bartender was a von Braun fan, and introduced the guidance M.I.T. professor to John and Robert Kennedy for timely conversation. It turned into an argument, according to Dr. Draper, with John Kennedy arguing that all rockets were a waste and their space use was even more so.[3] But then the Kennedys were known to pick arguments just for education of it or for entertainment. You will have to pardon these anecdotes but for the aerospace historian there seems something missing in all of the good books about John F. Kennedy to explain fully why he made the decision to go to the moon, why he made the stirring speeches that were written, and encouraged those in the space venture.

Another dialectical irony, perhaps, was that presidential candidate Kennedy's pivotal primary victory in West Virginia was aided by the downing of the U-2 and the collapse of the summit meeting in Paris in May 1960. Khrushchev's bragging about space exploits of the U.S.S.R. and the rattling of missiles on the top of the news provided ammunition for "a new frontier" and the so-called "missile gap" issues. "Space is our great new frontier," Kennedy said in West Virginia. Space was not to be a campaign issue of any importance against Richard Nixon, or any future elections. But Kennedy's "new frontier" theme meshed much of his rhetoric, and he never forgot his primary victory in West Virginia.[4]

President-Elect Kennedy got his administration ready while working at sunny Palm Beach, Florida. By mid-December, he had selected his cabinet and had established panels to work out recommendations on organization and policies as well as to identify persons useful in his administration. On missiles and space, Kennedy named campaigner Jerome B. Wiesner of M.I.T. Wiesner had served on advisory committees for many years, most recently as a member of the President's Science Advisory Committee (PSAC) and the Air Force's Air Research and Development Command's advisory committee on space chaired by Trevor Gardner. Kennedy also got his regular CIA briefings and lots of advice from job seekers.

If not urgent, two things became obvious about American space affairs in contrast to the forthcoming decline of the so-called "missile gap." The U.S. Air Force had just successfully attained operational tests of its instrumented satellites with recoverable data capsules for serving vital defense needs. Secondly, NASA's Project Mercury was in deep trouble technically after a succession of launch failures: Mercury Atlas blew up at 40,000-feet due to weak interface between capsule and booster; on election day, a Mercury-Redstone shut off at the moment of launch; there was a premature firing of the escape tower at Wallops; and on January 31, 1961, there was a 125-mile over-shoot on the landing of the chimpanzee "Ham." In contrast, the U.S.S.R. had launched a "zoo" (two dogs, rats, mice, flies, and seeds) into orbit and recovered them in late August 1960, and a dummy cosmonaut was orbited but not recovered on December 1, 1960. But at the very best, if the Air Force Atlas worked for NASA, it might orbit an American by the end of 1961.[5] For its part, the U.S. Air Force had moved on to the Titan as its ICBM until the Minuteman came in, and also preferred Titan for its manned booster. Its space champions argued that national defense supported the notion that it should become the operational space instrumentality, including manned flight to orbit, space stations, and the moon. A curious partnership, in effect, was joined between scientists opposed to any manned space flight whatsoever, or others aghast about the unprecedented dangers of launch stresses and weightlessness to astronauts, and the military space advocates seeking to diminish the mission of NASA in the national space program in the new Kennedy administration. In mid-December Administrator Glennan had complained to Chief of Staff Thomas D. White about "what NASA had done to deserve" the slander in a speech-preparation kit for general officers of the Air Force. It was withdrawn.[6] But the USAF network campaigned all out for space.

A preliminary organizational recommendation to the President-Elect was that the National Aeronautics and Space Council in the White House, one of Lyndon Johnson's inputs into the Space Act, and ignored by Eisenhower, should be abolished. This position was reversed after two visits of Vice President-Elect Johnson to Palm Beach, before and after Christmas. Mr. Kennedy was persuaded that the Vice President could well be utilized by chairing the Space Council. It was so announced in Florida. And, Mr. Johnson was asked to come up with the nomination for the post of NASA administrator, so he put his Senate Committee staff to work on it. Scientists plumped for a science administrator not a general or an industrialist. No one wanted the NASA job. Lyndon Johnson later stated that he interviewed 28 different persons for NASA Administrator. Congressman Albert Thomas of Texas recommended Deputy NASA Administrator Hugh Dryden to the President-Elect, but he was informed the choice was the Vice President's task.[7]

The chill that blizzarded Washington on Inauguration Day hit the National Aeronautics and Space Administration ten days earlier. NASA had prepared extensive briefings and source books to help provide for a smooth transition and explaining its ten year plan and its problems. No one asked for them. Then came the "Wiesner Report: of January 10, 1961.[8]

The report of the *Ad Hoc* Committee on Space with nine members, chaired by Jerome Wiesner, whom Kennedy announced as his Science Advisor, was released to the press. Its recommendations for the Pentagon were withheld so that it came out as a dissection of NASA. It blasted NASA at some length for giving Project Mercury, a marginal program, highest priority when perhaps it should have been cancelled. In particular, it criticized the leadership of NASA as being uninspired and poor managers. It submitted five principal objectives for the National Space Program, in the following order: (1) national prestige; (2) national security; (3) scientific observation and experiment; (4) practical non-military applications; and (5) international cooperation. If the Wiesner committee had not been fully briefed on NASA, the dedicated space committees of the Congress were, and they were as upset as NASA, which was to become evident later.[9] Administrator Glennan left Washington.

For what it was worth, President Kennedy's inaugural address called upon the Soviet Union, "Let both sides seek to invoke the wonders of science instead of its terrors. Together let us explore the stars, conquer the deserts, eradicate disease...." Robert Frost's poem was heroic.

Deputy NASA Administrator Hugh Dryden had submitted his resignation but it had not been accepted or rejected. He had been unable to make contact with Press Secretary Pierre Salinger and, in desperation, called Dr. Wiesner on January 25 to inform him that the Mercury-Redstone test with a chimp was scheduled for the 31st. It was likely this launching would arouse world-wide interest.[10] That afternoon, in his first press conference, President Kennedy said that if the Vice President did not come up with a NASA Administrator in five days he would select one. Not only did this demonstrate the new President's adroit use of a press conference to learn from and to lead public opinion, but also to shape up his own problems needing attention. Mr. Johnson called upon James E. Webb a second time, in fact it was Wiesner who was asked to call him in Oklahoma City to come to Washington as an indication the President's science advisor supported the idea. Mr. Webb did not want the NASA job and, when he met with Vice President Johnson, insisted that he was neither a scientist nor an engineer. He was told that President Kennedy wanted him. From LBJ and others Webb well knew NASA's dilemma.

When he met with President Kennedy on January 30, Webb repeated that he was not qualified as his science advisor and had publicly stated the NASA administrator should be a scientist or an engineer. To Webb's argument, President Kennedy replied (according to Webb): "The decisions, the policies that relate to the development of this nation's space capability run to the national and international arena requiring decisions of future capabilities that will be used in an unknown way. I would like you to undertake this job because you have had experience with national policy."[11] Webb asked if he would have to carry out decisions for NASA made by others, or if he would be able to fight for NASA's program? He was assured that Kennedy wanted Webb to work directly for him as the NASA Administrator. This Webb could not refuse and he also asked that Hugh L. Dryden be retained as NASA Deputy Administrator. The White House promptly dispatched the NASA nominations to the Senate Committee on Aeronautical and Space Sciences. It was now chaired by Lyndon B. Johnson's successor, Senator Robert Kerr of Oklahoma, who had been one of the sponsors of Mr. Webb. As was to be the NASA Administrator's practice, Webb never neglected to deal directly with President Kennedy on major policy, budgetary, or other questions.[12]

In his State of the Union message the same day, President Kennedy said, "I now invite all nations--including the Soviet Union--to join with us in developing a weather prediction program, in a new communications satellite program, and in preparation for probing the distant planets of Mars and Venus." He also said, "Today this country is ahead in the science and technology of space, while the Soviet Union is ahead in the capacity to lift large vehicles into orbit." The next day, NASA's vertical space flight of the chimpanzee "Ham" was a reasonable success.[13]

PRESIDENT JOHN F. KENNEDY AND JAMES E. WEBB

NASA Administrator in the White House in 1961 (NASA Photo No. VIP-10)

The change of administrations were never to be as uneven as the first one from Eisenhower to Kennedy. Because of subsequent space accomplishments NASA's apolitical and national mission became better known. Confirmed by the Senate, James E. Webb took charge of NASA on St. Valentine's Day. President Kennedy's choice of Lyndon B. Johnson as his Vice Presidential running mate, and his designation, once enabling legislation was passed, as chairman of the National Aeronautics and Space Council, along with his selection of James Webb as NASA Administrator--these two space-dedicated leaders were to shape the course of subsequent events for the Kennedy administration.[14] Both had savvy on the ways of Capitol Hill.

THE DECISIONS TO GO TO THE MOON

Forever into the future it will be asked why Americans in mid-1961 decided to place the first earthmen upon our nearby moon. It was to prove to work out as a series of historic episodes in the long story of mankind. Three Apollo astronauts voyaged around the moon in December 1968, and Armstrong and Aldrin walked on its surface on July 20, 1969, while ten more astronauts worked on the lunar surface by 1972. Many other explorations of our solar system by instrumented spacecraft as well as practical benefits for life on earth were to transpire. But the feats of Apollo were not to be repeated by anyone up to the present moment. In American history, Columbus sailed the ocean blue in 1492, while in 1969, Neil Armstrong "leaped for all mankind."[15]

Here space is only feasible to review again the involvement of President John F. Kennedy in space affairs in the wholesale spectrum of White House concerns, leading to his decision to recommend to the Congress on May 25, 1961, that the United States should fly men to the moon "before this decade is out." This recommendation became a commitment of the Congress and was endorsed by the American people without debate in 1961. One Congressman muttered that they had better find gold on the surface of the moon to pay for the effort, but H. R. Gross of Iowa said similar things about the cost of many Federal programs. There had been many decisions made before the Kennedy administration which made a lunar program recommendation feasible technologically. There were many subsequent decisions which made possible the carrying out of the Apollo enterprise--the "race to the moon" debates with budgetary cuts in 1963, the growing war tragedy in Southeast Asia, and social discontents all concurrent with the fulcrum of the thermonuclear balance of world power during the 1960's aroused second thoughts.

Much has been written about the "decision" to go to the moon in the memoirs of those close to President Kennedy after he was assassinated in Dallas, and the flood of journalistic accounts after the Apollo missions. Impartial and documented treatment of the historical sequence and context of President Kennedy's decision to recommend the lunar landing has been attempted in two broad-gauge historical essays: my "Historical Perspectives on Apollo," prepared after the shocking Apollo 204 fire in 1967;[16] and, Professor John M. Logsdon's now-classic, *The Decision to Go to the Moon.*[17] Logsdon interviewed most principals around President Kennedy just before *Apollo 11.* The basic history seems fairly but not wholly complete until the yet-sensitive documentation might become available that surely must exist in White House records. Historians can never assume that everything is fully known concerning space policy, technology, and politics of the past three decades. But much we know.

Here it is submitted that President Kennedy's innate sense of history in the making was the essential ingredient making for his decision to go to the moon after the world-circling space flight of Soviet Major Yuri Gagarin of April 12, 1961. It was a political catalyst for taking a responsive action offering an opportunity to help fashion the future. A technological and organizational feasibility seemed to exist for a lunar landing by 1970, if not sooner, and psycho-political sanction was estimated for its viability by his advisors including members of the Congress, some scientists, laymen, NASA and Department of Defense spokesmen. The success of the Mercury-Redstone vertical space flight of Alan Shepard on May 5, 1961, dispelled all major doubts as his address to the Congress presenting his Apollo recommendation set the entire Apollo endeavor into motion nation-wide.

No one has yet said for certain precisely when John F. Kennedy decided Americans would fly to the moon for it was a process of cumulative decisions and actions. Historically it was essentially made before the Shepard flight, and was long before the first American orbited in February 1962. It was made before the moon rocket had been designed and before the mode of its landing men on the moon had been determined. And in the process, even the debacle known as the "Bay of Pigs" [April 17-19, 1961] and the bone chilling confrontation with Soviet Chairman Nikita Khrushchev in Vienna [June 3-4, 1961] did not slow the momentum for Apollo in the White House. Eisenhower on his farm in Gettysburg voiced displeasure with its priority, and, after Dallas, it was a Kennedy promise which Lyndon B. Johnson was to see to it that it was fulfilled. But in a historic sense it turned out just as President Kennedy recommended to the Congress on May 25, 1961: "It will not be one man going to the moon--if we make this judgment affirmatively it will be an entire nation."

Because Kennedy's decision to race openly to the moon was inspired by his sense of history in the making, there was never any doubt that it was a NASA job. The Department of Defense had more immediate and more expensive requirements. Once the national commitment to go to the moon was made, however, it was to prove even more historically significant that the Apollo goal was attained eight years later. There were other commitments such as active involvement in Southeast Asia, and crises in Berlin and about the missiles in Cuba. But the national decision set in motion by Yuri Gagarin's space flight, and requested of Americans by John F. Kennedy, was fulfilled.

Unlike any previous new President, John F. Kennedy was perforce gusted by on-going perturbations of the Space Age. He chose to act, and to appear, as a full-time functioning chief executive. It became his style. But on space matters he had to rely on those more expert, as shown by the release of the Wiesner Report and the selection of the NASA Administrator. His

PRESIDENT KENNEDY ADDRESSES JOINT SESSION OF CONGRESS

May 25, 1961

(NASA Photo No. 70-H-1075)

promise to make Vice President Lyndon Johnson the chairman of the National Aeronautics and Space Council in the White House, however, was not submitted to the Congress for legislative action until April 10, 1961. In the meantime, external events related to space intruded. And Kennedy rarely neglected the morning papers.

On his inauguration day it was announced in Moscow that *Strelka*, one of two female dogs safely recovered from orbit in August 1960, had given birth to six puppies. One of the pups would eventually romp on the White House lawn. But President Kennedy became involved front and center on space in his weekly news conferences and formal exchanges with the Kremlin on opportunities for cooperation in space. At the same time, the Soviet Premier rattled his nuclear-tipped ballistic missiles for European and U.S.S.R. domestic consumption. Most important were the pending budgetary questions anchored on accepting or changing the Eisenhower space estimates for the Department of Defense and NASA for FY 1962. Such were prelude to a climax on consideration of the manned space program to follow Project Mercury before the Kennedy administration was 100-days old.[18]

The so-called "space race" trended disturbingly. It was never neglected by the news media. On February 4, 1961, the Soviet Union launched a dummy cosmonaut, which was not to be successfully recovered. At his next news conference, Kennedy was asked about the U.S. man-in-space plan. He replied: "We are very concerned that we do not put a man in space in order to gain some prestige and have that man take a disproportionate risk..., even if we should come in second in space." The President's Science Advisory Committee (PSAC) had already renewed its inquiry on the reliability of Project Mercury, including concerned life scientists for the unprecedented launch and weightlessness stress on astronauts. On February 12, Kennedy dispatched congratulations to Premier Khrushchev for the launching of a space probe to the planet Venus. Khrushchev replied with appreciation for Kennedy's words on "this outstanding achievement of peaceful science and for wishes for success in the new stage of the exploration of the cosmos." Khrushchev noted Kennedy's State of the Union call for applying science for the benefit of mankind and for the settlement of the problem of disarmament, welcoming further discussions.[19] In his news conference the same day, Kennedy repeated his hope for peaceful relations with the U.S.S.R. He added that on "boosters we [the United States] are behind and it is a matter of great concern... unless we are able to make a breakthrough before the Saturn booster comes into operation.... We have sufficiently large boosters to protect us militarily, but for the long, heavy exploration of space, which requires large boosters, the Soviet Union has been ahead and it is going to be a major task to surpass them."

By the time James Webb had been sworn in as NASA Administrator, Defense Secretary Robert McNamara had already been functioning almost a month. The so-called "missile gap" had been discounted as the mobile Polaris and the siloed Minuteman missiles were on schedule. It had been announced on January 25th that the USAF had selected the Titan II rocket as a booster for the Dyna-Soar manned orbital winged-reentry vehicle before Webb had been nominated. The Air Force had also achieved successful orbiting of a 4,100-lb test satellite carrying photographic equipment in January. Space detection and tracking systems were placed on operational status in Colorado Springs in February.

For NASA the big event was the successful flight test of the Mercury-Atlas (MA-2) on February 21. With the new Administrator in office just a week, NASA had to take full responsibility for the "belly band fix" for the mating of the Mercury capsule to the thin-skinned Air Force rocket.[20] Two days later, NASA and DOD signed a formal understanding which reconfirmed the National Launch Vehicle Program, one calling for an integrated development and procurement of booster rockets. Neither DOD nor NASA would initiate the development of a large launch vehicle or booster without the written authorization of the other. On March 6, Secretary McNamara assigned all "research, development, test, and engineering of Department of Defense space development projects to the U.S. Air Force. Three days later, the U.S.S.R. launched *Spacecraft IV* into orbit, and successfully recovered a dog and other animal passengers--the second time they had accomplished this feat. It was an operation repeated by the Soviet space program on March 25, 1961. "Space race" portends were animated in the public press as well as in official channels.

Chairman Overton Brooks of the House Committee on Science and Astronautics sent President Kennedy a blunt letter on March 9. His committee had requested NASA, during the Eisenhower administration's last months, to prepare a plan for landing men on the moon. Brooks said that the Wiesner Report released on January 11 was the only statement of space policy existent. The abridged version, with its Pentagon recommendations deleted, had been overly critical of NASA, which was soon to present its request for funds before his committee. Was it President Kennedy's intent to downgrade NASA while increasing the space role of the Department of Defense? This letter was not answered for almost two weeks as the White House had to develop a reply.[21] This required Kennedy's approval.

While we are waltzing here concerning the historical momentum a-building for consideration of a voyage to the moon, several basic realities facing the White House should be noted. The President's science advisor, Jerome Wiesner, had closest access to him. Wiesner's job, sometimes one of being the devil's advocate, was providing guidance on all science and technology issues. He was aided by PSAC, most members of which were physicists, and historians have yet to have access to their papers. In view of the general opposition of scientists to manned space flight--even a few in NASA--Administrator Webb professed being surprised and pleased with a report of the Space Science Board of the National Academy of Sciences on February 27. Chaired by Lloyd Berkner, the SSB report said:

> Scientific exploration of the Moon and the planets should be clearly stated as the ultimate objective of the U.S. space program for the foreseeable future. This objective should be promptly adopted as the official goal of the United States space program and clearly announced, discussed, and supported. In addition, it should be stressed that the United States will continue to press toward a thorough understanding of space, of solving problems of manned exploration, and of development of application of space science for man's welfare....[22]

A copy of this report was formally sent to the White House late in March, but its position must have been known by Academy members serving Kennedy.

Wernher von Braun, who had championed the rocket's objective for the goal of lunar flight since 1945, was now a familiar voice on the NASA team. He had unbelievable influence on Capitol Hill. The giant Saturn rocket, an Army orphan before becoming a NASA asset, was, with liquid-hydrogen upper stages, the first American large launch vehicle suitable for mounting a space station or a circumlunar manned mission. For its part, the U.S. Air Force and its contractors had prepared studies and plans for manned lunar missions. It also had priority instrumented observation satellite systems in operation. It had prime responsibility for strategic weapons development to keep the strategic nuclear deterrent viable with new generations of intercontinental bombers and intercontinental missiles. The U.S. Navy mainly had Polaris and a high interest in passive support space systems, and a few space cadets. But the military services were responsible directly to the Secretary of Defense for their funding, and their cause was voiced at the White House also by the chairman of the Joint Chiefs of Staff on the National Security Council. In contrast, as an independent and open agency, NASA, with its Space Act charter, could seek to mobilize resources nation-wide--military and civilian in government, industry and academe, and internationally--necessary for the achievement of space goals for their own sake. Pride in the United States and prestige of the United States in the eyes of the world presented unequal arguments on policy and goals between the advocates of a military or a civilian aegis for the space program. Even General Eisenhower had frequently said after the sputniks, "We have no enemies on the Moon." President Kennedy and NASA leaders were forced together to do whatever, if anything, should be done with Eisenhower's very minimal expansion funding for non-military space.[23] The "space race" did not go away any more than had the sputnik impact for Ike.

NASA had been upgrading its Ten Year Plan since the inter-regnum which would underwrite its presentations to the Congress on its FY authorization. It was a broad statement of missions in all classes, which need not be detailed here. In particular, the post-Mercury manned program, named Apollo since July 1960, now aimed for development of a lunar landing capability by the late 1960's as a "prime NASA goal." The Eisenhower budget for FY 1962, in brief, had crimped NASA's lunar mission technology development. NASA's testimony before congressional committees in February 1961 described Apollo both as an earth-orbiting laboratory and as a program for circumlunar flight that could lead to a manned lunar landing. Large launch vehicles were the pacing factor. The F-1 single-chamber million-pound thrust engine was key for a direct ascent, while the Saturn C-2 with liquid-hydrogen upper stages made feasible the circumlunar manned mission, an earth-orbital rendezvous of multiple launches for a lunar landing, and a space station potential. The bottom line was the FY 1962 budget, hopefully something beyond the Eisenhower level.[24]

In mid-March NASA Administrator Webb made his first public statement at the annual Goddard Memorial Dinner in Washington. He reported that NASA "is hard at work" on the on-going programs, and was "proceeding to a thorough examination of the present validity of the ten-year program worked out last year.... [This evaluation] will go forward without delay, and I feel sure that the President will submit any changes which he believes necessary in time for consideration by the House Committee on

Science and Astronautics during its present hearings on the Eisenhower budget." At the adjunct meeting of the American Astronautical Society, Webb added:

> Today in 1961 we stand before the frontier of space... and I have been giving much attention to the plans that should be made for continuing manned exploration of space beyond Project Mercury. We are giving careful consideration to the rate at which we should proceed.

On March 21, NASA administrators made extensive presentations to the director of the Bureau of the Budget and other members of the White House staff. The Bureau's director, David Bell, insisted that it was not at all certain that President Kennedy would support any additional monies for NASA's FY 1962 funding. Deputy NASA Administrator Dryden strongly stated that historical events would force presidential support for accelerating NASA's program, and also indicated that whether or not Mercury missions were launched depended upon the program directors who best knew the reliability of the equipment, and who would certainly not take undue risks. Webb, who had served as the first director of the Bureau of the Budget in the Truman administration, said that he would have to take NASA's case to the President. It was scheduled for the next day, March 22, 1961.

Before meeting with President Kennedy, NASA's James Webb, Hugh Dryden and Robert Seamans briefed Vice President Johnson on their eleven requests for $308 million supplemental funding. Requests stressed propulsion and Saturn development to "increase the rate of closure on the U.S.S.R. lead." It was followed by a rejoinder by BOB director Bell and his staff with reasons for opposing the NASA requests. After this, President Kennedy, McGeorge Bundy, Jerome Wiesner, and the AEC chairman Glenn Seaborg joined the group with the Vice President in the Cabinet Room. Dr. Edward C. Welsh, whom the Vice President had selected as executive secretary of the National Aeronautics and Space Council (NASC), was also present. It was an intensive two-hour conference.

As general manager of NASA, Dr. Seamans named the NASA programs for which the supplemental funding was requested: move the orbital date of Apollo from 1967 to 1965, accelerate development of the advanced Saturn C-2 so that circumlunar flight could be accomplished in 1967 instead of 1969, develop the 1.5 million-lb-thrust F-1 engine for Nova direct ascent to the moon for a landing by 1970 instead of 1973, and develop a prototype nuclear upper stage for lunar-base operations. Hugh Dryden outlined the flexibility of Apollo for a two-week crew mission in an orbital laboratory or a lunar landing flight by 1970. As was his practice for critical policy moments, Administrator Webb had personally invested many hours preparing his wholesale review of a full-spectrum role of space in the national interest, very much a strategic appraisal. Webb explained how the present situation had developed, and the rationale for implementation of NASA's revised ten year plan. The full text of the talking paper Webb used was first made available in Professor Logdon's *The Decision to Go to the Moon.*[25] Webb pointed out the distinctive, yet technologically meshing, NASA and DOD programs. "The first priority... should be to improve as rapidly as possible our capability for boosting large spacecraft." NASA needed the large boosters before the military "in order to achieve a number of major space exploration milestones." "Eisenhower had eliminated from his budget the preliminary design studies to begin the effort." A year had already been lost on Apollo.

President Kennedy listened intently, asked questions freely, and discussed what the decisions he was being asked to consider would involve. He made no decisions at this meeting. Kennedy called a meeting on March 23, the next day, which included Vice President Johnson, Welsh, Wiesner, and Bell. Johnson and Welsh urged approval of NASA's request, and particularly the large booster requirement, the latter also being supported by Wiesner. As it turned out, President Kennedy did not approve Project Apollo ($42.6 million requested) or any change in policy for which public and Congressional support need be calculated. He approved $56 million for Saturn development, $25.6 million for Centaur vehicle development (which included liquid-hydrogen engines), $9.3 million for propulsion R&D (F-1 engine), and $4 million for nuclear propulsion, and $19.2 million for launch and test facilities. This brought the Kennedy add-on of $125.7 million to the Eisenhower budget, to a total NASA figure of $1.235 billion. It did speed up the booster pacing, which for the Saturn C-2 did move up proposed Apollo milestones, including a lunar landing in the 1969-70 time period. In short, President Kennedy's March 23rd decision was one in principle for a post-Mercury manned flight program aimed at the moon. NASA's FY 1963 budget would permit the President to gain greater ken on the rate and the scale of the American space inheritance he now presided over, as well as greater confidence in the NASA push backed by the Vice President and the Congressional leaders. Kennedy had other crises.

The historical signficance of Kennedy's initial space decisions seems often overlooked in hindsight. The Soviet indicators were clear enough, and he had heard nothing favorable about Project Mercury although the breadth and thrust of NASA's presentations for its budget additions certainly had a sense of historic urgency to capture some of the future in space achievement. By endorsing the funding to accelerate the Saturn, C-2 development, NASA Leaders, perhaps optimistically, seemed to feel that Mr. Kennedy was "tending toward approval of Apollo" at this time.[26]

Two follow-on decisions were made after Kennedy's first session with NASA administrators on the FY 1962 supplemental. First, on March 23, the nomination of Dr. Edwin C. Welsh as executive secretary of the National Aeronautics and Space Council in the White House, went forward to the Senate. A seasoned Washington hand, Welsh had been an assistant to Senator Stuart Symington for the beginnings of space actions on Capitol Hill. Welsh's first task was to draft the proposed legislation authorizing that the Vice President preside over the NASC which had been promised in December 1960. As it turned out, this executive role for Lyndon Johnson became law on April 20, which proved most timely. An economist by background, Dr. Welsh meanwhile proved a wise space counsel who served two presidents.[27]

Also on March 23, President Kennedy answered the March 9 letter of Overton Brooks, chairman of the House Science and Astronautics Committee. He said: "It is not now nor has it ever been my intention to subordinate the activities of NASA to those of the Department of Defense.... [There are] legitimate missions in space and the application of space technology to the conduct of peaceful activities, which should be carried forward by the civilian space agency...." The House space committee was thus confirmed in its purview over NASA, and it accepted the White House add-on to the FY 1962 budget.[28] NASA witnesses were pressed to explain unfunded needs.

While the U.S.S.R. launched and recovered its third spacecraft populated with animal passengers on March 25, Project Mercury was driving to launch its first astronaut on a Mercury-Redstone as soon as feasible. PSAC's inquiry on Mercury, particularly the panel on life sciences, remained highly concerned about the lack of blood pressure sensors. Extra centrifuge runs by selected astronauts, Shepard, Grissom and Glenn, Soviet animal demonstrations, and X-15 pilot interviews were to no avail. By early April, the panel was recommending that 50 centrifuge runs with chimps be conducted prior to any Mercury space flight. It could not accept the norms and projections of the effects of space flight for conditioned astronauts by aerospace medicine specialists. A key NASA official said that if such an invalid report was submitted it should be published on the front page of the *New York Times* so PSAC would get full credit. Mercury director Robert Gilruth declared that they would have to move the entire project to Africa for enough chimps.[29]

President Kennedy was advised early in April that a Soviet man would make a space flight by April 15. On April 10, while rumors that a Soviet cosmonaut had or would be launched into space swept from the streets of Moscow to the Washington newspapers, the White House submitted to the Congress the draft amendments making the Vice President the chairman of the National Aeronautics and Space Council. The next day, press secretary Pierre Salinger was advised to prepare suitable statements, and that evening Kennedy was advised that the Soviet manned space flight would take place that night. It did.

On April 12, 1961, an unknown major of the Red Air Force circled the Earth in space. The Kremlin soon made known that his name was Yuri Gagarin.

KENNEDY'S PERSONAL DECISION: "THERE'S NOTHING MORE IMPORTANT"

The historic space flight of Yuri Gagarin on April 12, 1961, was no technological surprise in the White House. Nor was it an unpredicted blow to American prestige abroad as President Kennedy had warned at his press conferences repeatedly. The Kremlin's word warfare was instantly evident--"Let the capitalistic countries catch up with our country," Khrushchev said when congratulating Gagarin by telephone. A Soviet postage stamp commemorating the event was already on sale in Washington, D.C. Despite Kennedy's warnings about being second for some time, the American public and the Congress seemed shocked and ill-prepared for the dramatic implications of the first flight by a man in space. The sputnik syndrome was reincarnated in another news media orgy.[30]

What were John F. Kennedy's immediate reactions, officially in the White House as well as in his innermost thoughts? You will find very little indeed in all the memoirs and other books written after Dallas, and particularly in Arthur Schlesinger's "history" of *A Thousand Days* of the Kennedy administration.[31] In whom would John Kennedy confide any gutsy or heady reactions about Gagarin's symbolic feat right after it happened? His closest confident, Robert Kennedy, had no words for space, even long after Dallas. Of the later chain of events leading to the national decision process to go to the moon, much has been written. What did the CIA or the JCS have on space goals to suggest before both lost some White House clout at a Bay of Pigs, April 17-19, 1961? President Kennedy's science advisers on Mercury gave it a "high risk" go ahead, with medical doubts, on the very day of Gagarin's flight. "Shorty" Powers, the voice of Project Mercury, was a real patriot when asked about the Russian manned mission by a newsman, "We're all asleep down here." So Kennedy had questions, and undoubtedly resolved rather earlier than later that he had to get atop the space problem.

In the campaign to get elected, which candidate had used a "missile gap" and the need to get the country moving again, perhaps on the "space frontier" to win the nomination and, by a narrow margin, the White House? Who now sent the Eisenhower-sounding congratulations to Nikita Khrushchev on April 12? "The people of the United States share with the people of the Soviet Union their satisfaction for the safe flight of the astronaut in man's first venture into space.... It is my sincere desire that in the continuing quest for knowledge of outer space our nations can work together to obtain the greatest benefit to mankind." Who now issued a White House release for Americans?

> The achievement by the U.S.S.R. of orbiting a man and returning him safely to ground is an outstanding technical accomplishment.... The exploration of the solar system is an ambition which we and all mankind share with the Soviet Union and this is an important step to that goal. Our own Mercury man-in-space program is directed toward that same end.

Who had said before his election that when man reached the moon "his name will be Ivan"? John F. Kennedy's quick memory and his acute political instincts, his love of competition, debate, and winning, his intellectual and his managerial sense of history--these unsimple virtues were to be increasingly evident in the course of space history.

At his regular press conference on the afternoon of *Vostok 1*, President Kennedy responded to a Congressman's earlier reaction, when he said: "No one is more tired than I am [of seeing the United States second to Russia in the space field]. The news will be worse before it is better... We are, I hope, going to go in other areas where we can be first, and which will bring more long-range benefits to mankind. But we are behind." At this moment there was no projection of NASA's space flight milestones.

All bucks ended in the Kennedy White House. They were neither partisan nor personal as they had been for Eisenhower after the sputniks or after the U-2 episode in May 1960. NASA was apparently directed to hold a news conference on the Russian first, an unprecedented step to deflect some of the news media pressure. It was a sputnik crisis all over again except there was now a space agency. Let NASA field the questions, although there was very little known about the Vostok details. Administrators James Webb and Hugh Dryden dutifully explained to a packed news conference the significance of President Kennedy's supplemental to the Eisenhower FY 1962 budget--there was $28 million for large boosters serving future missions in the $126 million add-on, and advanced planning beyond Mercury was proceeding in NASA. With regard to the possible impact of the Gagarin mission on the Mercury program, Dryden laconically said he could not forecast what the ultimate impact would be: "As you know, our [NASA] programs are determined by the democratic process in this country, in very many forums, including the Executive Branch of the Government and the Congressional Branch." NASA spokesmen were very cautious in supporting its supplemental money now before the House Committee.

The next day, April 13, NASA witnesses were called to a special hearing of the House Committee on Science and Astronautics to satisfy its questions. It was not to be the only time that a Russian space spectacular boosted NASA's budgetary stock on Capitol Hill. But this hearing was, as this attendee can testify, electrified by intense concern about the nation's future history in space. It was focused on what to do rather than first find out what happened, as concern arose after sputniks. Chairman Brooks asked Administrator Webb whether NASA's long-range plan called for "catching up with the Russians?" Webb replied carefully that the President has made the fundamental decision to go ahead with the large rockets to carry vehicles including Apollo. Certainly you can proceed faster than the funds recommended by President Kennedy, if the Congress should wish to increase NASA funding. Without answering the question Webb stated: "I will take every step necessary to carry out the decision that is made and do everything possible to realize the maximum rate of progress from the resources provided." The NASA administrator carefully supported the White House position, and cautioned that the decision before the nation is "whether we now expect to proceed as we did in connection with the atomic bomb, with a substantial number of efforts going on in parallel, with all the resources that may be required to do this." Webb offered no false hope, saying in 1967 or 1968, with an all-out effort, the Soviet lead in large boosters might be matched.[32]

Congressman James Fulton specifically asked: "How much could the lunar landing be accelerated and how much would it cost?" Webb would not comment on the pace of acceleration. But on costs, he replied: "There are very large numbers that people use. Some people use a number of $40 billion to land a man on the moon. Others say half the amount." One can be certain that every word that NASA said on Capitol Hill was thoroughly digested at the White House, and front page and editorially in all newspapers.

The following day, April 14, when Yuri Gagarin was wildly greeted on Red Square in Moscow, NASA's budget hearings resumed before the House Committee. Associate NASA Administrator Robert C. Seamans, closest to program planning, became a star witness. The key question was asked by young Congressman David S. King: "I understand the Russians have indicated...their goal is to get a man on the moon and return him safely by 1967, the fiftieth anniversary of the Bolshevik Revolution. Now specifically, I would like to know, yes or no, are we making that specific target date to try to equal or surpass their achievement?" Seamans reply was that the U.S. target date was 1969 or 1970, about which King opined that "as things are now programmed we have lost." Could the 1967 date be met? Seamans replied: "This is really a major undertaking. To compress the program by three years means that greatly increased funding would be required.... It would cost $4 to $5 billion a year...." But he said "the goal may very well be achievable."

Congressman Edgar Chenowith of Colorado commented that such a decision would have to be made at a higher level than the House Committee. Seamans responded: "I think it is a decision to be made by the people of the United States." Chenowith disagreed: "The people of this country do not have technical knowledge.... When you talk about putting a man on the moon, they don't know what you are talking about.... I think this would have to be made at the higher level of the administration." NASA's Seamans had been a reluctant witness. He had held to his position that NASA had no plans to ask the White House for more money for FY 1962 after the Gagarin flight than approved on March 23. He had nonetheless been forced to suggest, Congress, the White House, and Americans willing, the landing of an American on the moon by 1967 was a goal that "may very well be achievable." There were many unstated assumptions. Meanwhile, at the same time, in Moscow, Nikita Khrushchev was uttering words honoring Gagarin without constraint on the glory of Soviet science and technology in the future: "In forty-three years of Soviet government, formerly illiterate Russia...has traveled a magnificent road.... This victory is another triumph of Lenin's idea.... This exploit marks a new upsurge of our nation in its onward movement toward communism." No U.S. Information Agency polls of the impact of the Gagarin flight on world opinion would be needed to reflect the new dimensions of the symbol of space achievement in world politics. NASA administrator Webb had already cleared with President Kennedy, on April 12, the first National Conference on Space to be held in Tulsa, Oklahoma, in May. Other portions of their White House talk are not yet fully known to this present day.

President Kennedy's decision to review the implications of the predicted, yet unprecedented, space flight of Yuri Gagarin led to a historic meeting with his key space advisors on April 14. He had concluded his discussions concerning the problem of the survival of a free Berlin with visiting Chancellor Konrad Adenauer. First, it was NASA spokesmen who had fielded the questions of newsmen and responded to the congressmen provoked to verbal action by the impact of the latest Soviet space feat. Yet the Gagarin flight which predominated in the outpourings of the news, now commanded his attention. Not yet ninety days in office, he had inherited a space problem not easily resolved. If he was not a leader and could find some way to ameliorate the Soviet space challenge, would he be plagued henceforward for appearing indecisive? Perhaps something other than NASA's technological milestones leading to a lunar landing could be offered? The Soviet's acknowledged lead in large rocket boosters for space was impressive. Mercury had been destined for a second-place start, which was now a fact. But there was an even chance to land an American on the moon in the 1967-68 period. This goal, if it could be afforded and achieved certainly would be a dramatic historic event. Kennedy had gone over this in some detail in the NASA budget session on March 23. And he had already approved NASA's Saturn C-2.

A meeting was set up for President Kennedy on the evening of April 14 with NASA administrator Webb, his deputy Dryden, science adviser Wiesner, budget director Bell and Theodore Sorensen. It proved historic not only that President Kennedy asked questions and reviewed once again a lunar landing target as a long-range strategic space goal. Additionally, this pivotal meeting was, in a recapitulation finale, to include a prominent newsman as an outside witness. Kennedy's calculated tactic of revealing an image of his personal concern for initiating consideration of a national decision for going to the moon, appeared in *Life* magazine on the newstands in a few days. It was a personal position that could not be erased for Hugh Sidey's column on the presidency in *Life* of April 21, 1961, was earlier available in hard type for millions of readers just as the CIA-managed invasion of Cuba began.[33] Once taken up with the political impetus of going for the moon on April 14, it appears that John F. Kennedy stayed with it until Dallas, despite some second thoughts along the way. And maybe on April 14 he merely wanted some diversionary cover?

In the Cabinet Room on the evening of April 14, Wiesner, Bell, and Sorensen met with Webb and Dryden to review "the next steps in the space race," according to Sorensen. While awaiting the President, they reviewed major parameters of acceleration. The pacing factor was propulsion and large booster development: the Saturn C-2 underway offered an orbital lab and a 1967 circum-lunar mission; the F-1 for a giant Nova rocket and the Air Force's large solid-propellent rocketry offered a new technology, and nuclear propulsion of the Rover type was not infeasible for the far distant future. PSAC's separate analysis of the large launch vehicle projection, even without reliable estimates of the Soviet development, would take three or four months to complete. The high $40 billion estimate to go for a lunar landing with parallel developments would be additional to all expanding defense estimates, now nearly at $50 billion for FY 1962 as compared to NASA's $1.3 billion. Project Mercury's up-and-down Redstone manned test flight looked likely in several weeks for what that was worth. Joined by President Kennedy, members of the group explained their estimates on how the gap in space technology appeared to them and what could be done. NASA had a plan with a timetable, with options dependent upon the extent to which parallel launch vehicle developments could be mobilized in a concerted effort. Hugh Dryden pointed out that even spending the high $40 billion to race for a lunar landing offered only an even chance, even if the attempt were made to leapfrog the Russians on launch vehicle development for a lunar landing by the 1967-68 period.[34]

Unbeknownst to Webb and Dryden, the President had also arranged for a not-for-attribution interview that evening with Hugh Sidey of *Life* magazine. Kennedy asked Sorensen to bring Sidey into the Cabinet Room. Kennedy reopened the discussion with a historical setting: "As I understand it, the problem goes back to 1948, when we learned how to make smaller warheads that could be carried with smaller boosters...."[35] Each of the participants was encouraged to restate his thoughts and recommendations with a reporter present concerning the space futures.

It appeared, as Sidey reported, "a discouraging picture of years and billions of dollars that separate the United States and Russia in space."

We may never catch up," Kennedy muttered, To Sidey, the President accepted the Soviet space challenge by his questions.

> Now let's look at this. Is there any place we can catch them? What can we do? Can we go around the moon before them? Can we put a man on the moon before them? What about Rover and Nova? When will Saturn be ready? Can we leapfrog?

Sidey reported: "The one hope, explained Dryden, lay in this country's launching a crash program similiar to the Manhattan Project. But such an effort might cost $40 billion, and even so there was only an even chance of beating the Soviets." In constrained words for *Life,* James Webb said: "We are doing everything we possibly can, Mr. President. And thanks to your leadership we are moving ahead now more rapidly than ever." But Kennedy said, according to Sidey: "The cost, that's what gets me," and Bell responded with his geometric projection of space costs. And, Wiesner cautioned, "now is not the time to make mistakes...."

Sidey concluded his reportage, as follows:

> Kennedy turned back to the men around him. He thought for a second. Then he spoke. "When we know more, I can decide if it's worth it or not. If somebody can just tell me how to catch up. Let's find somebody--anybody, I don't care if it's the Janitor over there, if he knows how."
>
> Kennedy stopped again a moment and glanced from face to face. Then he said quietly, "There's nothing more important."[36]

If President Kennedy's purpose of the second meeting including Hugh Sidey was to plant a story, he was successful. Sidey reported in *Life* that the President was "gravely concerned" about the implications of the Gagarin flight, and that he recognized "it was most urgent [--more] than ever to define U.S. space aims."[37] In his biography, cleared by Kennedy in 1962, Sidey wrote: "alone with Sorensen [after the meeting] he thought about the curious dilemna further. The cost was frightening. Yet the threat was there.... To Kennedy it was inconceivable that there was no way to accept the challenge and win the race if it was worth it and the country wanted to do it. 'I am determined to get an answer,' he said."[38]

Just recently, upon the tenth anniversary of *Apollo 11,* Hugh Sidey added a note to his earlier reflection of Kennedy's views on the evening of April 14, 1961, as follows:

> Out of his chair, Kennedy thanked the men for coming over, beckoning Ted Sorensen to follow him. In five minutes Sorensen came out [of the oval office]. "We are going to the moon," he said simply.[39]

As Sidey left Sorensen to depart, President Kennedy standing in the doorway of his office asked if he had his story. Sidey nodded affirmatively.

Of this meeting in mid-April 1961, Theodore Sorensen has also written that President Kennedy "was more convinced than any of his advisers that a second-rate, second-place space effort was inconsistent with this country's security, with its role as a world leader, and with the New Frontier spirit of discovery." He was as close to Kennedy as anyone.[40]

Hugh Dryden confided to one of his long-time associates that it appeared to him that James Webb's campaign to outline a $40 billion program as feasible for NASA was a little frightening: "He's talking about leveling out the [NASA] program at six billion dollars a year. And I think he's sold the President."[41] In NASA, Apollo was always a part of the whole.

And Professor John Logsdon in his careful tracing of the decision to to go the moon, concluded:

> He [Kennedy] reluctantly came to the conclusion that, if he wanted to enter the duel for prestige with the Soviets, he would have to do it with Russians' own weapon, space achievement. He found that there was no other means available to counter....[42]

What is more clear is that Kennedy's commitment to the lunar landing space goal was confirmed with each subsequent step in seeking some confirmation of its political feasibility with the Congress and the American public as well as seeking answers to his questions by way of gaining more confidence on NASA's projections. In fact, NASA itself would recommend that the lunar landing goal should be targeted for 1969 rather than 1967 because of naked chronology. President Kennedy's early personal response to the space challenge of the Soviet Union in a space endeavor survived a red-hot chain of strategic crises posed by Khrushchev's thermonuclear missile exchange threat in Vienna in June 1961 and the Cuban missile standoff in 1962. There was no linkage of the race to the moon posed by the Bay of Pigs several days after Kennedy's personal commitment. It was a CIA disaster and for others involved but, as McGeorge Bundy characterized it, it was "a brick through a window," no more.[43] The strategic implications of a space race derived from its openness, not a secret and closed effort like the Manhattan Project. Though some lunar technology developed might serve military as well as peaceful uses, the linkage of space technology to strategic purpose was already being served by the instrumented satellites of the U.S. Air Force.[44] This only enforced the rationale that going to the moon was clearly a job for NASA.

President Kennedy followed up on his personal space race decision of April 14, with a discussion on April 19 with Vice President Johnson and NASC secretary Edward Welsh on what to do next. Budget director Bell had already gone on record with Kennedy that, regardless of increases in the NASA budget, it appeared a "tail chase" in closing the "weight-lifting gap" with the Russians.[45] Science adviser Wiesner, who had been suggesting other science projects augmenting national prestige since January, now had PSAC studying the large booster problem involving both NASA and DOD programs. Vice President Johnson now recommended to President Kennedy that the National Aeronautics and Space Council function be activated to sort out the technological, managerial, and political feasibilities of an accelerated space program. There were basic questions to be answered and some consensus could be developed, should the NASC get into the action.[46]

On April 20, the day Congress passed the legislation authorizing the Vice President to chair the National Aeronautics and Space Council (NASC) in the White House. Kennedy submitted a memorandum with a series of questions on accelerating the space program to the Vice President.[47] These questions, reflecting the discussion the previous day, were apparently dictated by the President as no White House staffer claims their authorship.[48] Knowing Johnson was 1000% for space revealed Kennedy's commitment.

Vice President Johnson, who had embraced space issues fully since the sputniks, swung into action with his new White House task, He corralled all the most informed witnesses to answer formally the questions posed by President Kennedy, with both a sense of urgency as well as Lyndon Johnson's characteristic intensity. Professor Logsdon interviewed most of these witnesses in this interesting process.[49] By April 28, Vice President Johnson submitted a preliminary report to President Kennedy. Since it has only recently become available, our extended coverage of the decision to go the moon in 1961, may as well be served by the full text.[50] It should also be noted that it was not until April 25, that the President signed the legislation making the Vice President the chairman of the National Aeronautics and Space Council. And Mr. Johnson invited in many witnesses and monitors to his "hearings" including outside businessmen, select members of the Congress, and representatives of the Bureau of the Budget.

OFFICE OF THE VICE PRESIDENT

MEMORANDUM FOR THE PRESIDENT April 28, 1961

Subject: Evaluation of Space Program

Reference is to your April 20 memorandum asking certain questions regarding this country's space program.

A detailed survey has not been completed in this time period. The examination will continue. However, what we have obtained so far from knowledgeable and responsible persons makes this summary reply possible.

Among those who have participated in our deliberations have been the Secretary and Deputy Secretary of Defense; General Schriever (AF); Admiral Hayward (Navy); Dr. von Braun (NASA); the Administrator, Deputy Administrator, and other top officials of NASA; the Special Assistant to the President on Science and Technology; representatives of the Director of the Bureau of the Budget; and three outstanding non-Government citizens of the general public: Mr. George Brown (Brown & Root, Houston, Texas); Mr. Donald Cook (American Electric Power Service (New York, N.Y.); and Mr. Frank Stanton (Columbia Broadcasting System, New York, N.Y.).

The following general conclusions can be reported:

<u>a</u>. Largely due to their concentrated efforts and their earlier emphasis upon the development of large rocket engines, the Soviets are ahead of the United States in world prestige attained through impressive technological accomplishments in space.

b. The U.S. has greater resources than the U.S.S.R. for attaining space leadership but has failed to make the necessary hard decisions and to marshal those resources to achieve such leadership.

c. This country should be realistic and recognize that other nations, regardless of their appreciation of our idealistic values, will tend to align themselves with the country which they believe will be the world leader -- the winner in the long run. Dramatic accomplishments in space are being increasingly identified as a major indicator of world leadership.

d. The U.S. can, if it will, firm up its objectives and employ its resources with a reasonable chance of attaining world leadership in space during this decade. This will be difficult but can be made probable even recognizing the head start of the Soviets and the likelihood that they will continue to move forward with impressive successes. In certain areas, such as communications, navigation, weather, and mapping, the U.S. can and should exploit its existing advance position.

e. If we do not make the strong effort now, the time will soon be reached when the margin of control over space and over men's minds through space accomplishments will have swung so far on the Russian side that we will not be able to catch up, let alone assume leadership.

f. Even in those areas in which the Soviets already have the capability to be first and are likely to improve upon such capability, the United States should make aggressive efforts as the technological gains as well as the international rewards are essential steps in eventually gaining leadership. The danger of long lags or outright omissions by this country is substantial in view of the possibility of great technological breakthroughs obtained from space exploration.

g. Manned exploration of the moon, for example, is not only an achievement with great propaganda value, but it is essential as an objective whether or not we are first in its accomplishment -- and we may be able to be first. We cannot leapfrog such accomplishments, as they are essential sources of knowledge and experience for even greater successes in space. We cannot expect the Russians to transfer the benefits of their experiences or the advantages of their capabilities to us. We must do these things ourselves.

h. The American public should be given the facts as to how we stand in the space race, told of our determination to lead in that race, and advised of the importance of such leadership to our future.

i. More resources and more effort need to be put into our space program as soon as possible. We should move forward with a bold program, while at the same time taking every practical precaution for the safety of the persons actively participating in space flights.

As for the specific questions posed in your memorandum, the following brief answers develop from the studies made during the past few days. These conclusions are subject to expansion and more detailed examination as our survey continues.

Q.1 - Do we have a chance of beating the Soviets by putting a laboratory in space, or by a trip around the moon, or by a rocket to land on the moon, or by a rocket to go to the moon and back with a man? Is there any other space program which promises dramatic results in which we could win?

A.1 - The Soviets now have a rocket capability for putting a multi-manned laboratory into space and have already crash-landed a rocket on the moon. They also have the booster capability of making a soft landing on the moon with a payload of instruments, although we do not know how much preparation they have made for such a project. As for a manned trip around the moon or a safe landing and return by a man to the moon neither the U.S. nor the U.S.S.R. has such a capability at this time, so far as we know. The Russians have had more experience with large boosters and with flights of dogs and man. Hence they might be conceded a time advantage in circumnavigation of of the moon and also in a manned trip to the moon. However, with a strong effort, the United States could conceivably be first in those two accomplishments by 1966 or 1967.

There are a number of programs which the United States could pursue immediately and which promise significant world-wide advantage over the Soviets. Among these are communications satellites, meteorological and weather satellites, and navigation and mapping satellites. These are all areas in which we have already developed some competence. We have such programs and believe the Soviets do not. Moreover, they are programs which could be made operational and effective within reasonably short periods of time and could, if properly programmed with the interests of other nations, make useful strides toward world leadership.

Q.2 - How much additional would it cost?

A.2 - To start upon an accelerated program with the aforementioned objectives clearly in mind, NASA has submitted an analysis indicating that about $500 million would be needed for FY 1962 over and above the amount currently requested of the Congress. A program based upon NASA's analysis would, over a ten-year period, average approximately $1 billion a year above the current estimates of the existing NASA program.

While the Department of Defense plans to make a more detailed submission to me within a few days, the Secretary has taken the position that there is a need for a strong effort to develop a large solid-propellant booster and that his Department is interested in undertaking such a project. It was understood that this would be programmed in accord with the existing arrangement for close cooperation with NASA, which agency is undertaking some research in this field. He estimated that they would need to employ approximately $50 million during FY 1962 for this work but that this could be financed through management of funds already requested in the FY 1962 budget. Future defense budgets would include requests for additional funding for this purpose; a preliminary estimate indicates that about $500 million would be needed in total.

Q.3 - Are we working 24 hours a day on existing programs. If not, why not? If not, will you make recommendations to me as to how work can be speeded up?

A.3 - There is not a 24-hour-a-day work schedule on existing NASA space programs except for selected areas in Project Mercury, the Saturn C-1 booster, the Centaur engines and the final launching phases of most flight missions. They advise that their schedules have been geared to the availability of facilities and financial resources, and that hence their over-time and 3-shift arrangements exist only in those activities in which there are particular bottlenecks or which are holding up operations in other parts of the programs. For example, they have a 3-shift 7-day-week operation in certain work at Cape Canaveral; the contractor for Project Mercury has averaged a 54-hour week and employs two or three shifts in some areas; Saturn C-1 at Huntsville is working around the clock during critical test periods while the remaining work on this project averages a 47-hour week; the Centaur hydrogen engine is on a 3-shift basis in some portions of the contractor's plants.

This work can be speeded up through firm decisions to go ahead faster if accompanied by additional funds needed for the acceleration.

Q.4 - In building large boosters should we put our emphasis on nuclear, chemical or liquid fuel, or a combination of these three?

A.4 - It was the consensus that liquid, solid and nuclear boosters should all be accelerated. This conclusion is based not only upon the necessity for back-up methods, but also because of the advantages of the different types of boosters for different missions. A program of such emphasis would meet both so-called civilian needs and defense requirements.

Q.5 - Are we making maximum effort? Are we achieving necessary results?

A.5 - We are neither making maximum effort nor achieving results necessary if this country is to reach a position of leadership.

Lyndon B. Johnson (S)

This historic document, with its assessment and its preliminary recommendations, supported by a coherent rationale, was accepted by President Kennedy as a plan for action to gain space leadership in the 1960's. It was a triggering mechanism for subsequent steps: (1) confirm its logic with more detailed assurances of NASA and the Department of Defense that its technological projections were sound and institutionally feasible; and (2), confirm its political feasibility with the Congress and by the willingness of the American people to support a greatly accelerated program pointed toward landing men on the moon. Vice President Johnson would submit the NASA-DOD position paper to the President on May 8, the so-called Webb-McNamara paper.[51] President Kennedy would personally present his request for endorsement by the Congress and the American people to support the endeavor to land an American on the moon "before this decade is out" on May 25, 1961.[52]

In the meantime, also on April 28, President Kennedy was asked to make another space decision. It did represent a direct linkage to the low state of American prestige in Cold War management symbolized in the CIA-engineered debacle at the Bay of Pigs existent by April 19, for which the President accepted full responsibility. Kennedy was urged by some advisers to postpone or cancel the Mercury-Redstone flight scheduled for May 2, for which Astronaut Alan Shepard had been selected. Others, including members of the Congress urged that the news media buildup at Cape Canaveral dictated that the launching not be open to the press, although the beach watchers well knew neither a success nor a failure could be hidden. Several key advisers were still concerned about the likelihood of a launch failure with an astronaut, comparing it to the Vanguard disaster in the post-Sputnik era.[53] At one meeting with President Kennedy after he had been advised to postpone MR-3, secretary of the Space Council Edward Welsh spoke up: "Mr. President, why postpone a success?" Kennedy asked him if he had such confidence in the mission, and Welsh replied that it was no more dangerous than a flight from New York City to Hawaii in bad weather.[54] Kennedy made no space decision other than proceed with NASA's policy to launch only when ready, its estimate of the relative risks, and to do it in the open before the eyes of the world. He was but the first President not to interfere with NASA's launch schedule. He was also to learn to appreciate the imperatives of space pilot risks and of hardware reliability as well as the laurels for success that he himself would attain and share with the astronauts.[55] Mercury had begun under Ike.

As it turned out, Alan Shepard made his up-and-down Redstone flight on May 5, after several postponements for weather-stretched tensions. NASA was totally unprepared to wire in Kennedy's call from the White House to Shepard after his recovery. And, it was Kennedy who held the first reception for the space heroes in the Rose Garden of the White House on May 8, with thanks for the U.S. beginning. Though an orbital flight of a Mercury astronaut might happen by the end of the year, to match the Gagarin mission if NASA was fortunate, the space flight of Alan Shepard swept away any reservations concerning the space race to the moon that John F. Kennedy may have kept to himself.[56] Moreover, the first U.S. space flight received suprising world-wide acclaim, with only Khrushchev boasting that Americans had yet to accomplish orbital flight. It now remained for the President to ask the Congress for its blessings upon the historic "sporting chance," as von Braun termed it, to race to the moon.

FIRST WHITE HOUSE ASTRONAUT CEREMONY

Alan Shepard receives the NASA Distinguished Service Medal from President Kennedy, May 8, 1961 (NASA Photo No. 61-Adm-13)

The weekend following the Shepard flight, the staffs of the NASA administrator and of the secretary of Defense finalized their joint backup paper for the accelerated American space program. It had been accounted by representatives of the Bureau of the Budget, given a final editing by James Webb, and signed by the heads of both agencies. It was delivered to Vice President Johnson who, in turn, handed it to President Kennedy. The decision to go to the moon was now firmly on paper awaiting its transmittal to the Congress for its consent, and by the American people.[57] As it turned out on May 25, 1961, President Kennedy's call to land an American on the moon "before this decade is out" was the last of five requests made of the Congress. Before a joint session of both houses of the Congress, his speech on "Urgent National Needs" first outlined the need to meet Communist insurgency in developing nations, attack domestic socio-economic problems, and enhance military, intelligence, and civil defense capabilities. These are, Kennedy said, "extraordinary times. We face extraordinary challenge." The United States was "engaged in a long and exacting test of the future of freedom."[58]

Then, speaking with greater conviction, flair, and punch, Kennedy spoke last of the space challenge:

> I believe that this Nation should commit itself to achieving the goal, before this decade is out, of landing a man on the moon and returning him safely to earth. No single space project in this period will be more exciting, or more impressive to mankind, or more

> important for the long-range exploration of space, and none will be so difficult or expensive to accomplish. We propose to accelerate the development of the appropriate lunar spacecraft. We propose to develop alternate liquid and solid fuel boosters, much larger than any now being developed, until certain which is superior.... We propose additional funds...for unmanned explorations.... But in a very real sense, it will not be one man going to the moon--if we make this judgment affirmatively, it will be an entire nation. For all of us must work to put him there....
>
> It is a most important decision that we make as a nation. But all of you have lived through the last four years and have seen the significance of space and the adventures in space, and no one can predict with certainty what the ultimate meaning will be of mastery of space....
>
> It means a degree of dedication, organization and discipline which have not always characterized our research and development efforts.... New objectives and new money cannot solve these problems....[59]

NASA's $549 million increase for FY 1962, speeded launch vehicle development, additional money was included for global communication satellites and weather satellites. Defense was given $62 million for solid-fuel booster development. But President Kennedy made his decision a national commitment with this historic address calling for a goal of a lunar landing by 1970. Sustaining the commitment in subsequent years was not to be an assured thing by any means. Indeed, one of the unsung achievements of Apollo was that it was to be the largest single technological enterprise of a non-military nature ever undertaken in modern times.

The heroic initial phase to place man on the moon was concluded, moreover, by a democratic nation in an open enterprise over eight long years because its elected executive grasped the tiller of historical destiny in April 1961. John F. Kennedy set the goal and the course with a deadline.

SPACE ACCELERATION AND COLD WARFARE

Realities other than the "space race" led to a heating up of the Cold War during the take-off months of the Kennedy administration. The view of the bipolar world from the Oval Office perforce included space-related factors while problems, policies, public positions, and actions were orchestrated and control of future difficulties anticipated. As we have reviewed its impact, the first space flight by Yuri Gagarin was a premiere dramatic triumph in a sequence of Soviet spectaculars. More weight-lifting successes by the Soviets had to be assumed. However the more numerous but smaller American satellites had accomplished more science and demonstrated the value of weather, communication, and other application satellites. Kennedy's pivotal acceptance of the challenge to race openly to land an American on the moon--perhaps by 1968, which would be at the end of his second term, one fervent advocate pointed out--was only a declaration of intentions. Such was logical for the future frontier of mankind but the realities of national security, the world balance of power, and national leadership pressed daily perturbations. Apollo had no early Cold War clout.

HERBLOCK'S EDITORIAL CARTOON

"Fill 'Er Up—I'm In A Race"

----copyright 1961 by Herblock in The Washington Post

Used with permission of Mr. Herb Block, *Washington Post*

President Eisenhower's prehistoric hope for "open skies" in 1955, brusquely rejected by the Soviet Union, had only been accepted by the demonstration of the overflights of the sputniks in principle. Considerable effort to negotiate on the summit had been demolished by the Kremlin after the downing of the U-2 in May 1960. Yet, within three months after the last U-2 flight, the American *Discoverer 13* and *14* had ejected capsules, presumably containing photographic film exposed in orbit, which had been recovered. United States satellites giving early warning of missile launchings had been flown on test missions early in 1961. During the following months, the so-called "missile gap" campaign issue evaporated. The build-up of American ballistic missile systems also altered estimates of the strategic balance. In March 1961, President Kennedy had added ten more Polaris submarines and three more Minuteman squadrons to the schedule of keeping a thermonuclear deterrent to intercontinental warfare.[60]

Space achievements and their worldwide recognition, coupled with existence of operationally-ready ICBM's, buttressed Soviet Premier Khrushchev's pointy rattling of missile threats over Western Europe. International cooperation in non-military space benefits as well as seeking an accord with the Soviet Union on some limits to strategic weapons systems by means of a nuclear test ban treaty. These proved to be sustaining priorities of the Kennedy administration, and subsequently, during Apollo's race to the moon. Such White House concerns have been little evidenced in histories of the Apollo achievement. NASA administrators Webb and Dryden were, of course, most concerned with the Cold War context of Apollo while, at the same time, mobilizing the difficult and open endeavor to landing Americans on the moon.

Within ten days of President Kennedy's call before the Congress to go to the moon, a summit meeting with Premier Nikita Khrushchev took place in Vienna, Austria, on June 3-4, 1961. It picked up where Eisenhower's attempt had fallen with the U-2 in May 1960, except that Khrushchev, while loudly rattling Soviet missiles and space achievements, insisted that he would unilaterally make a separate peace treaty with Germany as the U.S. had done with Japan in 1945. The increased flood of émigrés to the West from East Germany, however, threatened the existence of the Iron Curtain. In Southeast Asia, the new American leadership was also being severely tested on the global Cold War battlefield. Kennedy had symbolically ordered American military advisers in Laos to put on their uniforms after the Bay of Pigs. If the Kremlin was testing the new American President, a summit meeting could not make the world situation much worse. Space cooperation and competition were included on Kennedy's agenda in seeking some accord on a nuclear test ban and an easing of tensions over Berlin. According to all accounts, Kennedy vowed to argue on every question--well prepared and with vigor.[61]

At the Vienna summit, through interpreters, President Kennedy verbally tackled Soviet Premier Nikita Khrushchev with the suggestion: "Let us go to the moon together?" This was not to become known to the American space community until Kennedy voiced it again at the United Nations in September 1963. Some background seems in order that the twin pillars of Kennedy's space policy of competition and cooperation be more clearly appreciated.

Vice President Lyndon Johnson had early voiced the space thrust of the Kennedy administration in public. On May 12, before the National Assembly in Saigon, Vietnam, Johnson said: "America is moving forward in the great adventure of outer space.... We are not afraid for the world to see our efforts because we have confidence in our ability to succeed." In his address at the U.S. Military Academy at West Point, on June 1, Johnson said: "Urgently, we must push on with all determination into space so as to win it for peace before some other nation conquer it for war and destruction."[62]

Khrushchev had intimated to Eisenhower during his visit to the U.S. in 1959, that the moon was the objective of the Soviet space program, which had been confirmed by the three "Luniks." President Kennedy, however, had not received any response from the Kremlin to his inaugural address invitation "to explore the stars together." Steps preliminary to the Vienna summit had resulted from the congratulatory exchanges between Kennedy and Khrushchev after the space flights of Gagarin and Shepard. Missile and space exploits had been Khrushchev's standard political stance, and he had not hesitated to threaten that four weapons could take care of England in a war.

The day after his call for approval of the lunar-landing goal before the joint session of the Congress, President Kennedy telephoned opening remarks for the first NASA Conference on the Peaceful Uses of Outer Space, in session at Tulsa, Oklahoma. This conference, said Kennedy, dealt "with the very heart of our national policy on space research and exploration."[63] Alan Shepard's Mercury spacecraft, *Freedom 7*, was already on exhibition at the Paris Air Show. It pointed up the cardinal fact that no picture had yet been released on the launching of Gagarin's *Vostok 1*. And the Soviet Union was thus persuaded to show general movies on the training of cosmonauts, and release more details of Gagarin's flight. In the meantime, the United States had already gone forward with England and France in preparation of a draft of a nuclear test ban treaty to which the Soviet Union had also been invited to consider at the discussions at Geneva.

In Vienna, only the social events were highly publicized. The pivotal point was reached, in the form of an oral confrontation threatening a nuclear exchange over the fate of Berlin. Khrushchev laid it on the line, by saying: "I want peace but if you want war, that is your problem."

It was only during mealtime conversations that informal summitry seemed to spawn exchanges of pleasantries and possibilities. Gagarin's automated flight, opined Khrushchev, raised serious questions about going to the moon. Perhaps, responded Kennedy, "we should go to the moon together?" Khrushchev responded negatively, but added in half jest--"All right, why not?" These rhetorical jousts were hardly symbolic, as well as Khrushchev's intimations that only the United States could afford to go to the moon.[64]

One space consequence of the Vienna summit was a surprise to President Kennedy. In an exchange of pleasantries seated together at the Austrian state dinner, Jacqueline Kennedy said to the Soviet Premier that she had noted that Strelka, the Soviet space dog recovered from orbit, had recently had puppies--"Why don't you send me one of the puppies?" This was duly communicated via the translator. Two months later, Soviet Ambassador Menshikov delivered a little Russian puppy to the Oval office. Kennedy asked: "How did this dog get in here?" Later, Mrs. Kennedy confessed: "I'm afraid that I asked Khrushchev for it in Vienna."[65]

While still in Vienna, Kennedy gave a secret off-the-record interview to James Reston, correspondent of the *New York Times*, in which he recounted the grim standoff which had transpired, shades of a crisis similar to Europe in 1939, and the global Cold War problems, particularly that of Berlin. One of his commentaries, which Reston did not reveal until much later, was: "Now we have a problem in trying to make our power credible, and Vietnam looks like the place."[66] When he returned to Washington, Kennedy ordered more troops for Europe. He also nominated the commander of the Strategic Air Command, General Curtis E. LeMay, as chief of staff of the U.S. Air Force.

Additional to formal letters and diplomatic exchanges, however, Kennedy and Khrushchev were to maintain correspondence via personal letters thereafter.[67] While the race for strategic weapons systems and the Space Race were to be sustained in earnest, President Kennedy was not to be dissuaded from his pursuit of a test ban treaty holding an olive branch for increased international cooperation in space. In his address before the United Nations on September 1961, he said:

> ... As we extend the rule of law on earth, so must we extend it to man's new domain--outer space.
>
> All of us salute the brave cosmonauts of the Soviet Union. The new horizons of outer space must not be driven by the old bitter concepts of imperialism and sovereign claims. The cold reaches of the universe must not become the new arena of an even colder war.
>
> To this end, we shall urge proposals extending the United Nations Charter to the limits of man's exploration in the universe, reserving outer space for peaceful use, prohibiting weapons of mass destruction in space or on celestial bodies, and opening the mysteries and benefits of space to every nation. We shall propose further cooperative efforts between all nations in weather prediction and eventually in weather control. We shall propose, finally, a global system of communications satellites linking the whole world in telegraph and telephone and radio and television. The day need not be far away when such a system will televise the proceedings of this body to every corner of the world for the benefit of peace....[68]

The Kennedy administration was to have ended before technology and international law were to make his words at least partial realities. Another consequence of the Vienna summit was the resumption of nuclear testing by the Soviet Union in August 1961, one shot being a 58 megaton shocker. And the erection of the Berlin wall cemented the East-West standoff.

In the meantime, American and Soviet scientists maintained very informal contacts at international meetings. Gus Grissom made his Mercury sub-orbital flight in July while Gherman Titov in *Vostok 2* made his 17-orbit flight on August 6. Titov's flight demonstrated twenty-five hours of weightless space flight. When NASA and the Department of Commerce held an International Weather Satellite Workshop in November 1961, to encourage use of Tiros photographs, Soviet participation did not transpire.[69] If the summit at Vienna had nourished Khrushchev's inclination for brinksmanship games with Kennedy, the ultimate crisis was to be erected in the Cuban missile confrontation in 1962. American missile power-in-being, backed by satellite evidence that the United States did not possess a missile gap, helped ordain changes in the Kremlin leadership. Such was of utmost import to the course of events, which served as the backdrop for the environment in which the Apollo thrust to the moon meshed with sustained efforts to increase cooperation in the non-military benefits of space exploration and application.[70] The Kennedy administration understood the double-barreled space thrust of racing to the moon and also seeking international cooperation, but the Apollo team, their avid supporters on Capitol Hill, and much of the general public became a little perplexed by 1963.

Apart from the problem of communist insurgency in Southeast Asia, Robert Kennedy's specter of "five hundred World War II's in less than a day" dominated White House concerns during the Cold War crises over Berlin in 1961, and particularly during the Cuban missile crisis in 1962. NASA's non-military program dominated by the Apollo commitment assumed almost a historical trajectory of its own. Ranger, Lunar Orbiter, and Surveyor were now essential for aiding the lunar landing. While not neglected, scientific and application satellites got a modest share of money or attention. Weather and communications satellites proved both useful on an international scale, yet for Cold War reasons, did not enhance space cooperation with the Soviet Union. The first formal task of the National Aeronautics and Space Council under Vice President Johnson was to generate recommendations for national policy on communications satellites. To this end a report was issued on June 15, 1961, which sorted out the issues on public, commercial, and R&D aspects. Relay, Telstar, and Syncom got moving and President Kennedy, as we have seen, called for international cooperation in the U.N. in September. By July 1962, *Telstar I* demonstrated a revolution in global communications. The legislation for the Communications Satellite Corporation proved controversial, one provoking a unique filibuster by liberals in the Senate before its passage.[71]

After the Rose Garden ceremony for Scott Carpenter in May 1962, President Kennedy offered his appreciation of the space program in a conversation with Carpenter and Walter C. Williams. Kennedy said he believed that the United States could successfully engage openly and peacefully the enemies of freedom, and win, by utilizing fully scientific and technological superiority in which the space frontier was a part. In the battle for men's minds, an open democratic society had the advantage. No one wanted nuclear war, unless in retaliation, because everyone would lose. Economic war was a slow process. He wanted the men of Mercury to appreciate how important their contribution had been.[72]

THE MOON RACE AND SPACE PREEMINENCE

There had been no debate or roll call vote in the Congress on the Apollo commitment as it started. The buildup of NASA facilities involved regional politics; for example, Merritt Island in Florida became the launch site, the Space Task Group in Virginia became the Manned Spacecraft Center in Houston, and the Mississippi Test Facility and the Michoud plant outside New Orleans were instituted. These matters directly involved NASA with the White House as well as with the space committees on the Hill, which demanded an accounting from NASA and, in turn, controlled the purse strings.[7] The opening of the NASA Electronics Research Center in Cambridge, Mass., was coupled with the election of Senator Edward Kennedy.

D. Brainerd Holmes had been named to head NASA's Office of Manned Space Flight, billed as the "Moon Czar" in *Time* magazine. Holmes was a driving force mobilizing the Apollo team and getting it moving, along with the definition of the interim Gemini program using the Titan booster. Mobilizing industry as the Apollo configuration shaped up was levered by budgetary matters. Administrator Webb's province was the White House and the Congress as well as energizing nation-wide interest and support for preeminence in space. NASA would require support of universities because NASA centers were not like Army camps in the southland.

At all times, President Kennedy encouraged debate, discussion, and interplay, and used the telephone frequently. When the necessary and productive debate took place on the basic decision by 1962 of how to go to the moon, which configured the Saturn V and the three Apollo spacecraft, the debate took place near a live microphone when President Kennedy visited the Marshall Space Flight Center in Alabama. Science adviser Jerome Wiesner was encouraged to play a devil's advocate role. Later, when Kennedy visited Cape Canaveral, he was given an Air Force tour and a NASA tour. He had jocular questions and comments as he costed the large Air Force Titan launch complex rising out of the Banana River while NASA was proud of its less-expensive, but costly, sonic pile-driving on Merritt Island for the VAB and launch complexes 37 and 39.

President Kennedy's greatest space testament, now classic, was made at Rice University in Houston on September 12, 1962. He declared that the United States should "become the world's leading spacefaring nation."

> We sail on this new sea because there is new knowledge to be gained, and new rights to be won, and they must be won and used for the progress of all people....
>
> We choose to go to the moon... in this decade and do the other things, not because they are easy, but because they are hard, because that goal will serve to organize and measure the best of our energies and skills, because that challenge is one that we are willing to accept, one we are unwilling to postpone, and one which we intend to win, and the others too.

It is for these reasons that I regard the decision last year to shift our efforts in space from low to high gear as among the most important decisions that will be made during my incumbency in the office of the President....[74]

There were a few acute problems between NASA and DOD not worked out in the Astronautics and Aeronautics Coordinating Board (AACB), one particularly on the control of the Gemini program.[75] It ended up with the Air Force Manned Orbital Laboratory program intended to exploit near space while NASA voyaged to the moon. All major program decisions involved the White House. The NASC under the Vice President was not about to get between Secretary McNamara and Administrator Webb and President Kennedy.

The "moon race" urgency prompted Brainerd Holmes to propose a $400 million supplemental for FY 1963. It was supported by Wernher von Braun of the Marshall Center and Robert R. Gilruth of MSC. Administrator Webb knew that neither the White House nor the Congress would support such an

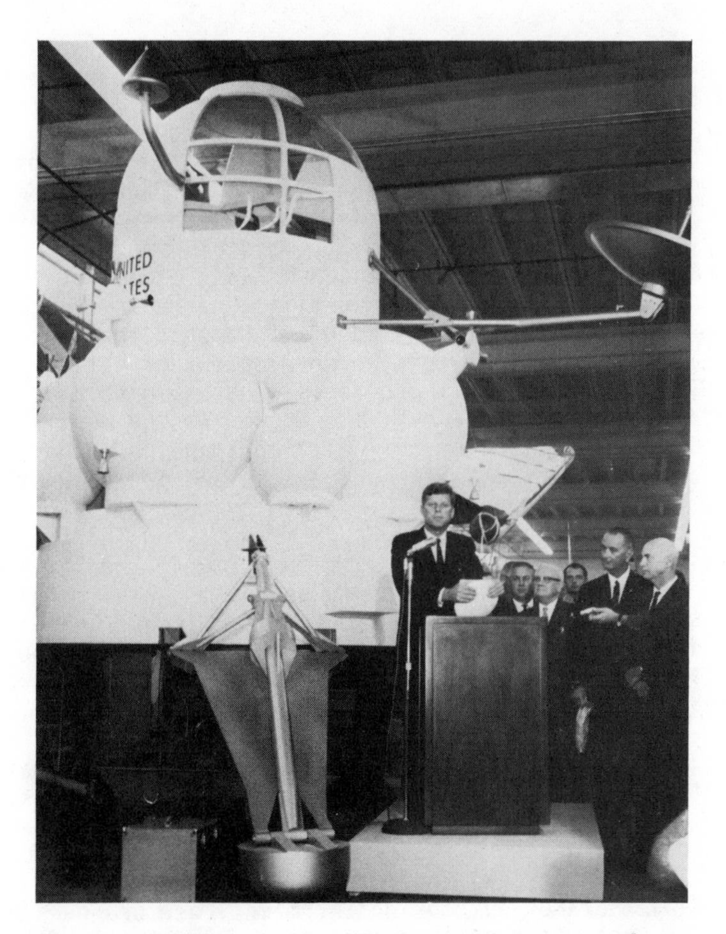

President John F. Kennedy at the NASA Manned Spacecraft Center after his speech at Rice University on September 12, 1962, with Vice President Johnson and NASA officials (NASA Photo No. 62-ADM-16)

asking, besides evidencing a lack of orderly program execution crimping the normal budget cycle. In a session with the President, Holmes suggested other portions of NASA be cut. Webb stood firm. Kennedy decided against the supplemental, but stated that the manned lunar landing was all important.

Astronauts Kennedy enjoyed, starting with John Glenn in early 1962. After the Mercury flight of Gordon Cooper in May 1963, NASA headquarters cancelled MA-10 so energies could be concentrated on Apollo. The astronauts objected, and took their argument, which was no secret, to President Kennedy. Kennedy then advised Administrator Webb that it was his decision not that of the White House. Though Brainerd Holmes did not agree with the astronauts, when the end of Project Mercury was announced it was also made known that Brainerd Holmes would return to industry. Behind this sparse account is a complex story leading the nationwide dialogue on going to the moon. In public it had been sparked by inferences that there was no evidence that the Soviet Union was racing to the moon. It also involved continued discussion begun November 1962 between President Kennedy and the NASA Administrator.

President Kennedy considered that landing Americans on the moon was the major purpose of NASA's priorities, while Administrator Webb disagreed saying that the national objective was to become preeminent in space, and he would not take responsibility for a program that was not a balanced one. Kennedy requested that Webb put his argument on paper so that there would be clearer understanding. Webb's lengthy memorandum of November 1962, declared: "The objective of our national space program is to become preeminent in all important aspects of this endeavor and to conduct the program in such a manner that our emerging scientific, technological, and operational competence in space is clearly evident." He said that the manned lunar landing is "a natural focus for the development of national capability in space, and, in addition, will provide a clear demonstration to the world of our accomplishment in space." The answer was evident in the President's budget request for FY 1964 of $5.7 billion spending authority for NASA.[76] Dialogue continued within NASA headquarters and in the White House. Kennedy asked the Space Council to elaborate on what military value had derived from the Gemini and Apollo programs. It replied that the space program was an expensive but "a solid investment" and "will give ample returns in security, prestige, knowledge and material benefits."[77] NASA set about structuring a single headquarters. It had been a shock when it was learned, also in May 1963, that the estimated total cost of Gemini had shot over a billion dollars. The resignation of D. Brainerd Holmes prompted a reorganization.

The end of Project Mercury also prompted a wholesale nationwide dialogue about going to the moon. Sir Bernard Lovell's suggestion that he did not believe the Russians had a manned lunar program, later in the year to be stated by Khrushchev, was a fuse that exploded the Moon Race. If there was no race, what was the reason for an expensive or crash Apollo program? A plethora of outpourings from many sectors questioned whether Apollo was a stunt while some scientists argued about whether the surface of the moon would support a landing machine. The questioning reached Capitol Hill, not that it became a single issue of political import but it

Ceremonial unveiling of "Mona Lisa" at the National Gallery in Washington on January 9, 1963. Top picture was transmitted via *Relay* satellite to Agence France-Presse in Paris, bottom picture showed quality of picture received in France. Shown with President Kennedy (left to right) are Madame Malraux, French Minister of Cultural Affairs André Malraux, Mrs. Kennedy, and Vice President Lyndon B. Johnson (NASA Photo No. 63-ERelay-2)

tested the existence of the Apollo priority and shaved its margins. The advocates for curing cancer, poverty, and hunger were heard, while there was no question the Soviet manned and unmanned space missions were impressive. Some social historian should assay this 1963 phenomenon, which in many ways was a belated questioning which had not taken place when the lunar-landing goal had been set in 1961.[78] Space had to earn its own way, then as now, and President Kennedy stayed with his decision for the decade.

The debate was certainly boosted by President Kennedy's address to the United Nations on September 20, 1963. It was a call to the Soviet Union to get with the nuclear test ban treaty. The space portion of his speech had apparently been cleared over the telephone with James Webb. Kennedy made a strong argument for cooperation in space with the Soviet Union: "I include among these possibilities a joint expedition to the moon." While he had named Hugh L. Dryden of NASA as ambassador to initiate conversations with the Russians on possible cooperation in 1962, not much had resulted. Ameliorating the arms race had produced a gigantic file, starting with Eisenhower and continued by Kennedy, on correspondence with the Kremlin on disarmament and space cooperation. Some of the Apollo managers were not fully informed on this--neither was the public--so that Kennedy's U.N. speech erected false indicators that his support was weakening on the Moon Race. Several NASA speakers, questioning in public a joint lunar program, provoked calls from the White House to NASA.[79] The "moondoggle" critics had their say. But on October 10, the House of Representatives passed the

President Kennedy being briefed on Apollo at Blockhouse 37 at Cape Canaveral by Dr. George Mueller. Front row (left to right) George M. Low, Kurt Debus, Robert Seamans, James Webb, the President, Hugh L. Dryden, Wernher von Braun, General Lee Davis, and Senator George Smathers (NASA Photo No. 63-Adm-60)

Independent Offices Appropriations for FY 1964, which included $5.1 for NASA. Among the amendments accepted in H.R. 8747, one required Senate approval of any agreement that would use NASA funds in support of joint lunar exploration with Communist countries. It was Congressman Albert Thomas (D.-Texas) who led the support for the NASA budget.[80] That same day, President Kennedy presented the Collier Trophy for 1962 to the seven Project Mercury astronauts in the Rose Garden, saying the award "will be a stimulus to them and to the other astronauts who will carry our flag to the moon and, perhaps, some day, beyond."[80] A month later he directed NASA Administrator Webb to develop technical proposals for "substantive cooperation with the Soviet Union in the field of outer space." And, on November 16, 1963, Kennedy visited Cape Canaveral, the third time in two years. When a model of the Atlas D was compared with the one of the giant Saturn V moon rockets he said, "This is fantastic." He also got a full briefing on Apollo, as seen above.

And so it was in late November 1963, a combination of earlybird political fence-mending in a key state for 1964, with an opportunity to upraise the space community. First President Kennedy had been scheduled to appear at a testimonial dinner in Houston for Congressman Albert Thomas who, as chairman of the Independent Offices Subcommittee of the Committee on Appropriations, had brought many Federal dollars to Texas. Then he was scheduled for appearances at the USAF School of Aerospace Medicine, and at Fort Worth and Dallas.[81]

On November 22, in a breakfast talk in Fort Worth, President Kennedy said that the space program was expensive, "but it pays its own way, for freedom and for America." It was the speech he was likely to give in Dallas.

Kennedy invited Congressman Albert Thomas, and Congressman Olin Teague of the House Science and Astronautics Committee, to sit with him on the short flight on *Air Force One* from Fort Worth to Dallas. The President told them that he very much wanted to get to the Cape to see the first launching of a Saturn in December. Kennedy said that he knew the space program needed a boost and he wanted to help.[82]

Until the shots rang out in Dallas, John F. Kennedy was in full control of the Nation's affairs, including the space program. Those in the space program were to miss him a lot. Some of them wonder why it is that there are no space quotations anywhere nearby his gravesite in Arlington Cemetery.

JOHN F. KENNEDY

REFERENCE NOTES

1. This quotation was one of the very few, if any others, made by a member of the Kennedy family during the *Apollo 11* mission which attained the lunar-landing goal called for by John F. Kennedy. It was offered by Sargent Schriver at the launching of *Apollo 11* at Cape Kennedy. "The Best View of All, From a Rocking Chair," AP dispatch of 7/16/69; *New York Times*, 7/17/69; NASA "Current News," 7/31/69, p. 16.

2. Interview of Wernher von Braun, by W. Sohier and the author, Huntsville, AL, 1964, for the JFK Library; cf. Th. C. Sorensen, *Kennedy* (NY: 1965), pp. 747-48.

3. Interview of C. Stark Draper, by the author, Constance, West Germany, 10/13/71; and also by the author and J. M. Grimwood, JSC, Houston, TX, 6/2/74, taped.

4. Th. H. White, *The Making of the President in 1960* (NY: 1961), pp. 115-19.

5. See Loyd Swenson, James Grimwood, and C. C. Alexander, *This New Ocean: A History of Project Mercury* (NASA SP-4201, 1966), pp. 304-364.

6. Author's *pre-Apollo 8* historical review of the coming of President Kennedy's decision to go to the moon is broadly documented in "Perspectives on Apollo," *Journal of Spacecraft and Rockets* (April 1968), pp. 369-82 (also AIAA Paper No. 67-839 revised). It was splendidly amplified by additional research and interviews by John Logsdon, *Decision to Go to the Moon* (Cambridge: 1970), *passim*. Most of the new information in this terse account is from excepted interviews by the author which will be fully documented in forthcoming publications.

7. White House, *Compilation of Presidential Documents*, 4 (December 9, 1968), pp. 1658-59.

8. T. Keith Glennan's personnel memoir in the Eisenhower Library (used by permission), plus personal presence on the scene and subsequent interviews by the author.

9. *New York Times*, 1/12/61, story and text; cf. Emme, "Perspectives," pp. 375-76.

10. Excepted interviews; cf. Emme, "Astronautical Biography: Hugh L. Dryden, 1898-1965," *Journal of the Astronautical Sciences* (AAS), 25 (April-June 1977), pp. 151-71.

11. James E. Webb's remarks at farewell dinner for Thomas O. Paine; Excepted interviews in NASA History Office and other public statements in which he repeated the substance of his session with President Kennedy.

12. Since this paper focuses upon presidents, readers should be cognizant of the genuine breadth and depth of James E. Webb's background and experience which he brought to his management of NASA. His answers to questions of Congressmen were sometimes lengthy in explanation to make a point, his energy endless, and his "fight" for NASA openly and behind the scenes not yet well fully documented. A biography of James Edwin Webb would indeed provide a valuable insight into the history of aerospace technology, and particularly in the Federal government from the Truman administration (he served as first director of the Bureau of the Budget and as Under-Secretary of State) to the landing of Apollo astronauts upon the surface of the moon. He learned the ropes of the Congress as an assistant to the chairman of the House Rules Committee, graduated from the law school of George Washington University, had a successful career in industry at Sperry Gyroscope and in the Kerr-McGee organization. He was a Marine Corps Reserve aviator, attached to a ready reserve unit under Lloyd V. Berkner, and served during World War II. He was also very active in various social study and educational activities such as the Frontiers of Science Foundation in Oklahoma, as president of Sciences Services, Inc., in Cambridge, Mass., and was interested in urban, international and management problems. When he became NASA Administrator he had to resign as a director of the McDonnell Aircraft Company, which had the prime contract on the Mercury spacecraft. Few people today realize that he became space conscious early, because of his contacts with McDonnell. In 1956, Webb gave an address at Colorado College on "Space: The New Frontier." But his view of not wanting the NASA job, despite the support of Jerome Wiesner and others close to the President, seems well founded. One witness saw Webb leave the President's office after the meeting on assuming NASA, who said: "He really was shook. I guess it was more than he had planned, perhaps actually talking Kennedy out of it." Administrator Webb, however, proved to be a pretty tough in-fighter for "my NASA program" throughout the corridors of Washington politics, and from January 30, 1961 onward.

13. The plethora of space-related events, which in breadth and sequence are necessary for historical understanding, should not be neglected by the serious student. Ill-informed hindsight should not be the trade mark of future historians. Cf. *Aeronautical and Astronautical Event of 1961*, NASA Report prepared by the History Office for the House Committee on Science and Astronautics, Committee Print, June 7, 1962, 113pp. Subsequent annual chronologies in the NASA *Astronautics and Aeronautics* series, as well as program chronologies, have served usefully as prerequisites for the writing of histories.

14. Historians may find it difficult to document any informal relations of Kennedy's closest advisers with either Lyndon B. Johnson or James E. Webb on space affairs. Both were serious-minded servants of the President at all times--it was "Mr. President" and to others "President Kennedy not "JFK." Free wheeling discussions were always preceded or followed by summaries or recommendations in writing. It was NASA people who

had to learn that its new administrator had nation-wide and personal sources for information concerning NASA activities, and no detail of public or Congressional relations escaped his surveillance. All major aspects of NASA policy, decision, and action, Webb insisted, was a collective product of joint consideration with Deputy Administrator Hugh Dryden and Associate Administrator Robert C. Seamans--which became called a "troika." Cf. J. E. Webb, "Foreword," *Administrative History of NASA, 1959-63,* by R. L. Rosholt (NASA SP-4101, 1966), pp. iii-vii. Jay Holmes, *America on the Moon* (Philadelphia: 1962), pp. 193-210.

15. Historian Melvin Kranzberg, by way of a couplet, described the impact of *Apollo 11,* as follows:

> "In fourteen hundred and ninety-two
> Columbus sailed the ocean blue.
> "In nineteen hundred and sixty-nine
> Neil Armstrong leaped for all mankind."

See Kranzberg's "Historical Perspectives on the Space Program," *Georgia Tech* Alumnus, 51 (Spring 1973), p. 8, and "A Message from the President," *American Scientist* (July 1979).

16. E. M. Emme, "Historical Perspectives on Apollo," AIAA Paper No. 67-839, revised January 17, 1968, published in *Journal of Spacecraft and Rockets* (AIAA), 5 (April 1968), pp. 362-89.

17. John M. Logsdon, *The Decision to Go to the Moon* (Boston: MIT Press, 1970) and reprinted paperback (Chicago: University of Chicago Press, 1977); and, "The Apollo Decision in Historical Perspective," 4th Annual AIAA History Lecture (AIAA Paper 79-0261), 19pp.

18. Specific references for most details will not be made here, which are contained in references in notes nos. 67 and 68. Additional references will. Also helpful are J. Holmes, *The Race to the Moon* (Philadelphia: 1962), pp. 189-205; Vernon Van Dyke, *Pride and Power: The Rationale of the Space Program* (Urbana, Ill.: 1964), pp. 20-29. Journalistic books on Apollo most frequently fail to treat the sequence of events with regard to their interfaces for leading to the national decision to go to the moon. It is to be hoped that future historians are not so gross, particularly when the rest of the evidence is in.

19. House Committee on Science and Astronautics, *Documents on International Aspects of the Exploration of Outer Space,* Staff Report, Document No. 18, 5/9/63, pp. 189-91.

20. L. Swenson, J. Grimwood, A. Alexander, *This New Ocean: A History of Project Mercury* (NASA SP-4202, 1966), pp. 318-22, C. Brooks, J. Grimwood, L. Swenson, *Chariots of Apollo* (NASA SP-4205), pp. 22-25. Most detailed source on Apollo program development is, I. Ertel, *et al., Apollo Spacecraft: A Chronology* (NASA SP-4009, four vols., 1969f), and D. Akens, *Saturn Illustrated Chronology* (NASA MHR-5: 1971).

21. The House Committee on Science and Astronautics did not have oversight on military space affairs. See Ken Hechler, "A History of the House Committee on Science and Technology, 1958-1979," in press.

22. Lloyd V. Berkner had been involved in space science since World War II, was known as "the father" of the International Geophysical Year, and had created the Space Science Board before NASA was officially in being. He was also a friend of James E. Webb from Navy reserve days in the 1930's. Cf. Holmes, p. 194; C. Atkins, "NASA and the SSB" (NASA Historical Note No. 62: 1966), pp. 43-46.

23. Regretfully, other than a few memoirs, there are no pertinent histories on space affairs available from the military services, the science and industrial establishments, or the sensitive policy elements of the White House at the highest levels. For their part, NASA's professional histories deal only with those relationships documentable in NASA files, hearings of the Congress, some oral history, and reports and releases of other agencies and institutions. NASA's annual chronologies, *Astronautics and Aeronautics: A Chronology of Science, Technology, and Policy,* 1963-present, rely on available documentation including select news media sources. Most but not all NASA historical works are cited in A. Roland, compiler, "A Guide to Research in NASA History" (NASA History Office HHR-50, February 1979), pp. 15-17.

24. See *Chariots for Apollo,* pp. 22-25.

25. Logsdon, *Decision,* pp. 98-99; R. Rosholt, *Administrative History of NASA, 1958-63* (NASA SP-4101, 1966), pp. 190-91; *Chariots of Apollo,* pp. 25-26; and excepted interviews by the author.

26. Logsdon, p. 99: *Chariots for Apollo,* p. 26.

27. Welsh had served on the Senate staff for the Air Power hearings on the "bomber gap" in 1956, and as Symington's liaison to Lyndon Johnson's Investigating Subcommittee, Special Committee, and standing space committee, and had chaired a space panel for President-elect Kennedy. Cf. Holmes, p. 195; Rosholt, pp. 189-90.

28. House hearings of NASA witnesses were concurrent to deliberations in the White House. They were front page news in the Washington press so Kennedy also read them, along with the world's "space race: items.

29. Excepted interviews by the author. Cf. *This New Ocean,* pp. 330-31; M. Link, "The Season of Crisis: 1961," *Space Medicine in Project Mercury* (NASA SP-4003, 1965), pp. 112-23.

30. Cf. Emme, "Perspectives on Apollo," pp. 378-79; Logsdon, *Decision,* pp. 104-105; Rosholt, p. 191, Holmes, pp. 196-97.

31. A. M. Schlesinger, Jr., *A Thousand Days: John F. Kennedy in the White House* (Boston: 1965). In conversation with the author, Professor Schlesinger indicated that he had "nothing to do with the space decision," as he was out of the country." Washington, D.C., 9/16/67.

32. U.S. Congress, House Committee on Science and Astronautics, *Discussion of Soviet Man In Space Shot,* April 13, 1961, pp. 1-5, 31.

33. Prominence of the six-million-plus copies of the weekly *Life* on space affairs cannot be assayed here. Each issue was dated to reach all subscribers via mail before that date so that it reached the news stands at least three or more days before its issue date. According to Robert Sherrod, *Life* claimed that each copy was read or scanned by more readers than any other weekly. Hugh Sidey, "How the News Hit Washington," *Life* (April 21, 1961), pp. 26-29, and *John F. Kennedy, President* (New York: Atheneum, 1963), "Space Challenge," pp. 110-123.

34. Based, in part, on excepted interviews by the author. On purpose of the meeting, Theodore Sorensen, *Kennedy* (New York: 1965), pp. 590; cf. Logsdon, *Decision*, p. 106.

35. Sidey, *Kennedy*, pp. 121-23; cf. Emme, "Perspectives," p. 378; Logsdon, pp. 106-107.

36. Hugh Sidey, "How the News Hit Washington," *Life*, 4/21/61, pp. 126-29.

37. Sidey, *Life*, p. 129; Sidey letters to author, 2/13/80, 4/80.

38. There is almost nothing of Sidey's rationale explaining Kennedy's sense of this historical moment that offends any historical evidence. That Kennedy reconfirmed Sidey's account before published in book form, which is not mentioned, makes it more authoritative than it might otherwise appear.

39. Hugh Sidey, "Adversity Made JFK Aim Us for the Moon," *Washington Star*, 7/15/79, p. D-3. For a dramatic rendering of this meeting, see William Manchester, *The Glory and the Dream: A Narrative History of America, 1932-1972* (Boston: 1974), pp. 927-29.

40. Sorensen, *Kennedy*, p. 525. Of all the participants in the April 14 meeting, only Sorensen would not interpret Kennedy's views from a predetermined relationship, whether BOB, PSAC, or NASA. And this paper is intended to focus on the president's role.

41. Interviews of Paul A. Dembling, NASA General Counsel, by the author, with Richard Balderston, 6/26/72, and 8/9/72. In 1961, NASA headquarters was still in the Dolley Madison House, a block from the White House.

42. Logsdon, *Decision*, pp. 105-7.

43. As quoted in the historical context of the descent of the U.S. into a conflict in Vietnam during the Kennedy administration, in David Halberstam, *The Best and the Brightest* (Greenwich, Conn.: 1972), pp. 94-95.

44. Little has yet appeared in the way of documented history on the role of reconnaissance and early warning satellites in U.S. estimates of Soviet missile and space capabilities. Best discussions are in Herbert York and Alan Greb, "Strategic Reconnaissance," *Bulletin of the Atomic Scientists*, v. 33 (April 1977), pp. 33-42; R. Cargill Hall, "Instrumented Utilization of Space," in *Two Hundred Years of Flight in America* (AAS History Series, v. 1, 1977), pp. 183-212; Philip Klass, *Secret Sentries in Space* (New York:

1971), pp. 102-122; Harlan B. Moulton, *From Superiority to Parity: The U.S. and the Strategic Arms Race, 1961-1971* (Westport, Conn.: 1973).

45. Bell's memorandum to the President in late March 1961 (actually undated) stressed the priority of problems on earth instead of dashing off into space. As a BOB policy position not specifically tied to budgetary details its existence was unknown to NASA Administrator Webb, who requested a copy from the History Office when it was published. Cf. National Archives, *A Documented History of America's Venture Into Space* (Washington: 1970), p. 37.

46. From his Capitol Hill office, Vice President Johnson maintained fully his senatorial contacts, and particularly his keen interest in the functioning of the Senate Committee on Aeronautical and Space Sciences, his replacement as chairman being Senator Robert Kerr. Johnson often attended committee meetings *ex officio*. As chairman of the National Aeronautics and Space Council in the White House, Mr. Johnson held meetings much like Congressional committee hearings, and Senator Kerr participated in several sessions. Johnson did not call the secretaries or the chiefs of the armed services, but he did attend meetings of the National Security Council. Cf. Logsdon, *Decision*, pp. 112-18, a splendid documentation of the NASC "hearings." No historian has yet seen all White House records.

47. On April 19-20, 1961, Kennedy also attempted to brook political criticism of the Bay of Pigs, contacting Republican leaders including General Eisenhower and Barry Goldwater, and taking full responsibility. Space matters were not discussed according to those interviewed by the author. Without exception, those savants declaring that the Bay of Pigs disaster was responsible for the decision to go to the moon are those who were opposed to such an acceleration. This historian's view of no direct linkage of the Bay of Pigs to the decision to go to the moon, remains as stated in 1967. Cf. "Perspectives on Apollo," p. 379.

The most direct White House response to the Bay of Pigs debacle was the creation of the Defense Intelligence Agency (DIA) in the DOD, the development of the "Green Berets," and, most immediately, Kennedy's resolve to be firm at his upcoming summit meeting with Khrushchev.

48. Point on authorship of Kennedy's questions was recently verified at a luncheon of principals (Sorensen, Welsh, and Webb) at the Library of Congress on the anniversary of *Apollo 11* 7/19/79. Tape recording, Library of Congress, Music Division, Recorded Sound, Nos. 17-18.

49. Cf. Logsdon, *Decision*, pp. 108-9.

50. Office of the Vice President, "Memorandum for the President: Evaluation of the Space Program," 4/28/61, original in the John F. Kennedy Library (Confidential, declassified in 1974). Cf. Leonard Bruno, *We Have a Sporting Chance: The Decision to Go to the Moon* (Washington: 1979), a catalogue for an exhibit at the Library of Congress, July 16-September 16, 1979, 23pp. Kennedy's memorandum to the Vice President of 4/20/61, with its questions, is reproduced on p. 15.

51. This secret NASA-DOD paper was entitled: "Recommendations for Our National Space Program: Changes, Policies, Goals." It was the backbone of the accelerated U.S. Space Program, one that delineated the tasks of NASA and of DOD, and the funding. NASA would develop the 12-million-pound-thrust booster for which greatest experience had been gained while DOD would develop the equivalent 260-inch-diameter solid-fuel equivalent. President Kennedy approved the recommendations on May 10. Cf. John Logsdon, *Decision*, pp. 124-27.

52. Delay in the President's call upon the Congress to support the acceleration of the space program was that other urgent "national needs" were yet to be worked out to be combined in the same message.

53. One former Science Adviser to the President watched the launching of Alan Shepard on a television set in the Executive Office Building. As the countdown came to the moment of launch, he sat stooped down with his head in his hands. As the Redstone rose, he muttered repeatedly: "That poor-sonna-bitch, that poor-sonna-bitch, that poor-sonna-bitch..." Excepted eye-witness source interviews by the author.

54. Oral history interviews of Dr. Welsh in the NASA History Office, and in the John F. Kennedy Library.

55. President Kennedy's identification with the astronauts grew with each space mission. Once they had flown in space they were "astronauts."

56. Various White House staffers have mentioned that there were preliminary efforts to develop a rationale for the lunar-landing decision, one comparing it to FDR's "50,000 airplanes" call in May 1940 and the Manhattan Project, and Truman's decision to develop the hydrogen bomb, but the work load to support the other expanded requests to be made of the Congress made it a still-born exercise.

57. Mr. Johnson handed the Webb-McNamera report to the President at the press conference for Astronaut Shepard at the State Department on May 8, 1961. The Vice President then left for the Far East. Cf. Logsdon, *Decision*, pp. 125-27.

58. Full text of Kennedy's address to a joint session of the Congress on "Urgent National Needs,on May 25, 1961, *Public Papers, Kennedy, 1961*, pp. 403-405. His final approval on May 10, was delayed in being presented to the Congress for two weeks in order to clump a number of supplemental funding requests. In the meantime, word of his decision leaked into the press, and NASA management and planners had been in high gear for months.

59. President Kennedy further edited his speech before delivery, and departed from it in the latter portion of his space challenge when he called for "dedication, organization, and discipline." Theodore Sorensen submits that Kennedy was not at all certain that sufficient support would be forthcoming to fulfill the lunar-landing commitment. When Kennedy and Sorensen rode in the limosine back to the White House after the address, Kennedy indicated that he had been a little disappointed by the limited enthusiasm accorded his space remarks. But the Congress as a whole was

not nearly space-dedicated as were the Senate and House space committee chairmen and members who constantly interacted with the White House. Sorensen, *Kennedy*, p. 592. Actually, as Logsdon has pointed out, members of the Congress space committees were perhaps a little annoyed that they had not been able to play a greater role in the White House program decisions. Logsdon, p.129. It was to be two years before some hostility to the accelerated Space Program became significant on Capitol Hill.

60. "Space race" histories without sufficient regard of White House problems of maintaining peace seem as prevailing as memoirs and accounts of the Kennedy administration insufficiently treating the President's comprehensive appreciation of space-related aspects of both NASA's and DOD's programs. The best treatise on Kennedy's political rationale and strategy accorded space remains Theodore Sorensen, *Kennedy*, pp. 523-29.

61. Evolution of U.S. space policies for international cooperation and open competition are treated in Arnold Frutkin, *International Cooperation in Space* (New York: 1965), Vernon Van Dyke, *Pride and Power: The Rationale of the Space Program* (Urbana: 1964). NASA histories on Mercury, Gemini, and Apollo programs, focused on technological progress, should not be ignored, while DOD histories of space activities are not available. On U.S. military policy related to space, see R. Cargill Hall, "Instrumented Exploration and Utilization of Space," in *Two Hundred Years of Flight in America* (San Diego: 1977), pp. 198-213; Harland B. Moulton, *The United States and the Strategic Arms Race, 1961-1971* (Westport, CN: 1972).

62. Lyndon B. Johnson, "Statements on Space, 1959-1961," MS in LBJ Library (1961), p. 9.

63. NASA, *Proceedings of the Conference on the Peaceful Uses of Outer Space* (NASA SP-7, 1961), p. 3.

64. T. Sorensen, *Kennedy*, pp. 543-50; A. Schlesinger, *A Thousand Days*, p. 364.

65. Schlesinger, pp. 366-67.

66. Cf. interview of James Reston by D. Halberstam, *The Best and the Brightest*, pp. 95-98.

67. Most useful references: Senate Committee on Aeronautical and Space Sciences, *Statements by Presidents of the United States on International Cooperation in Space--A Chronology: October 1957-August 1971*, prepared by The NASA Historical Staff (Document No. 92-40, 9/24/71), 126pp.; and, Senate Committee on Aeronautical and Space Sciences, *Documents on International Aspects of the Exploration and Use of Outer Space, 1954-1962* (Senate Document No. 18, Staff Report, 5/9/63), 405pp.

68. *Public Papers, Kennedy*, 1961, pp. 622-23.

69. Edward C. and Linda N. Ezell, *The Partnership: A History of the Apollo-Soyuz Test Project* (NASA SP-4209: 1978), pp. 33-35.

70. A full-fledged history has yet to be available on the so-called "missile gap" estimates, and particularly with regard to the near war crises of the early 1960's. Kennedy's issue of his political campaign in 1960 was apparently based upon the obsolete projection for 1963 time period made in 1957-58. The actual buildup of U.S. and Soviet strategic nuclear weapons, revalidated early in the Kennedy administration, found that a "missile gap" did not exist. The estimate of three to one Soviet advantage was apparently found in error by the U-2 missions in May 1960. All Republican protests during the campaign were dismissed as political propaganda, and perhaps because CIA briefings of candidate Kennedy were not updated on missiles, and did not include U.S. bomber force capabilities. In mid-1961, the balance greatly favored the U.S. Best general references are: Th. Sorensen, pp. 608-13; Schlesinger, pp. 317, 498-500; Moulton, pp. 60-71, Klass, pp. 59-71, 96-122; and, Halberstam, pp. 296-303.

On the Soviet side, the tapes that N. Khrushchev recorded after his downfall in 1964 and smuggled to the West, should not be ignored. Voiceprint verified, his thoughts and recollections are in *Khrushchev Remembers: His Last Testament*, trans. and edited by Strobe Talbott (Boston: 1974). This volume is more detailed and seems helpful in contrast to an earlier edition hastened out.

71. See White House, *United States Aeronautics and Space Activities, Report to the Congress from the President of the United States, 1961* and *1962*. These were prepared by NASC staff during the Kennedy and Johnson administrations. Cf. Delbert D. Smith, *Communications Via Satellite: A Vision in Retrospect* (Boston: 1976); R. C. Hall, *Lunar Impact: A History of Project Ranger* (NASA SP-4210, 1977); Bruce Byers, *Destination Moon: A History of the Lunar Orbiter Program* (NASA TM X3487, 1977).

72. Interview of Walter C. Williams by the author, 3/25/64.

73. Ken Hechler, *Towards the Endless Frontier: A History of the House Committee on Science and Technology* (Washington: 1980), *passim*. It flushes out many of the details unstated in hearings and reports in the House Committee on Science and Astronautics, 1958-1978.

74. NASA often provided rough draft material for White House speeches, which was the case with Kennedy's speech in Houston in 1962. The President added some thoughts in the oral presentation also.

75. Cf. Barton Hacker and James Grimwood, *On Shoulders of Titans: A History of Project Gemini* (NASA SP-4203, 1977), pp. 117-128.

76. Cf. Jay Holmes, "A Preliminary History of NASA, 1947-1963," NASA Historical Note, August 1968; and "Challenge and Response, 1947-1963," in *Preliminary History of NASA, 1963-1969* (1969), pp. I-49 - 65; C. Brooks, J. Grimwood, L. Swenson, *Chariots for Apollo* (NASA SP-4205, 1979).

77. Lyndon B. Johnson, *Vantage Point* (New York: 1972), pp. 281-82; Jay Holmes, pp. I-58-59.

78. NASA, *Astronautics and Aeronautics: Chronology on Science, Technology, and Policy - 1963* (NASA SP-4004, 1964), *passim;* Vernon Van Dyke, *Pride and Power: The Rationale of the Space Program* (Urbana, IL: 1964).

79. Robert R. Gilruth made some off-hand remarks in a speech at the National Press Club, while Associate Administrator Robert Seamans gave a background talk on Capitol Hill which leaked to the press. Both had pointed to the difficulty in getting the Apollo program integrated to attain its goal without adding the Russians. The NASA History Office had a call from the White House as to documentary holdings on Russian cooperation, which received a negative reply at that time.

80. Cf. NASA, *Astronautics and Aeronautics - 1963,* pp. 384f.

81. Laurence F. O'Brien, *No Final Victories: A Life in Politics from JFK to Watergate* (NY: 1974), p. 155.

82. Olin Teague, letter to the author, 1/24/79, 1p.

Lyndon B. Johnson

No politician had a more sustained concern or official interest in American space affairs than Lyndon B. Johnson. It began when he viewed *Sputnik I* or its booster rocket in the first evening orbit in Texas skies over the banks of the Pedernales River adjoining his ranch. It ended when he left the White House in January 1969. He had announced in March 1968, that he would not run for a second elected term as President. Problems on Earth overwhelmed his daily life despite his genuine enthusiasm for man's conquest of outer space.

While Majority Leader of the Senate, Mr. Johnson was chairman of three committees concerned with space: the Investigating Subcommittee of the Committee on the Armed Services (Truman's old committee); the Special Committee on Space and Astronautics (which shaped the Space Act); and the Senate Committee on Aeronautical and Space Sciences (the standing committee). As Vice President in the Kennedy administration he played a pivotal role in helping shape the initial commitment to accelerate the space program in response to the orbital flight of Yuri Gagarin. President Kennedy made him chairman of the National Aeronautics and Space Council in the White House and requested its policy recommendations, as we have seen.

Lyndon Johnson came from modest origins and was, by all accounts, of very high intellect only tempered to many by his accent and vocabulary. He read swiftly to discern pivotal points on staff papers, wrote little but listened to those who could inform him. He was perhaps a singular President in knowing when he moved into the White House much about what most pieces of the Federal establishment were supposed to be doing. He looked for help and support from the most informed or connected people available to advance his causes--the "little man," the Nation, Texas, and Lyndon Johnson. When he said "let us reason together," it often appeared to mean that he knew the problem, he had a solution, and he would twist an arm or there would be "some head knocking" to gain a consensus. As a politician and leader he was impatient and ruthless according to all of his associates that have written or have been interviewed about him. But his imperial residence was not the White House. It was his fiefdom on his ranch where he entertained astronauts, kings, prime ministers, scholars, and friends.

Lyndon Johnson's tragedy was his being a wartime President of the longest and most unsuccessful war in American history. But without doubt he was, however, an essential personage in the shaping of the American space program and its success. Fortunately, less needs to be said here for there are shelves of relevant Senate hearings, documents, and reports. His memoirs, *The Vantage Point*, do not neglect some of his space connections, otherwise hardly mentioned by political historians to date.[1] In the NASA History Office and in the Lyndon B. Johnson Library you will find an unclassified "Preliminary History of NASA, 1963-1969."[2] And, Mr. Johnson did not neglect to be interviewed by Walter Cronkite on the eve of the launching of *Apollo 11*.[3] Throughout his presidency he was also served by Vice President Hubert H. Humphrey, who was Chairman of the National Aeronautics and Space Council with its able staff under Edward C. Welsh. He also maintained close working relations with NASA Administrator James Webb.

ON THE HILL

After seeing Sputnik arch over on the evening of October 4, 1957, Senator Johnson had phoned his Senate Subcommittee colleagues of both parties to get their support for investigative hearings on missiles and space. He could hardly have been surprised by the Soviet satellite, as Moscow had announced completion of ICBM tests in August. Senator Johnson had longstanding friends at the School of Aerospace Medicine, one of the first space-minded sectors of the Air Force, and General Bernard Schriever also came from his part of Texas. If Senator Johnson had presidential aspirations, he could be encouraged by the fact that he had fully recovered from his heart attack in 1955. Senator Stuart Symington had held the "Air Power" hearings in 1956 after the appearance of the Soviet heavy jet bombers. With the Soviet ICBM and the sputniks the Democratic Congress perforce should help President Eisenhower. In Texas, Senator Johnson had made stirring speeches about the need for a new Manhattan Project for the missile crisis and for making certain that the moon was not conquered by the Russians. In Washington, the Investigating Subcommittee of the Senate Committee on the Armed Services made preparations for an inquiry on missiles and satellites, the primary concern first being to assure national security. After *Sputnik II* had orbited a little Russian dog early in November 1957, the Investigating Subcommittee began its hearings on the 25th. Chairman Johnson carefully insured that they were conducted in bipartisan fashion, not to place blame for what happened but to assist in fashioning a response to the challenges. To some it appeared as if he was acting presidentially. Later, as President he was also to strive for "a broad national consensus" to work for the future.

The Subcommittee hearings were thorough, open sessions appearing on television, but worth the reading now for the initial impact of the sputniks on Capitol Hill. Most witnesses were from the Department of Defense. Director of the IGY Project Vanguard, John P. Hagen, was able to get his organization chart on record, one showing the tangle of support and coordinating interfaces helping to explain how the "missile mess" had not served the IGY satellite program. There were only two all-out space advocates as witnesses: Dr. Edward Teller and Dr. Wernher von Braun. On December 17, Senator Johnson asked von Braun twice whether the U.S. space program should have a military or civilian aegis to best serve the future. Twice von Braun declined to respond. He was, of course, working to get a satellite in orbit under Army auspices with a Redstone, and he was also a member of the civil-oriented Rocket and Satellite Research Panel (RSRP) which had proposed a billion-dollar-a-year space agency separate from DOD.[4] In the 17 recommendations made by Johnson's Subcommittee at the turn of the year, the possible organization of the national space program was not one of them. It did recommend highest priority for the million-pound-thrust clustered rocket (Juno IV) for space. The rest had to do with national defense systems.[5]

When President Eisenhower made it known that a space organization would be recommended for legislation (a flood of bills already being submitted by Senators), Lyndon Johnson secured authority and selected a blue-ribbon Special Committee on Space and Astronautics. All members were chairmen or ranking minority members of Senate committees, plus Stuart Symington, to prepare the space legislation. Johnson secured a very talented staff, including a Hollywood-seasoned lawyer, Edwin L. Weisl, Cyrus Vance, Homer J. Stewart of Caltech, and others.[6] From the start, however, Senator Johnson and President Eisenhower were in complete agreement that the peaceful exploration and exploitation of outer space should reside in a non-military space agency. It was also Johnson's determination that what would become the National Aeronautics and Space Administration should be spared the fate of being overwhelmed by the Department of Defense as had happened to Vanguard. It should also have an international role, which was included in the process of beefing up the bill sent to the Hill by the White House. Denial of defense applications in space was not at all intended. Space was too important to be placed entirely in either the hands of the military or the scientists, which very few witnesses proposed. It was the Space Council in the White House, derived from the Harvard Business School, which replaced the proposed advisory board to perpetuate the function of the National Advisory Committee for Aeronautics. Over a drink in the White House, Senator Johnson proposed to his fellow Texan that President Eisenhower himself chair the National Aeronautics and Space Council. Ike agreed so that the space act would get out. Mr. Johnson also traded off some Senate committee positions, such as the Joint Congressional Committee for space, with the House Committee under John McCormack, with the help of Speaker Sam Rayburn. With the passage of the National Aeronautics and Space Act of 1958, Senator Johnson became chairman of the standing Senate Committee on Aeronautical and Space Sciences. To give his committee power, its first action was to require authority to review NASA's budget requests. Needless to detail, the Senate space committee was indispensable in the take-off and development of NASA. Its prints became educational textbooks on the space program.[6]

AS VICE PRESIDENT

John F. Kennedy's selection of Lyndon B. Johnson as his running mate proved helpful in a close election. When President Kennedy suggested General James M. Gavin as NASA administrator, Johnson stated that he said: "It would be a mistake to appoint any military man to head the organization." After interviewing "about twenty men" he persuaded James E. Webb to be considered for the job. It was to be fortuitous, as we have previously seen, that Kennedy turned to Johnson as chairman of the National Aeronautics and Space Council to fashion the feasibility of going to the moon. Vice President Johnson's interim report to President Kennedy on April 28, in 1961, became national space policy: "The U.S. can, if it

President Johnson examines book of Astronaut Edward White's "space walk" at NASA's Manned Spacecraft Center in Houston, June 12, 1965. With the President were (left to right) NASA Administrator James Webb, Astronaut James McDivitt, Assoc. Administrator Robert Seamans, President Johnson, and Astronaut White. (NASA Photo No. 65-H-972)

President Lyndon B. Johnson and German Chancellor Ludwig Erhard, and their associates, listen to an Apollo briefing at NASA Kennedy Space Center, September 27, 1966. (NASA Photo No. 66-H-1318)

will, firm up its objectives and employ its resources with a reasonable chance of attaining world leadership in space during the decade." In the space dialogue in the summer of 1963, discussions in the White House revolved around the national goal of "space preeminence" vs. "beating the Russians to the moon." Vice President Johnson began to get memoranda of inquiry from President Kennedy: How can the large amount of money for space be justified? How does the moon program serve military control in space? Indeed the U.S.S.R. had racked up a sequence of space spectaculars while Mercury had just come to a conclusion. Soviet "Cosmonette" Tereshkova had more time in orbit than all the Mercury astronauts combined, and the Cosmos series was becoming a long list.

AS PRESIDENT

The shocking death of John F. Kennedy in Texas made the beginning of Lyndon Johnson's presidency most difficult. At the request of Mrs. Kennedy, President Johnson rammed through the change of the historic name of Cape Canaveral to Cape Kennedy. He was obligated to fulfill Kennedy's legislative goals on civil rights, in Southeast Asia, and in space. His election over Barry Goldwater in 1964 was decisive, although the space program with its Apollo priority was not a campaign issue.[7] Champions of Air Force space requirements, however, pressed for increased augmentation of reconnaissance requirements made evident by the Cuban Missile Crisis in 1962. In December 1963, the Department of Defense announced the Manned Orbital Lab program, which was ultimately to prove uneconomic compared to the Air Force's instrumented satellites.[8] In his inaugural address, President Lyndon Johnson declared the need to "assure our preeminence in the peaceful exploration of outer space--in cooperation with other powers, if possible, alone if necessary."[9]

In 1965, the war in Vietnam began to crimp the space program. NASA budgets began to get very tight. It was NASA Administrator Webb who fought the fight on Capitol Hill to avoid the threat to the attainment of the lunar landing itself, with the buildup peak to be reached in 1967. By 1966, came the "Great Society" theme of the Johnson administration. New NASA programs were scaled down. A modest Mariner for Mars replaced a then more expensive but larger Voyager aimed for a landing on Mars in the 1970's. Gemini missions perked public interest. Johnson hailed the astronauts.

While NASA's milestones toward the lunar mission began to appear in Ranger and Lunar Orbiter missions, the deepening of the Vietnam involvement and discontent on the home front coincided. One of the cartoonists linked craters on the surface of Mars with the war.

On January 27, 1967, a fateful day, came the Apollo 204 fire on the pad which snuffed out the lives of Astronauts Virgil Grissom, Ed White, and Roger Chaffee at the Kennedy Space Center. It happened within hundreds of minutes of a crowning achievement in the White House--the signing of the International Space Treaty. It had involved almost a decade of considerable effort on Mr. Johnson's part, ever since he had been asked by President Eisenhower to present the U.S. position on the internationalization of space before the United Nations in November 1958. The tragedy

THE WASTELANDS

Joaquin De Alba in THE WASHINGTON DAILY NEWS, July 30, 1965
(Courtesy of THE WASHINGTON STAR)

challenged the integrity of Apollo and NASA. Beyond the shock, it was NASA's lengthy explanations to the Congressional space committees that helped to restore confidence in NASA's revalidation of the Apollo spacecraft, which cost almost a year's delay in the flight schedule. Brevity here is permissible only because of available works.[10]

In a speech to an audience of educators in Nashville, Tennessee on March 15, 1967, President Johnson stated--and this had leaked out to the press--that the contribution of military space operations were equal to ten times everything else which had been spent on space, including Apollo.[11] It was first mention by any President that more precise information about military activities elsewhere in the world was a vital part of the national space program. But the costs and the pressures of Vietnam were to force him to cancel the Air Force MOL program.[12]

The eighteen-month hiatus in U.S. manned space flights was ended with *Apollo 7*, launched on a Saturn IB, in October 1968. While it restored integrity to Apollo, President Johnson had already announced that he was not going to stand for reelection. James E. Webb had just resigned from NASA, having been told earlier by the President that his "number one government agency was no longer NASA." Inroads on the NASA budget dictated, Mr. Webb said, that "the United States is not pursuing for the time being at least, its goal of 'preeminence in space.'"[13] NASA would succeed in virtually ending the "Space Race" with the circumlunar voyage by *Apollo 8* at Christmas time in 1968. Deputy Administrator Thomas O. Paine had become Administrator. But *Apollo 8* was to bring about the attainment of an age-old dream of Jules Verne before Lyndon B. Johnson left the White House.

Apollo 8 gave mankind the precious view of the Earth as the "blue marble" when seen from the vicinity of the moon. Yet Earth seemed inescapable in a year of riots and assassinations, with the belated release of the crew of the Pueblo, the continued conflict during the annual truce in South Vietnam, and the fourth explosion of a thermonuclear device in mainland China. Yet the man-rating flight of Saturn V carried men to the moon.

One tends to agree with Henry Kissinger that Lyndon B. Johnson was the chief casualty of the Vietnam War. He did not decide to be a war President. He later saw *Apollo 11* achieve John Kennedy's space goal, a symbol of what he had helped fashion, and which had helped to boost his trajectory into the White House. In his memoirs, Mr. Johnson ended on space with this basic thought: "The new adventures in space that lie ahead will bring with them excitement and new accomplishments as great as anything we have witnessed in the epic period just past, when we proved ourselves once more to be the sons of the pioneers who tamed a broad continent and built the mightiest nation in the history of the world."[14] One of President Johnson's final tributes to space was a memorial honoring the crews of *Apollo 7* and *Apollo 8* in the White House. Lyndon B. Johnson soon left Washington for his ranch.

Signing of a memorial document to be hung in the Treaty Room of the White House on December 3, 1968. Signing the document are (left to right): *Apollo 7* Astronauts Walter Cunningham, Donn Eisele, Walter Schirra, and *Apollo 8* Astronauts William Anders, James Lovell, and Frank Borman. Standing are Charles A. Lindbergh (also a signer), Mrs. Johnson, the President, James E. Webb, and Vice President Humphrey. (NASA Photo No. 68-H-1300)

LYNDON B. JOHNSON

REFERENCE NOTES

1. Lyndon B. Johnson, "Space," *The Vantage Point: Perspectives of the Presidency, 1963-1969* (New York: 1971), pp. 270-86. Cf. book of resident historian in the Johnson White House, Eric F. Goldman, *The Tragedy of Lyndon Johnson* (New York: 1969), which does not even mention Apollo or NASA.

2. At the request of President Johnson, all governmental departments and agencies were asked to prepare histories for the Lyndon B. Johnson Library. To date, space papers have not to my knowledge been opened at the Johnson Library. On space, the following references are useful: *Aeronautics and Space Report of the President*, 1963-1969, annual reports prepared by the staff of the National Aeronautics and Space Council during the administrations of John F. Kennedy and Lyndon B. Johnson. On general coverage NASA's annual chronology should be consulted, *Astronautics and Aeronautics: Chronology on Science, Technology and Policy*, 1964-1969 NASA SP's 4005-8, 4010, 4014. It goes without saying that the *Public Papers of the President* Series for Mr. Johnson are indispensable.

3. CBS, Interview of Lyndon B. Johnson by Walter Cronkite, Cape Kennedy, 7/5/69. Transcript in NASA History Office.

4. Preparedness Investigating Subcommittee of the Committee on Armed Services, U.S. Senate, *Hearings: Inquiry Into Satellite and Missile Programs*, Part I, December 17, 1957, pp. 579-606.

5. U.S. Senate, Special Committee on Space and Astronautics, *Compilation of Materials*, No. 2, April 14, 1958, p. 207. The Preparedness Investigating Subcommittee was noncommittal on space organization, though the only pending possibility was the creation of the Advanced Research Projects Agency in DOD. An attempt was made to record the history of the Special Committee, not too helpful, in Alison Griffith, *The National Aeronautics and Space Act* (Washington: 1962), with a foreword by Senator Johnson, pp. iii-iv.

6. The summary document is *Committee on Aeronautical and Space Sciences, United States Senate, 1958-1976* (Committee Print, December 30, 1976), 87 pp.

7. In a speech to the National Space Club in Washington, July 17, 1962, Senator Barry Goldwater came out with full endorsement for "space preeminence for the United States," later published in the *Airpower Historian*, X (April 1963). This was in contrast to former President Eisenhower's dim view of Apollo and most other Republicans of the cost of the Kennedy Space Program. Space was not an issue in the 1964 political campaigns.

8. Useful discussion of relationship of DOD-NASA Space Programs is in R. Cargill Hall, "Instrumented Exploration of Space: The American Experience," *Two Hundred Years of Flight in America* (AAS History Series, vol. 1: pp. 198-212.

9. Emme (ed.), *Statements of Presidents*, "Lyndon B. Johnson," p. 69.

10. See list of NASA published histories in *A Guide to Research in NASA History* (NASA 1979), pp. 15-18.

11. *Astronautics and Aeronautics, 1967* (NASA SP-4008), p. 76.

12. Cf Curtis Peebles, "The Manned Orbital Laboratory," *Spaceflight*, 22 (April 1980), pp. 155-160.

13. James E. Webb, in his Diebold Lecture at Harvard University, September 30, 1968. Mr. Webb would not criticize either President Johnson's grappling "in a period in which attention has been focused on other serious needs and problems," or the "phenomenon" of "decision by crisis" which gave space high priority after Sputnik and the Gagarin flight. But in his retirement statements he did not hesitate to state that the U.S. might face a new space crisis one day.

14. *The Vantage Point*, p, 286.

Richard M. Nixon

Forever it is recorded that Richard M. Nixon was President of the United States when two Apollo astronauts first landed on the moon of the Earth in July 1969. It was a historic moment in the long history of mankind. The *Eagle* lander yet stands as a monument at Tranquility Base as live-television-boosted memories dim with time. It was President Nixon who approved the development of the Space Shuttle. It will soon provide for revolutionary earth-orbital capabilities for a new era of space mobility and utility. Yet today all this contrasts starkly with the tragedy of the Watergate uncoverings. Such becomes a trilogy of White House tragedies when Watergate is coupled with John F. Kennedy's rendezvous at Dallas and Lyndon Johnson's defeat in Vietnam and on the home front.

Mr. Nixon made his tracings on space history before he moved into the White House as 37th President. His eight years as Vice President with President Eisenhower provided him with global perspectives and surrogate partisan political responsibilities. He lost Cook County and barely the election as President in 1960. And in 1968, Mr. Nixon became the first Vice President to win the White House in his own right since Martin Van Buren. He won reelection in 1972 with a landslide victory, losing only the state of Massachusetts and the District of Columbia.

Unfortunately for space historians, the Watergate crisis leading to the unprecedented resignation of President Nixon and the march of legions of lawyers have not facilitated the availability of non-Watergate presidential papers, or even the papers of Vice President Nixon. These gaps are not filled for historians by the flood of opinionated memoirs, commercial or journalistic gleanings, and all the rest of the outpourings since Watergate. Perspectives and time are needed. Therefore only a brief examination of Mr. Nixon's relationships with the space program can be provided here.

AS VICE PRESIDENT

Congressman Richard Nixon came to Washington in 1947, a product of the Duke University Law School and U.S. Navy service during the war. It was the moment when Soviet aggrandizement in Europe challenged the peace and resulted in the genesis of the Cold War. In the process of the evolution of a new foreign policy from the blockade of Berlin to the Korean War, Nixon became a Senator, a master of hyperbolic debate, which his early biographer, Earl Mazo, described as "the bayonet of the Republican offenses."[1] When General Eisenhower finally decided to run for the White

House, his selection of Nixon as his running-mate was based on Nixon's anti-communist crusades rather than any particular new dispensations of understandings about the future of the United States in a nuclear and jet-age world. Nixon's political partisanship meshed well as a surrogate; he had tens of thousands of invitations to speak one year, and he had political instincts which Ike sometimes consulted. It was Nixon who suggested, for example, that Eisenhower visit Senator Lyndon Johnson after his heart attack in 1955, and that Sherman Adams' position in a scandal was "untenable."[2]

In the well-ordered management staff of the Eisenhower White House Vice President Nixon gave his constitutional responsibility as President of the U.S. Senate the traditional minimal attention. Particularly in Eisenhower's second term, after his heart attack, Nixon was more a part of the operations in that he was in meetings of the National Security Council and conducted assignments on legislative matters. Eisenhower signed the document giving Nixon "acting President" powers, and after his "minor stroke" after Sputnik, Nixon presided over NSC meetings, which later included the National Aeronautics and Space Council sessions thereof.

When Sputnik came on October 4, 1957, Vice President Nixon became the first administration spokesman not to belittle the Soviet accomplishment. In a speech in Oklahoma City on October 12, Nixon supported the Eisenhower position: "It is obvious we are behind as far as the ability to launch a satellite is concerned... but there is a tendency to overestimate what the satellite will do in military power. Russia is not one iota stronger than it was before it put it up. As far as the missile field is concerned, we intend to keep the Soviet Union from gaining an advantage, and keeping our advantage."[3] Sherman Adams, in San Francisco on October 14, made his infamous remark that "the serving of science, not high score in an outer space basketball game, has been and still is our country's goal." The next day in San Francisco, the Vice President stated :

> ... We could make no greater mistake than to brush off this event [of Sputnik] as a scientific stunt of more significance to the man on the moon than to men on earth. We have had a grim warning and a timely reminder of a truth; we must never overlook that the Soviet Union has developed a scientific and industrial capacity of great magnitude.[4]

The next day's press conference of Secretary of State Foster Dulles voiced the Nixon line that "Sputnik was a very useful thing, as it had created unity of purpose between the administration, the Congress, and the people."[5] As we have previously noted, Eisenhower had earlier assigned PSAC with the task of coming up with the administration's responses on missiles and space, first attention being given to the ICBM race which the sputniks' launchers had demonstrated in the Soviet Union.

It was Wernher von Braun who arranged a briefing for Vice President Nixon on January 22, 1958, on the "National Space Establishment" proposal of the Rocket and Satellite Research Panel (RSRP). The RSRP had grown out of the V-2 Research Panel in 1946, coordinating scientific experiments by all governmental and university laboratories. Chaired by James A. Van Allan, chief experimenter on the Army satellite project, the RSRP called

Vice President Nixon revisited the Jet Propulsion Laboratory of Cal Tech on February 17, 1958, after the launching of *Explorer 1*, the first U.S. satellite. Examining the spacecraft model with him are (left to right): Clark B. Millikan and President Lee A. DuBridge of Cal Tech; Director William H. Pickering, Jack Froehlich, V. C. Larsen, and Robert J. Parks of the Jet Propulsion Laboratory. (NASA Photo No. 80-H-116)

for a billion-dollar-a-year non-military space organization, a concept developed before Sputnik, and submitted to the White House in December 1957. Nixon was briefed in his Capitol Hill office. He recommended that the Atomic Energy Commission hear this briefing, and directed Colonel Cushman, his aide, to arrange it. The next day the RSRP members briefed the AEC, and later the Joint Committee on Atomic Energy on Capitol Hill. The result was that Senator Clinton Anderson submitted his bill for instituting the civilian scientific portion of the U.S. space program in the AEC. Historically, the campaign of the AEC to incorporate space within its mission prodded the White House to study the alternatives in the national space organization.[6] Indeed the first White House decision on space organization was made quickly, that the AEC would not assume responsibility for the civilian space effort. The AEC was so advised. With the launching of *Explorer I* on January 31, the Sputnik crisis was sufficiently cleared to proceed on getting the space program organized. Vice President Nixon's views were politically sensitive and timely.

Visiting the Jet Propulsion Laboratory in Pasadena on February 2, Nixon said to the press that JPL "had not had the credit it deserves for its part in the development of the satellite, *Explorer*. Insofar as the public is concerned, the part played by the Army and its arsenal in Huntsville, Alabama, is well known. I have followed the work at Cal Tech with interest."[7] The *New York Times* reported that Nixon had said that *Explorer* is "only the first down in the first quarter of the game.... The U.S. must continue a stepped-up effort in the field of basic scientific research and applied research as it applies to outer space.[8]

Two days later, during a meeting with the President and Congressional leaders in the White House, Nixon spoke up for a civilian space agency. It requires mention mainly because it is generally not mentioned even in some otherwise fair histories. President Eisenhower told the Capitol Hill Republicans that rocket projects would probably remain under ARPA in the Pentagon because space projects used missile boosters. Dr. James R. Killian, Ike's Science Adviser, charged with coming up with a recommended space organization, spoke up. Killian said that he was "doubtful about this arrangement" as science and exploration would likely not be served under the military. This reflected PSAC's views. Vice President Nixon also spoke up in support of Killian, saying that space research and exploration should be conducted by an agency with no connections with the military. According to "Assistant President" Sherman Adams, Eisenhower stuck to his position, remarking that the government "was in no position to pour unlimited funds into expensive scientific projects," and he would rather "have one good Redstone nuclear-armed than a rocket that could hit the moon." Eisenhower said that he was not going to set up another competition between the Army, the Navy, and the Air Force for a lunar probe. Nixon spoke up again in favor of a non-military space agency. Eisenhower finally allowed that: "I don't rule out that eventually there might be a Department of Space."[9]

When Vice President Nixon returned to Cal Tech on February 17, he addressed an audience of 2,000 on the campus and held a news conference. He said: "It is vitally important, that we continue to emphasize that our effort[s] in this [space field] are for peaceful purpose.... Once we limit scientists to what military planners think possible, we place a limit that would be destructive in a contest in other areas. It cannot be under military control. We cannot tie down scientists to specific objectives which military men or political leaders may deem possible."[10] Eisenhower approved the recommendation for the creation of a NASA on March 5, 1958.

When the National Aeronautics and Space Act of 1958 was sent by the Congress to the White House, the top signature of the Congressional leaders was that of Richard Nixon as President of the U.S. Senate. With Eisenhower's signature it became law on July 29, 1958.

Vice President Nixon was dispatched to Moscow in 1959 just before Nikita Khrushchev's scheduled summit meeting with Eisenhower in Washington. The famous "kitchen debate" in the American exhibit at a Moscow trade fair was a public discussion. In private talks, Khrushchev confided that a Soviet missile had headed for Alaska but fortunately dropped into the ocean. Nixon gave a television and radio talk heard by millions of Russians, and *Isvestia* carried the full text. In it, Nixon stressed coexistence, saying:

> Let us expand the concept of 'open skies.' What the world needs are open cities, open minds, and open hearts. Let us have peaceful competition, not only in producing the best factories but providing better lives for our people. Let us cooperate in our exploration

of outer space. As a worker said to me at Novosibirsk, let us go to the moon together.[11]

Before he left Russia for Poland, Nixon thanked the Soviet press for carrying his speeches in full. He added that "many speeches I make in the United States do not receive such coverage."[12]

Vice President Nixon campaigned unflaggingly for the nomination for the presidency. Early in 1960, he was briefed on NASA's "MOM" concept, a closet term used sparingly by some NASA planners for getting a man on the moon quicker than the 10-Year Plan which retiring Eisenhower did not fund.[13] It called for a manned lunar landing early in the 1970's, which was about as far out as Nixon's campaign space rhetoric went. There were no votes to be gained or lost in debating on space as an issue, although John Kennedy meshed the space frontier with his "getting the country moving again."[14] The closeness of the election made it a bitter defeat for Mr. Nixon in 1960. Yet Richard Nixon's comeback and election as President in November 1968 became all the more remarkable.

THE APOLLO CLIMAXES

President-Elect Nixon wasted no time forming his administration before *Apollo 8*, to be man's first voyage around the moon at Christmas time. He had campaigned that the United States should "never be second in space." Early in December 1968, Nixon announced that President Lee A. DuBridge of the California Institute of Technology would be his Science Adviser. He also announced that Dr. Charles Townes, a key adviser on Apollo for NASA and a Nobelist, would head a task force to make recommendations on space. He named Dr. H. Guyford Stever as head of a task force to make recommendations on the needs of general science and technology.[15]

By December 11, President-Elect Nixon introduced his Cabinet in person on television. In the meantime, Vice President-Elect Spiro Agnew would inherit Lyndon Johnson's and Hubert H. Humphrey's job of being the chairman of the National Aeronautics and Space Council. Except in the opinion of some Baltimore friends, Mr. Agnew had no particular ken for his statutory responsibility--it had been created for Lyndon Johnson in 1961. There was no great wisdom in changing NASA administrators, or even announcing one until after *Apollo 8* and the change of administrations on January 20, although Astronaut Frank Borman was to become a confidant. Nixon named Robert C. Seamans, Jr., as Secretary of the Air Force. This was just days before Lyndon Johnson announced the termination of the B-70 bomber program.[16]

In his inaugural address, President Richard Nixon sounded a space theme, saying:[16]

> ... Those who would be our adversaries, we invite to a peaceful competition--not in conquering territory or extending domination, but in enriching the life of man. As we explore the reaches of space, let us go to the new worlds together--not as new worlds to conquer but as a new adventure of mankind....

He quoted Archibald MacLeish's poem inspired by *Apollo 8:*

> To see the Earth as it truly is, small and blue and beautiful in that eternal silence where it floats, is to see ourselves as riders on the Earth together....

A climax in space exploration was timed for a new President. Little wonder that the *Apollo 7* Astronauts rode on a NASA float in the Inaugural Parade. And, in his first press conference President Nixon announced that Frank Borman of *Apollo 8* would make a goodwill trip to Europe.

Thomas O. Paine was not actually reaffirmed as NASA Administrator until the Goddard Dinner in early March, when Science Adviser DuBridge gave a statesmanlike review of the historical evolution of the space program. It was not until April that *Apollo 8* Astronaut William Anders was named as Executive Secretary of the National Aeronautics and Space Council in the White House. Meanwhile, of course, the Nixon FY 1970 budget for NASA was reduced $45 million, but monies were provided for additional Saturn V's to continue manned exploration of the moon. *Apollo 9*, and then *Apollo 10* permitted the President to honor astronauts. Indeed, *Apollo 10* performed all of the mission of the lunar landing except touchdown (LEM flown to within 9.6 miles of the lunar surface). While the glories of manned lunar flight were shared by the world and the White House, it was also becoming apparent that the post-Apollo space program could not command dollar resources prompted by the space race begun with the sputniks. On May 19, Senator Edward Kennedy, speaking at the dedication of the new Goddard Library at Clark University, called for a slow-down of the space program.[17] That was a profanity to the memory of John F. Kennedy who had set Apollo in motion, if not also to the memory of Robert H. Goddard. The landing of *Apollo 11* was just sixty-odd days away, as also was a personal tragedy at a bridge at Chappaquiddick.[18]

President Nixon had early revealed his intent to exploit fully the increasingly dramatic successes of Apollo in augmenting his personal initiatives to "de-Americanize" the Vietnam War and to seek a détente in tri-polar geopolitical world with Moscow and Peking. On January 30, just ten days in the White House, while explaining Astronaut Frank Borman's goodwill trip abroad, Nixon said: "We do want to band together with other people on this earth in the high adventure of exploring the new areas of space." Borman's warm reception in Europe in February, including Czechoslovakia, was prelude to sending the *Apollo 9* crew to the Paris Air Show in May, and then sending Borman to Moscow in June 1969. NASA's information plan notwithstanding, Nixon pulled out all stops on *Apollo 11*. He was targeted for criticism when the plaque on the *Eagle* lander was publicized. It had President Nixon's signature below the autographs of the three astronauts, and the message--"We Came in Peace for All Mankind." A *Washington Post* editorial, entitled "Our Mark on the Moon," traced all those before President Nixon, from Jules Verne on, who had made contributions to the concepts and technology making a lunar landing possible. It even cited a so-called "history" by the NASA Historian (actually only an AIAA Paper) which had not directly referred to Mr. Nixon. But the flap was immortalized in the Herb Block cartoon which follows:[19]

HERBLOCK'S CARTOON

Lunar Hitchhiker

----copyright 1969 by Herblock in The Washington Post

Used with permission of Mr. Herb Block, *Washington Post*

On the eve of the launching of *Apollo 11*, President Nixon telegraphed the three astronauts at the Cape (i.e., hard copy) on the forthcoming "triumph all men will share," and also telephoned them just before lift-off. He then issued a proclamation naming July 21, as a "National Day of Participation." He announced that *Apollo 11* was carrying symbols commemorating earlier Russian and American space heroes who had died in the evolution of space flight.

Everyone here today probably recalls his conversation via communications satellites to the astronauts on Tranquility Base. But when he greeted the astronauts in their quarantine trailer on the carrier *Hornet*, President Nixon spoke nervously, saying: "I think I am the luckiest man in the world, and I say this not only because I have the honor to be President of the United States, but particularly because I have the privilege of speaking for so many in welcoming you back to earth.... I was thinking, as you know, as you came down, as we knew it was a success, and it had only been eight days, just a week, that this is the greatest week in the history of the world since the Creation, because as a result of what happened in this week, the world is bigger, infinitely..., as a result of what you have done, the world has never been closer together."[20]

Apollo 11 astronauts are greeted by President Nixon from inside their Mobile Quarantine Facility aboard the *USS Hornet* after returning from the Moon. (Left to right): Neil A. Armstrong, Michael Collins, and Edwin E. Aldrin, Jr. (NASA Photo No. 69-H-1196 or 108-KSC-69P-658)

If Nixon was the first President to witness a spashdown live, he was also just launching a diplomatic voyage around the world. He carried the "spirit of Apollo" on *Air Force One* to the Philippines, Indonesia, Thailand, Vietnam, India, Pakistan, Romania, and England, before returning to Washington. He became the first President to visit behind the Iron Curtain. When in Bucharest, he said: "I believe that if human beings can reach the moon, human beings can reach an understanding with each other."[21] A decade later Henry Kissinger was to write in his memoirs of this global trip that Nixon had a "moon glow" and had a "moon walk." To be sure it became tangy old stuff that the moon-landing achievement bode solutions for international problems and lots of other things. Even Henry Kissinger was moved to a recognition of some significance to space demonstrations. In the company of his children, he visited the gigantic Apollo-Saturn on its launching pad at the Cape just before the take-off of *Apollo 12*.[22]

President Nixon's travels abroad did almost become legendary--three in 1970, and three in the election year of 1972, including his historic visits to China and the Soviet Union.[23] But the greatest space splashdown party was the gigantic bash, the Apollo dinner held in the Century Plaza Hotel in Los Angeles, sponsored by the President, "to honor the first manned landing on the moon." The bulk of aerospace leaders attended, plus the widows of deceased astronauts, Mrs. Robert H. Goddard, aviation pioneers, the entire Cabinet, the Joint Chiefs of Staff, the governors of forty-four states, former Vice President Humphrey, and others. *Apollo 11* changed history without social trappings. It was a world-wide event, and it concluded the "Space Race" accelerated by President Kennedy.

The significance of President Nixon's post-splashdown diplomacy was noted. Nixon's round the world trip, from splashdown to Southeast Asia to Romania could be seen; a *Life* magazine editorial commented: "a half moon journey around the great Communist land mass of the Soviet Union and China. The Communist powers were not on his [Nixon's] itinerary, but the violent ideological quarrel was much on his mind in all the political business he conducted... a kind of symbolic signalling that diplomacy frequently employs these days, that the U.S. had no intention to play a favorite in the rift between China and Russia...."[24]

POST-APOLLO DECISIONS

Winding down the inherited Vietnam conflict with honor and winding up détente until the Watergate uncoverings were companion with the completion of five more Apollo explorations on the surface of the moon, three missions to the *Skylab* station, and the decision for the U.S.-U.S.S.R. Apollo-Soyuz Test Project (ASTP) during the Nixon administration. The decision to develop the Space Shuttle for the 1980's began within the first months, along with interplanetary and satellite applications.

President Nixon established a Space Task Group as it "was necessary for me to have in the near future recommendation on the direction which the U.S. space program should take in the post-Apollo period." His memorandum of February 13, 1969, formally sanctioned an interagency mechanism involving NASA and DOD as principals under the National Aeronautics and Space Council, chaired by Vice President Spiro Agnew. Science Adviser DuBridge was to come up with a long-sought post-Apollo plan before the climax of the lunar landing which had dominated the 1960's. Space momentum should not be lost because of a lack of a firm plan, if only to buttress budget requests. Nothing public was stated about DOD projects which would buttress strategic arms limitation negotiations with the U.S.S.R. historians being dependent on sparse official and industry reports.

Congressional budget cuts to the lowest level for NASA since 1963 in FY 1969 had indeed been a strong negative trend. Moreover, the White House faced both costly and hence contradictory projections from NASA with its near-term intent to explore the lunar surface in subsequent Apollo missions. Project Skylab, as well as the Air Force's Manned Orbital Laboratory, was also additional to development of a recoverable Space Shuttle, of a space station ultimately for 100 inhabitants in lunar orbit, and the longer-term goal of landing men on the planet Mars. What made the best sense on NASA's costly menu?[25]

President Nixon's own advisers appeared to have differing rationales and goals, at least in their public remarks concerning "new directions" and "balancing that enterprise." The initial space panel under Townes had submitted a recommendation for a $6 billion per year effort ($2 for DOD, rest for NASA), opposing the large orbital space station but endorsing Space Shuttle development for NASA and DOD, and a commitment to unmanned planetary exploration coupled with higher priority for practical benefits and international cooperation. Critics for ending Apollo "soon after" the first lunar landing was achieved, such as Ralph Lapp in the *New York Times*, voiced their priorities. Nixon requested that DuBridge assess the recommendation to create the Space Task Group to develop the Post-Apollo program, and to report on the "possibility of significant cost reductions in the launching and boosting operation of the space program."[26]

The Space Task Group submitted its report to President Nixon on September 15, 1969, the preparation of which under NASA's Paine and the Air Force's Seamans is beyond our scope here.[27] It was entitled "The Post-Apollo Space Program: Directions for the Future." Optimistically, it recommended NASA's long-range goal of a manned Mars landing, a focus to give order to exploiting existing capabilities and developing coherent interim capabilities in a balanced program--enhancing space applications and defense needs, a strong program of manned lunar exploration and instrumented planetary exploration for science, and the development of new operational technology, some involving international participation, and with a view to reduced costs of space operations. It offered options, with a decision for the Space Shuttle request in FY 1971, for the manned mission to Mars in 1980, in 1986, or before the close of the century. The earliest date involved $100 billion, probably reflecting optimistic conversations with President Nixon while awaiting the splashdown of *Apollo 11* on the carrier *Hornet*, and including the notion that an international aegis be mobilized to support the Mars landing venture. It did explain the origin of Vice President Agnew's call for a Mars landing at the launching of *Apollo 11*.[28] It was an idea before its time.

On receipt of the Space Task Group report, White House Press Secretary Ron Ziegler stated that President Nixon concurred in the STG's rejection of the extreme space opportunities: (1) to land men on Mars as soon as possible regardless of cost, and (2) to eliminate manned space flight after the completion of Apollo. Nixon's decisions would be made in the budget.

President Nixon's budget message of February 2, 1970, said: "Man has ventured to the moon and returned--an awesome achievement. In determining the proper place for future space activities we must carefully weigh the potential benefits.... I have received many exciting alternatives for the future. Consistent with other national priorities we shall seek to extend our capabilities--both manned and unmanned." NASA's budget was $4.148 billion for FY 1971. The boom was busted for the Apollo-dominated era. In March 1970, the White House released the first statement on national space objectives since Eisenhower and Kennedy, the preamble of which pointed to the lack of "a comprehensive plan" for a post-Apollo space program. The "Apollo hangover," as Professor John Logsdon termed it in his preliminary analysis of the Space Shuttle decision, was quite evident in the optimistic space futures evident in the manned Mars landing goal, and which was to be lowered to a more economic and long-term development of the Space Shuttle for the 1980's.[29]

The pivotal decision on the Space Shuttle Program, from all available evidence, was made by President Nixon through negotiations with the new NASA Administrator James C. Fletcher, who replaced Paine in May 1971. After much discussion with White House staffer Peter Flanagan and OMB Director George Shultz, it was clear to Fletcher and Deputy Administator George M. Low that President Nixon wanted a manned space program with the following constraints:

- No major increase in NASA's budget level.

- Any manned program which would be developed in the billions of dollars category would have to satisfy OMB's "cost effectiveness criteria."

- The Shuttle would have to advance the state of the art in space vehicles, but use as much of the Apollo technology as possible.

- The program would have to be achievable in a reasonably short time (hopefully by the end of 1976, which turned out not possible), but not a "crash program" in the sense that Apollo was.

"Cost effectiveness" gave NASA a chance to think through its entire program for the next thirty years.

Fletcher and Low had their own criterion which required that NASA proceed along the basis of a balanced program between manned space, space science, aeronautics and space applications. This meant not permitting the Space Shuttle to dominate NASA's budget for the next several years as the Apollo program had done in the 1960's.

After much negotiation with middle management in OMB, a recommendation was made to President Nixon that NASA develop a single-stage Space Shuttle using two low-altitude boosters (using either solid fuel or storable propellants, the decision to be made at a later date). This program was to be completed in early 1978 at a cost of no more than 5.2 billion dollars (with a 20% allowance for overrun) measured in 1971 dollars. Fletcher and Low also demanded a commitment that NASA's budget would remain at least at the current level of 3.2 billion dollars in outlays (in 1971 dollars). This latter was necessary in order to maintain a balanced program within NASA and still bring the Shuttle in on schedule. Other lower-cost alternatives would not be anything like the cost effectiveness of the Shuttle.

When all options were presented to George Shultz, Peter Flanagan, and finally to President Nixon, the large version with a 65,000-pound payload capability, with cargo dimensions of 60-feet in length and 15-feet in diameter, was easily the most attractive and the best buy. Middle management in OMB, also various other White House staffers, tried in vain to sell the President on a much cheaper program using a Titan upgrade as a booster. It was clear this had no economic advantages over the Space Shuttle design submitted.

The decision for the Space Shuttle was made by President Nixon late in December, and Low and Fletcher flew to San Clemente early in January 1972, at which time the President made the announcement to proceed. During these discussions, Nixon mentioned that international participation was greatly encouraged but as the U.S. Air Force would undoubtedly require use of the Shuttle for many of its missions, military considerations should also be taken into account. This made it difficult for international participation in the design of the Shuttle itself, but opened the door for an international payload which ultimately led to the European "Spacelab" program. Ten European nations built the manned laboratory that would be used for future manned space experiments.

President Nixon with NASA Administrator James C. Fletcher at San Clemente, California, on January 5, 1972, the day it was announced that development of the Space Shuttle would be undertaken for the 1980's and 1990's. (NASA Photo No. 72-H-31)

President Nixon examines model of the Apollo-Soyuz spacecraft with the *Apollo 16* astronauts and NASA Administrator on June 15, 1972. Shown (left to right) are Charles M. Duke, Administrator Fletcher, the President, John W. Young, and Thomas K. Mattingly in the Oval Office. (NASA Photo No. 72-H-826)

Due to severe budget constraints in 1972, NASA's budget was cut by almost a half-billion dollars. The first flight date of the Shuttle slipped by one full year. Later cuts, though less severe, reduced confidence even in the 1979 date, which undoubtedly had some impact on the schedule delays in 1978-1980. Thus, although President Nixon (and subsequently President Ford) continued to support the Shuttle Program in principle, the budgetary process with its cuts did not allow the orderly development that the Apollo Program had enjoyed.[30]

The transition from the Apollo conclusion to the undertaking of the Shuttle proved a most difficult period in NASA's history. There were also increased Department of Defense requirements in the national space budget supporting President Nixon's drive for détente and reelection. He never neglected to participate in the honors for the Skylab achievements while the perturbations of winding down Vietnam were to be followed, after his landslide re-election, by Watergate.

Historians normally avoid generalizing from specific moments or events in time, unless they illustrate a meaningful fabric of past realities in focus. President Nixon's farewell to a NASA audience was to be brilliantly reported in Elisabeth Drew's *Washington Journal--The Events of 1973-1974*, a collection of her contemporary writings. With her generous tolerance for this non-commercial paper, her observation of the awards ceremony for the *Skylab 3* astronauts at the NASA Lyndon B. Johnson Space Center in Houston perforce is pertinent here. It was Mr. Nixon's last NASA appearance before a captive and a select audience on a government reservation during the Watergate siege. Miss Drew wrote:[31]

> We hear "Ruffles and Flourishes" and Hail to the Chief.... Citations of awards are read.... His [Mr. Nixon's) eyes are barely visible. But he poses with each astronaut, and reworks the tired face into a broad smile. And when he begins to speak he can pump the emphasis into his voice and his gestures. He can make the effort. He is a pro, remarkably resilient. The President speaks without notes. He refers to the astronaut medals, "which I hope they will never have to hock." He says that he invited the astronauts and their wives to Camp David, and then goes into an almost mystical description of the place; the President rarely speaks of nature, beauty, physical things. "The clouds are right around you; when it is like that, you may think you were in space." He says he would like to volunteer for a space voyage. He tells the audience about his blood pressure.... "A great people must always explore the unknown. Once a great people gives up or bugs out, drops out of competition for exploring the unknown, that people ceases to be great." He talks about the Spanish and French and British explorers and about the American pioneers .... I have seldom heard him so discursive...
>
> Then he talks about the nation's being at peace "for the first time in twelve years," and about the fact that "all prisoners of war are home," and about the negotiations with Russia and China.... Praising the astronauts of the aborted *Apollo 13* mission, he says, "they didn't fail. The men and the women on the ground didn't fail, because you are only a failure when you give up, and they didn't give up." ....

The President again seems reluctant to leave the platform, he waves to the crowd, and then he waves at the band to stop it from playing the exit music.... As Nixon finally departs, the Clear Lake High School Band plays "The Washington Post" march.

RICHARD M. NIXON

REFERENCE NOTES

1. Best early biography of Richard Nixon remains Earl Mazo, *Richard Nixon: A Political and Personal Portrait* (NY: 1960). The only pre-Watergate work by a historian is Bruce Mazlish, *In Search of Nixon: A Psychohistorical Inquiry* (Boston: 1973). Nixon books are, *Six Crises* (NY: 1968), part of his comeback campaign, and *RN: The Memoirs of Richard Nixon* (NY: 1977), additional to the official *Presidential Papers* volumes.

Post-Watergate works have not been assessed here; cf. commentary by Edwin M. Yoder, Jr., "Instant Historians and Mr. Nixon," *Washington Star*, 12/18/75, p. A-24. Yoder submits a "quotation" of Sir Walter Raleigh in 1614: "Whoever, in writing modern history, shall follow truth too near at the heels, it may haply strike out all the teeth." Non-political historians will eventually have their say after greater perspectives and much more of the full documentation are available, and, on space material, beyond the NASA histories and chronologies.

2. Mazo, pp. 173-77; R. Cutler, *No Time for Rest* (Boston: 1965), p. 307.

3. *Chicago Tribune*, 10/13/57. This and other primary and secondary items on the impact of the sputniks may be found in E. Emme, "From Sputnik to Apollo: A Chronological Source Book on the Creation and Early Years of NASA, 1957-1961"(MS, NASA Historical Report No. 99, 1977), copies in NASA History Office, and Library of the National Air and Space Museum.

4. *New York Times*, 10/16/57; cf. Sherman Adams, *Firsthand Report* (NY: 1961), p. 412.

5. *Facts on File, 1957*, p. 338.

6. "Notes" of RSRP Meeting, 2/14/58; interviews by the author of W. von Braun, H. E. Newell, Wm. Stroud, N. Spencer, *et als.* Nixon's recollection in *Pres. Documents*, 3/1/71. Cf.H. E. Newell's forthcoming *Beyond the Atmosphere: The Early Years of Space Science* (NASA SP-4211).

7. Jet Propulsion Laboratory, *Lab-Oratory* (March 1958), p. 3.

8. *New York Times*, 2/3/58.

9. Sherman Adams, *Firsthand Report*, pp. 17-19; Excepted interviews of participants. Adams' book was cleared by Eisenhower before publication.

10. Gladwin Hill, "Nixon Asks that Space be Civilian Zone," *New York Times*, 2/18/58; *Lab-Oratory* (JPL) (March 1958), pp. 1-3.

11. William Safire, *Before the Fall* (NY: 1975), p. 3; Mazo, pp. 219-20; Interview of Herman Mark, NASA Lewis Research Center (5/21/77), who was at the NASA-USIA exhibit, which had 120,000 visitors daily; "Explorer VI is Hit of U.S. Fair in Moscow," *New York Times*, 8/21/59.

12. *Keesing's Contemporary Archives, 1959-1960* (London), August 1-8, 1959, pp. 16933-35; Time-Life Books, *Time Capsule/1959* (NY: 1960), pp. 27-32.

13. Excepted interviews.

14. John Logsdon, *Decision*, pp. 66-67; T. H. White, *The Making of the President in 1960* (NY: 1961), pp. 115-19; Cf. E. M. Emme (ed.), "Richard M. Nixon," *Statements by Presidents on International Cooperation in Space: A Chronology - Oct. 1957-Aug. 1971* (Senate Document No. 92-40: 9/24/71), pp. 91-126.

15. Most subsequent details are found in *Astronautics and Aeronautics: A Chronology on Science, Technology and Public Policy, 1969* (NASA SP-4014, 1970). It is a most complete annual edition.

16. *A/A* 1969, pp. 20-21.

17. *Congressional Record*, Senate, 3/13/69, p. 522755.

18. The aerospace community had generously contributed the bulk of the money for the outstanding modular Goddard Library at Clark University, the dedication of which was not considered by most attendees as a political event. See *A/A, 1969*, p. 147.

19. "Our Mark on the Moon," editorial in *Washington Post*, 7/3/69, p. A-14. The "NASA's official history of the Apollo project," to which it referred, was the author's "Perspectives on Apollo," AIAA Paper No. 67-839, later published in the *Journal of Spacecraft and Rockets* (AIAA), April 1968, pp. 369-82. Herb Block's cartoon appeared in the *Washington Post*, 7/4/69. We are greatly appreciative of his permission to publish it in this paper.

20. "Since the Creation" statement did appear to top all post-*Apollo 11* comments. It seems also true that President Nixon had the largest television audience to date, roughly estimated at over 700 million persons worldwide. Cf. *Fourth Colloquium on Oral History*, ed. G. P. Colman (NY: 1970), pp. 8-9.

21. "Statements of Presidents on ... Space," pp. 105-106.

22. Henry Kissinger, *White House Years* (Boston: 1979), pp. 223, 808-809; preliminary article in *New York Times Magazine*, 8/17/69, pp. 26-29, 76-80.

23. Interesting tidbits in J. F. ter Horst and R. Albertazzie, *The Flying White House: The Story of Air Force One* (NY: 1979), *passim*.

24. "Editorials: Journey round a new landscape," *Life* (July 1969), p. 30.

25. NASA's basic input was a massive outline submitted to the Space Task Group entitled *Goals and Objectives for America's Next Decades in Space* (Washington: September 1969), 108 pp.

26. *Space Task Group Report to the President, the Post-Apollo Space Program: Directions for the Future* (Washington: September 1969), 29 pp.

27. Paul R. Brockman, "The Space Task Group--1969: Catalyst for the Redirection of American Space Policy," Honorable Mention in the NSC Goddard Historical Essay Competition," 1975, 28 pp.

28. Excepted interviews. Cf. Charles J. Donlon, "Space Shuttle Systems Definition Evolution," NASA paper submitted to the Acting Director, Space Shuttle Program, 7/11/72, copies later sent to White House, 6 pp.

29. John M. Logsdon, "The Space Shuttle Decision: Technology and Political Choice," *Journal of Contemporary Business*, 7 (August 1978), pp. 13-29; and "The Policy Process and Large-Scale Space Efforts," *Space Humanization Review* (to appear in initial issue in 1979), 23 pp.

30. Excepted interviews. Cf. H. A. Scott, "Space Shuttle," *A/A* (6/79), pp. 54-59.

31. Elizabeth Drew, *Washington Journal--The Events of 1973-1979* (NY: 1975), pp. 209-211.

Gerald R. Ford

When Vice President Gerald R. Ford became the 38th President in August 1974, he faced extraordinary problems. It was, he said, a job to which he had never aspired. For thirteen times he had been elected as a Congressman from Grand Rapids, Michigan, and had a diligent career. Early Mr. Ford had become a member of the powerful House Appropriations Committee, and in 1964, was elected as Minority Leader. Congressman Ford had not been elected as Vice President in December 1973; he was nominated by President Nixon and confirmed by the Congress to replace Spiro T. Agnew. In his Inaugural Address eight months later, President Ford said: "Truth is the glue that holds government together."

When President Ford's 30-month term had ended, President Jimmy Carter first said: "For myself and for our nation, I want to thank my predecessor for all he has done to heal our land." While the institution of the presidency survived Watergate and the decade-long Vietnam debacle, the concurrent history of space affairs should not be neglected.

In Mr. Ford's subsequent autobiography, *A Time to Heal*, sadly he does not refer--by index subject, name, word, or picture on any page--to American space affairs before or during his presidency.[1] It is as much testimony to ghosted drafts and publication commerce as well as unhelpfulness to genuine history or public memory.

President Ford supported NASA and the Shuttle development, NASA's role in solar energy research, and perhaps the pinnacle achievement of détente in the Apollo-Soyuz mission with the U.S.S.R. The last Apollo mission was *ASTP*, and it demonstrated that Americans and Russians could get along in the hostile environment of space.[2] President Ford had his summit meetings in Vladivostok and Helsinki, resulting in the SALT I accords. He also welcomed the first test flight of the B-1 bomber, and supported the Trident and CM systems. President Ford restored a Science Adviser in the White House, Congressional action which he did not veto, and retained able James Fletcher as NASA administrator. The Department of Defense portion of the space program was funded as it made SALT I and its follow-on viable.

And President Ford dedicated the opening of what quickly became the world's most popular museum, the National Air and Space Museum on the Washington Mall. It was a national Bicentennial Event in July 1976, which was augmented by the landing of a sensible spacecraft on another planet--the two historic Viking landers and two orbiters probing Mars, the sister planet of the Earth.[3]

President Ford deserves more public memory than his pardon of Richard Nixon to de-orbit Watergate into the past. He fought for his party's nomination in the Bicentennial Year. And he failed to get elected by a narrow margin of 2% of the total vote, or less than 8,000 votes in Mississippi and 7,000 more in Ohio. Might-have-beens are not the stuff of genuine history. Some additional notes on American space history during the Ford administration might at least stir some future historians.

BACKGROUND

When the so-called Space Age took off on the fire-tail of Soviet ICBM rockets in 1957, the House Committee on Appropriations knew where the funding had gone for missiles and space. Few could argue that national defense was not of paramount importance, which included Republican Congressman Gerald R. Ford. When Majority Leader John W. McCormack was named as chairman of the House Select Committee on Astronautics and Space Exploration on March 5, 1958, junior but able Ford was a logical member. This was the time when Congressional leaders had been informed that President Eisenhower had decided to create a National Aeronautics and Space Agency atop a reconstituted National Advisory Committee for Aeronautics. The White House bill was not transmitted to the Hill until April 2. But the House Select Committee held lengthy hearings on space organization legislation before the Senate Special Committee on Space and Astronautics, authorized on February 5, began its detailed hearings during May 1958.[4]

Mr. Ford attended all House Select Committee hearings, confronted every witness in the process which was to result in the National Aeronautics and Space Act of 1958. It was Congressman Ford, however, who insisted that NASA's annual budget authorization should not be placed with the standing space committees of both Houses of the Congress. This was Senator Lyndon Johnson's capstone for giving his Senate standing space committee the power of authorizing NASA's budget. Ford argued that the authorization process by standing committees would merely prolong NASA's obligations on the Hill additional to the functions of the Appropriations Committees. But the views of the Senate Majority Leader Johnson prevailed. He also could function, and did, through House Speaker Sam Rayburn of Texas, and had his way.[5]

Congressman Ford did not become a member of the standing House Committee on Science and Astronautics. The fact that he became Minority Leader in 1964 was testimony enough on his effectiveness as well as his involvement in all sectors of the work of the House of Representatives.

IN THE WHITE HOUSE

Vice President Ford had no space duties like his predecessors. President Nixon had abolished the National Aeronautics and Space Council as set up by President Kennedy with the Vice President as chairman. Mr. Ford, throughout his White House tours, it is said, continued to keep in touch with the people he knew best. He read his home town newspaper daily. This was in contrast to one of his predecessors who had one of his staffers write his weekly letters to his mother. Vice President Ford's value to the Nixon administration, as well as his own, was his knowledge of and

House Select Committee on Astronautics and Space Exploration members and staff before hearings on April 16, 1958. Member Gerald Ford (standing to right) and Chairman John W. McCormack (seated to right). Witness was H.L. Dryden, NACA Director (seated center), with Echo communications satellite models as the backdrop. (NASA Photo No. 76-H-550)

intimate contact with Capitol Hill. He could relate to his long-time friends on both sides of the aisle. He once referred to Congressman Olin Teague, the space stalwart of the House: "He is your tiger in your tank."[6] Everyone knew precisely what he meant. In turn, Congressmen Teague and others always had an open ear in the White House.

Budgetwise, the U.S. space program fared somewhat better under the Ford administration, not counting inflation and help on Capitol Hill. The Space Budget for FY 1974 had been $4.6 billion ($2.9 billion for NASA), the lowest dollar level since 1963. Under President Ford, FY 1975 total rose to $4.9 billion, $5.3 billion for FY 1976, and $6 billion ($3.4 billion for NASA) for FY 1977.[7] Costs for the Space Shuttle rose with fabrication and launch facilities, which increasingly involved Air Force space futures.

President Ford was to be spared the excitement of the reentry of *Skylab*, but a problem was identified. Early in 1975, the problem for the 79-ton *Skylab* was initiated by the return to earth of its S-II stage. NORAD had advised NASA Administrator Fletcher on January 11, that the S-II would re-enter on Los Angeles on its next orbit. It did not. Its following orbit would take it over the heavily-populated northeastern U.S. Two days later, NORAD verified that the S-II had disappeared. Its splashdown was never fully confirmed except by a flash of light south of the Azores seen by an airline pilot. Increased solar activity, however, shortened the reentry date of Skylab to 1980 rather than 1983. In the meantime, the Shuttle was scheduled to begin its flight tests in 1979, and possibly could get to Skylab, if anyone thought it should be done.[8]

The *Apollo-Soyuz Test Project (ASTP)* in July 1975 directly involved the Ford White House. For NASA, *ASTP* filled the manned space flight gap between *Skylab* and the Shuttle. It was born of Nixon's "spirit of Apollo" applied to his tri-polar world concept of detente. President Ford was briefed by NASA's Fletcher, and persuaded to participate personally in the *ASTP* demonstation. International cooperation in manned flight was an advancement in space affairs involving the Soviet Union and the United States, though *ASTP* was strictly a one-shot venture. President Ford watched the launching of the *Soyuz* in the U.S.S.R. on television in the auditorium of the State Department, and in the company of Soviet Ambassador and Mrs. Dobrynin. Later, Ford talked nine minutes instead of the five scheduled, with the American and Soviet crews assembled in orbit. He said that it had "taken us many years to open this door to useful cooperation in space between our countries." This was also the gist of Soviet Chairman Brezhnev's remarks to the *ASTP* crews. It had also been NASA's job to pacify Senator William Proxmire and others regarding the safety of the Soyuz (the first launch attempt failed), as well as answer critics claiming unnecessary technology transfer of Apollo know-how.[9] After the successful mission, President Ford entertained the two Soviet cosmonauts and the three American astronauts at the White House. They later trooped to the United Nations in New York to deliver a U.N. flag orbited on the *ASTP*.

President Ford watching *ASTP* crew members Stafford, Slayton, and Kubasov on television, as he talks to them while they orbited the earth on July 18, 1975. (NASA Photo No. 75-H-804)

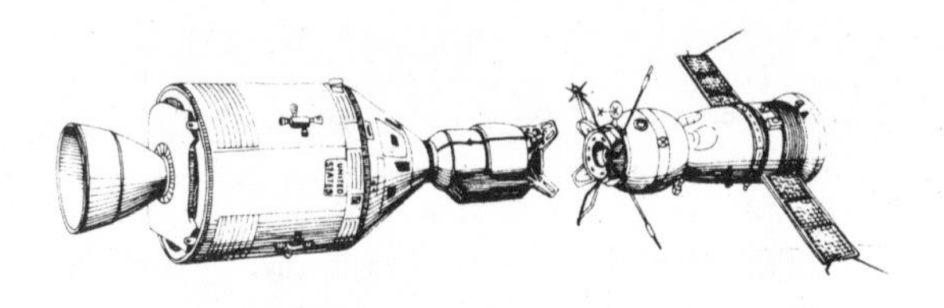

Drawing of Apollo-Soyuz Mating *(ASTP)*

Apollo-Soyuz crew meets President Ford in White House. (Left to right) Gen. V. Shatalov, Cosmonauts V. N. Kubasov and Alexei A. Leonov, Ambassador A. Dobrynin, President Ford, NASA Deputy Administrator George M. Low, and Astronauts Thomas P. Stafford, Donald Slayton, and Vance D. Brand. (NASA Photo No. 74-H-807)

Apollo-Soyuz at long last terminated the 14-year-old Apollo. NASA Administrator James C. Fletcher was provoked to vent a most candid strategic outlook for the future of the space program in an address to the National Academy of Engineering on November 10, 1975. It was neither input to a White House staff paper nor testimony for a Congressional committee. Dr. Fletcher said:

> Today our space programs are heavily oriented in the 'problem-solving' direction, and have produced notable accomplishments... [i.e., meteorological, communications, navigation, and resources satellites, and the re-usable Space Shuttle will add man's flexibility to the near-Earth space environment and open the way for manufacturing and processing of new materials that cannot be duplicated on Earth....]
>
> These programs all have a common denominator--they are all space programs structured to provide direct service to mankind... and will be expanded....

In concentrating on the 'now' problems we are forced to ask questions about the future: Are we losing sight of the 'dream'? As one Congressman expressed it: Are we sacrificing our destiny in order to satisfy our desire for immediacy in everything....

The grave problems that confront our times should not force us to hang a price tag on everything we do and then haggle over the prices as though we were shopping in some ancient Eastern bazaar. The danger of this mercenary approach is that we may lose sight of the incalculable rewards beyond the innermost fringe of our space goals.... It was narrow vision that prompted the Vikings to overlook the potential of America five hundred years before Columbus.

Dwindling resources and contamination of the planet Earth's environment are more recent examples of man's tunnel vision. And the blinders are still on. Our answer to the looming energy crisis is apathy. We should have been concerned about energy a quarter of a century ago. Instead, we recklessly plundered a leftover treasure from the Sun as if it was unlimited.... With uranium and fossil fuels headed towards depletion we should be giving serious attention to solar energy.... If we had placed the same emphasis years ago on ways to utilize solar energy as we have put into the development of a nuclear generating capacity, we might already be well along the road of solving the energy crisis. We should begin to think seriously about putting up a permanent, manned space station, an engineering feat well within the limits of current technology....

We should also expand our exploration of the planets. From these undisturbed worlds we can gain a better understanding of the forces that shape our own planet and its evolution. It is not at all unlikely that someday we may look upon some of these distant worlds and their satellites as havens from a ravaged and teeming Earth--long range alternatives to the orbiting space colony....

Is NASA, itself, becoming shortsighted? Our expenditures are weighted heavily in favor of contemporary needs.... It represents an accommodation with current constraints.... The 'now' attitude of the public leads to a second time horizon--the political term of office. A legislator must be responsive to the demands of his constituents. As a result, he seldom is afforded the luxury of thinking beyond his present term and thereby perpetuates the 'now' philosophy....

The Office of Management and Budget, which controls the government's pursestrings, rarely plans beyond one or two years at a time. It is responsive to the dictates of political and economic pressures which more often than not reflect only the day-to-day needs of society. NASA's Space Shuttle program is an excellent example of the effects of year-to-year budget cycles. The program has never been funded in its entirety, but has been piecemealed together out of the Agency's overall yearly budget. Yet, if NASA did not proceed with the development of the Shuttle, the nation would be without a major new space program for the 1980's....

Space offers us an alternative for the future. Our race can squander its potential and continue our unchecked momentum down the slopes of time toward the shore of the primeval sea to join the great reptiles and Nature's other unsuccessful experiments. Or, can we accept the challenge of the great spaces between the worlds and establish our citadels among the stars.

The choice, as the historian Wells once said: 'It's the Universe or nothing.'[10]

The NASA Administrator's long-range view of space missions is not known to have provoked any White House reaction. But such perspectives on the eve of an election year might engage some future historian, at least one familiar with the impact of Copernicus, or Darwin, or Einstein, or von Braun on futures of the past.

On May 11, 1976, President Ford signed the National Science and Technology Policy, Organization and Priorities Act of 1976. It prescribed a White House Office of Science and Technology Policy, whose director would be science adviser to the President. It was a post-Nixon action and mainly called for long-range R&D planning. But it championed the study of a cabinet-level department for fuels, energy, and material urged by its Democratic sponsors. When the director of the National Science Foundation, Dr. H. Guyford Stever, was nominated by Ford as the director of OSTP, it again aroused Congressional concern on presumed duplicity. Stever, in fact, had been acting as White House science adviser since 1973. As such, this legislation, highlighting energy problems, did not address future goals in space nor most of the broader spectrum of interests captivating the space-responsible Congressional committees. Election-year posturing was also evident.[11]

President Ford's most comprehensive statement on space was made in his dedication of the new National Air and Space Museum as a Bicentennial Event on July 1, 1976: He said, in part:

> This beautiful new museum and its exciting exhibits of the mastery of air and space is a perfect birthday present from the American people to themselves. Although it is impolite to boast, perhaps we can say with patriotic pride that the flying machines we see here, from the Wright brothers' 12-horsepower engine to the latest space vehicle, were mostly 'Made in U.S.A.' How many of us vividly recall the thrill of the first take-off? How many recall the first news of Lindbergh's safe landing in Paris? How many saw man's first giant step that planted the American flag on the Moon....
>
> The hallmark of the American venture has been a willingness--even an eagerness--to reach for the unknown....
>
> In the next 100 years, the American spirit of adventure can find out even more about the forces of nature, how to harness them, preserve them; explore the great reaches of the ocean, still an uncharted frontier; turn space into a partner for controlling pollution and instant communication to every corner of the world; learn how to make

our energy resources renewable and draw new energy from Sun and Earth; develop new agricultural technologies so all the deserts of the Earth can bloom; conquer many more of humanity's deadly enemies, such as cancer and heart disease.

As Thoreau reminded us, long before the age of air and space, 'The frontiers are not east or west, north or south, but wherever man fronts a fact.' The American venture is driven forward by challenge, competition, and creativity. It demands of us sweat and sacrifice and gives us substance and satisfaction. Our country must never cease to be a place where men and women try the untried, test the impossible, and take uncertain paths into the unknown.

Our Bicentennial commemorates the beginning of such a quest, a daring attempt to build a new order in which free people govern themselves and fulfill their individual destinies. But the best of the American adventure lies ahead....[12]

President Gerald Ford and Vice President Nelson Rockefeller greet one another at ribbon-cutting ceremonies that marked opening by the President of Smithsonian's National Air and Space Museum on July 1, 1976. (Left to right) Chief Justice Burger, Vice President Rockefeller, President Ford, Michael Collins, Director of the Museum, and Dillon Ripley, Secretary of the Smithsonian Institution. (NASA Photo No. 76-H-641)

On July 19, 1976, President Ford issued a proclamation for "Space Exploration Day, 1976." It commemorated the seventh anniversary of the landing of *Apollo 11's Eagle* on the surface of the moon--"a dream of thousands of years realized.[13] It fit the Bicentennial spirit and there were to be no more manned space flights until the Space Shuttle got going. On September 8, it was announced by the White House that the first Space Shuttle would be named *Enterprise*. This was not a campaign divot. The *Enter-*

prise had its factory roll-out on September 17. It was the first tangible public appearance of the "workhorse" for space missions in the 1980's.

President Ford had honored the death of Charles A. Lindbergh in 1974. He had welcomed two Soviet fliers who had flown over the North Pole from Russia to the United States before World War II. One of his last official acts, in early 1977, was to award the National Medal of Science to NASA's Wernher von Braun, on his death bed.[14]

Gerald Ford's presidency was anything but imperial. He worked the problems and made decisions. After he accepted his party's nomination for re-election, he survived two assassination attempts as well as the carping of the news media and the butts of Johnny Carson for stumblings on the steps of *Air Force One*.[15] If he regarded the House of Representatives as his "second home" after the White House, he had 28-years of public service. His had been a no-nonsense government career. He never visited Cape Canaveral to witness a space mission launching, although he had toured the facility with the House Committee in 1958. As Hugh Sidey, *Time-Life's* watcher of presidents, once observed, President Ford "has done his homework."[16] This was, in time, to become more broadly appreciated as the so-called "technocratic approach" of his successor, sans any Washington experience, was applied to the management of the Federal establishment.

President Ford and Administrator James C. Fletcher examine model of the Space Shuttle in the White House, September 8, 1976.
(NASA Photo No. 76-H-721)

GERALD R. FORD

REFERENCE NOTES

1. Gerald R. Ford, *A Time to Heal: The Autobiography of Gerald R. Ford* (NY: 1979), 454pp.

2. E. C. and L. N. Ezell, *The Partnership: A History of the Apollo-Soyuz Test Project* (NASA SP-4209: 1978).

3. Cf. *Aeronautics and Space Report of the President - 1974, 1975, 1976* (Washington: NASA, 1975-1977); NASA, *Astronautics and Aeronautics: A Chronology of Science, Technology, and Policy - 1974* (Washington: NASA SP-4019, 1977).

4. Ken Hechler, *Toward the Endless Frontier: History of the Committee on Science and Technology, 1959-79* (Washington: House Committee Print, 1980), pp. 1-27.

5. Hechler, pp. 24-25; cf. U.S. Senate, *Committee on Aeronautical and Space Sciences, 1958-1976* (Washington: Senate Committee Print, December 12, 1976), pp. 21-23.

6. Hechler, p. 196.

7. *Report of the President, 1977,* pp. 96-97.

8. Cf. Everly Driscoll, "Skylab Nears Its Re-Entry from Orbit," *Smithsonian Magazine* (May 1979), pp. 81-89.

9. Ezell and Ezell, pp. 328-29.

10. James C. Fletcher, "The Outlook for the Space Program," address to the National Academy of Engineering," Washington, D.C., November 10, 1975, 14pp (Text in NASA Historical Archives) Unfortunately, scholars are still awaiting the appearance of the annual NASA volume, *Astronautics and Aeronautics: A Chronology on Science, Technology and Policy,* for the years 1975 and thereafter.

11. Cf. Hechler, "Science in the White House," pp. 624-50.

12. *Presidential Documents, Ford, 1976,* vol. 12, pp. 1103-04.

13. White House Release (unnumbered), "Space Exploration Day, 1976, A Proclamation," January 19, 1976, 1p.

14. Cf. F. I. Ordway and M. R. Sharpe, *The Rocket Team* (NY: 1979), p. 404.

15. Cf. J. F. ter Horst and R. Albertazzie, *The Flying White House* (NY: 1979), pp. 72-73.

16. Hugh Sidey, "He Has Done His Homework," *Time*, 2/3/75, p. 12; Cf. J. A. Kilpatrick, "The Ex-President No One Hates," *Washington Star*, 12/26/76, p. A-11.

James E. Carter

Perspectives on the over-half-completed administration of President Carter and the American space program are historically quite skimpy. He became the second engineer graduate in the White House, and the first from the U.S. Naval Academy and the post-World War II generation. The former one-term Governor of Georgia conducted an unflagging primary and election campaign overcoming the "Jimmy who?" tag and upsetting the pollsters. Additional to his being the first chief executive coming from the Deep South, he won in an election noted for its poorest turnout at the polls in twenty-eight years with a promissory theme for cleaning up the mess in Washington. He touched enough voters to deny by 2% margin the election of Gerald Ford.[1]

As it had been since the first U.S. satellite program in 1955, space affairs did not become a singular political issue in the presidential contest during the Bicentennial Year. During the campaign, Mr. Carter did share with a reporter that he had once seen an unidentified flying object in 1969. He thought perhaps after he was elected that an explanation of this experience might be forthcoming.[2] If the aerospace vote was at all identifiable, it was in the State of California, where President Ford carried that state.

President Jimmy Carter requested and publicly issued a formal and public statment on U.S. space policy in 1978. It did arouse considerable comment for its limited goals in outer space. On Capitol Hill international events were also to conspire to increase defense aspects of the national space enterprise. He identified with the Apollo astronauts, the twenty-year anniversary of NASA, and the tenth anniversary of *Apollo 11*. But until the Space Shuttle got flying in space, even the splendid interplanetary missions of the Vikings, Pioneers, and Voyagers, as well as the profuse utility of communications, weather, and earth-sensing satellites, did not provide supreme media events as those when Americans were in space flight, the piggy-back test flights of the *Enterprise* notwithstanding.

The peanut became the first lapel symbol of the Carter administration, much as the PT-109 tie clasps had been the marker for the John F. Kennedy clan. And as the new President exhibited his strangeness concerning the ways of Washington, the peanut became the cartoonists' delight.

Space concerns ranked low in President Carter's action list if the selection of an administrator for the National Aeroanutics and Space Administration was any measure. The reason was the delay in selecting a Science Adviser to the President. This appointment was normally an early

"Up, Up and Away!"

(Courtesy of Francis Brennan of *Newsweek Magazine*, as published in the *Washington Post*, 11/14/76)

one to assist recruitment of relevant agency heads. Not until March 1977, did Carter select his Science Adviser in Dr. Frank Press of M.I.T. A geophysicist of high reputation, Dr. Press was an international specialist on earthquakes and had engaged in governmental seismic experiments on the moon for the Apollo program, and on underground nuclear tests. Press had made a trip to China in 1974 to participate in a conference on earthquake prediction.[3] The science community was relieved when, despite Mr. Carter's campaign promise of cutting down the White House staff, Frank Press was also named director of an Office of Science and Technology Policy. NASA's administrator for six years, Dr. James Fletcher, waited for his resignation from the new administration to be accepted. It became effective on May 1. For NASA it had been an uneasy interregnum.

It was Dr. Press who recommended Dr. Robert A. Frosch as the new administrator of NASA. They had known each other professionally for over twenty-five years.[4] President Carter talked to Frosch in April 1977, but did not nominate him until June. This timing permitted the concurrent nominations for the directors of the National Science Foundation, the National Oceanic and Atmospheric Agency, and the National Bureau of Standards. The Congress approved all nominees promptly; none of them were sworn in by the President, and Dr. Alan Lovelace remained as deputy administrator of NASA. From all indications, Dr. Frosch understood the ways of Washington in his difficult chores for NASA with the White House, the Congress, the

Pentagon, and all obvious space hustings during these years of sustenance budgets, the technical crunch getting the stretched Space Shuttle development into its test-flight stages. A seasoned research administrator, Frosch also displayed a grasp of the intellectual challenge of space futures, muted with a sardonic sense of humor. Cost overruns on the Shuttle development, the cancellation of the fifth orbiter, and all the other demands on NASA budgeting gave NASA management little respite in its OMB or Congressional obligations. Importantly, Frosch was to deal directly with President Carter upon critical occasions.[5]

For its part, the new Congress had to become accustomed to dealing with a President from the majority party, one with no Capitol Hill experience like all his predecessors since Eisenhower. The veto-prone President Ford had been predictable, but not Mr. Carter. Budgetwise, the Carter research and development budget for FY 1978 was more or less that which had been submitted by President Ford.[6]

Traditionally the first months of a new administration, beyond the party-switch management changes, was one of trying to fit campaign stances into Federal operations. An open administration, zero-base budgeting, human rights, and détente not Cold War--these were not controversial in principle. Early Carter challenges involved the urgent energy crisis, which became a "moral equivalent of war" leading to the creation of a new Department of Energy. Cancellation of the B-1 bomber and the Clinch River breeder reactor gave rise to strong Congressional arguments, as well as the controversial inquiry leading to the resignation of OMB director Lance. Steps toward a Strategic Arms Limitation treaty with the U.S.S.R., the Panama Canal treaties, opening the China connection, and the Egypt-Israel peace treaty were all part of the geopolitical backdrop around space policy and programs during the early Carter administration. Newly authorized Congressional committees, with purview of future NASA programs and oversight of all space activities, were concerned with the utilization of space technology. The House Committee on Science and Technology, while supporting NASA budget requests for FY 1978, pressed for expanding space applications and more imaginative long-range planning.[7] In February 1977, the Senate Committee on Commerce, Science, and Transportation assumed the functions of Lyndon B. Johnson's initial standing committee on the Aeronautical and Space Sciences. It had been the stalwart of international activities in space, and was not unmindful of national defense applications.

The Space Shuttle program assumed a new nomenclature as the Space Transportation System (STS) subsequent to a series of approach and landing tests of the *Enterprise*--a piggy-back landing in February 1977, and a separated Orbiter landing in August at Edwards, California. The modified 747 airliner would subsequently carry the *Enterprise* to the Marshall Center in Huntsville for dynamic tests, and thence on to Kennedy Space Center for testing of stacking and launch facilities. The U.S. Air Force continued developing facilities and making plans for ultimate use of the STS as it provided capabilities long desired for its mission in space.

In the spirit of détente, the United States and the U.S.S.R. signed a five-year accord covering possible space cooperation in manned flights and for an international space station, on May 6, 1977. Discussions in November considered experiments feasible in a long-duration mission involving the Shuttle and the Soviet Salyut space station. This was additional to the "Intercosmos" program for scientific exchanges.[8]

In the public sector a vast array of space systems proposals were propagated from large-scale structures for colonies, telescopes, solar energy plants, and whatever, to industrialization, mining, and manned colonies on the moon or nearby planets. Feasibilities were merely constrained by resources, time, and proving with the STS in the 1980's.[9]

NATIONAL SPACE POLICY

The dearth of any commitment of the Carter administration on space beyond inherited programs, was ameliorated on June 20, 1978.[10] A White House Press Release was issued, entitled, "Directive on National Space Policy." It came out strongly for freedom in space, and for recommitting "the use of outer space by all nations for peaceful purposes and for the benefit of all mankind." It seems wise here to quote this document at some length. Its meaning and implications serve sustained discussions for many future space opportunities shaped by national needs and by the resources committed thereto by the Government and the private sectors.

The White House release stated that President Carter had directed the Policy Review Committee of the National Security Council (NSC) to "thoroughly review existing policy and overall principles which guide our space activities." It established "ground rules for the balance and interaction of our space programs to insure achievement of the interrelated national security, economic, political, and arms limitation goals of the United States." It established a "NSC Policy Review Committee to provide a forum for all Federal agencies... to advise on proposed changes to national space policy... and to provide for rapid referral of issues to the President for decision as necessary." The Policy Review Committee (PRC) would be chaired by the Director of the Office of Science and Technology Policy, Dr. Frank Press.

President Carter's Directive on National Space Policy, did not, "in its classified portions," concentrate on "overlap questions" or "deal in detail with the long-term objectives of our defense, commericial, and civil programs. Determining our civil space policy... will be the next step." The White House release did state clearly, "as a result of this in-depth review," that the following National Policies and Space Principles would prevail to guide the conduct of American "activities in and related to space programs":

National Policies

. To advance the interests of the United States through the exploration and use of space.

. To cooperate with other nations in maintaining the freedom of space for all activities which enhance the security and welfare of mankind.

Space Principles [numbering added]

1. The United States will pursue space activities to increase scientific knowledge, develop useful commercial and government applications of space technology, and maintain United States leadership in space technology.

2. The United States is committed to the principles of the exploration and use of outer space by all nations for peaceful purposes and for the benefit of all mankind.

3. The United States is committed to the exploration and use of outer space in support of its national well-being.

4. The United States rejects any claims to sovereignty over outer space or over celestial bodies, or any portion thereof, and rejects any limitations on the fundamental right to acquire data from space.

5. The United States holds that the space systems of any nation are national property and have the right of passage through and operations in space without interference. Purposeful interference with space systems shall be viewed as infringement upon sovereign rights.

6. The United States will pursue activities in space in support of its right of self-defense and thereby strengthen national security, the deterrence of attack, and arms control agreements.

7. The United States will conduct international cooperative space activities that are beneficial to the United States scientifically, politically, economically, and/or militarily.

8. The United States will develop and operate on a global basis active and passive remote sensing operations in support of national objectives.

9. The United States will maintain current responsibility and management relationships among the various space programs, and, as such, close coordination and information exchange will be maintained among the space sectors to avoid unnecessary duplication and to allow maximum cross-utilization of all capabilities.

Elaboration of the National Policies and Space Principles in President Carter's Directive discussed some guidelines to provide added thrust pending future issuances. These also had best be included here, despite their length, since single issue space savants lose breadth:

Our civil space programs will be conducted to increase the body of scientific knowledge about the earth and the universe; to develop and operate civil applications of space technology; to maintain United States leadership in space science, applications, and technology; and to further United States domestic and foreign policy objectives within the following guidelines:

- The United States will encourage domestic commercial exploitation of space capabilities and systems for economic benefit and to promote the technological position of the United States; however, all United States earth-oriented remote sensing satellites will require United States government authorization and supervision or regulation.

- Advances in earth imaging from space will be permitted under controls and when such needs are justified and assessed in relation to civil benefits, national security, and foreign policy. Controls, as appropriate, on other forms of remote earth sensing will be established.

- Data and results from the civil space programs will be provided the widest practical dissemination to improve the condition of human beings on earth and to provide improved space services for the United States and other nations of the world.

The United States will develop, manage and operate a fully operational Space Transportation System (STS) through NASA, in cooperation with the Department of Defense. The STS will service all authorized space users--domestic and foreign commercial and governmental--and will provide launch priority and necessary security to national security missions while recognizing the essentially open character of the civil space program.

Our national security related space programs will conduct those activities in space which are necessary to our support of such functions as command and control, communications, navigation, environmental monitoring, warning and surveillance, and space defense as well as to support the planning for and conduct of military operations. These programs will be conducted within the following guidelines:

- Security, including dissemination of data, shall be conducted in accordance with Executive Orders and applicable directives for protection of national security information. Space-related products and technology shall be afforded lower or no classification where possible to permit wider use of our total national space capability.

- The Secretary of Defense will establish a program for identifying and integrating, as appropriate, civil and commercial resources into military operations during national emergencies declared by the President.

- Survivability of space systems will be pursued commensurate with the planned need in crisis and war and the availability of other assets to perform the mission. Identified deficiencies will be eliminated and an aggressive long-term program will be applied to provide more assured survivability through evolutionary changes to space systems.

- The United States finds itself under increasing pressure to field an anti-satellite capability of its own in response to Soviet activities in this area. By exercising mutual restraint, the United States and the Soviet Union have an opportunity at this early juncture to stop an unhealthy arms competition in space before the competition develops a momentum of its own. The two countries have commenced bilateral discussions on limiting certain activities directed against space objects, which we anticipate will be consistent with the overall U.S. goal of maintaining any nation's right of passage through and operations in space without interference.

- While the United States seeks verifiable, comprehensive limits on anti-satellite capabilities and use, in the absence of such an agreement, the United States will vigorously pursue development of its own capabilities. The U.S. space defense program shall include an integrated attack warning, notification, and contingency reaction capability which can effectively detect and react to threats to U.S. space systems.

President Carter's issuance of a comprehensive space policy document, with NSC aegis, on June 20, 1978, one apparently approved on May 11, was an historic step. It had been prompted by some urgency to prevent the escape of anti-satellite technology from the on-going SALT II negotiations under way with the Soviet Union. It also served to explain more clearly the DOD experimental use of the STS with NASA, or until the Air Force got an Orbiter of its very own. In principle, this U.S. Space Policy was consistent with long-standing NASA-DOD roles and missions first instituted under President Eisenhower, enforced as national policy by the National Aeronautics and Space Act of 1958, and its amendments. In practice technical potentials had become operational realities for the near future, and particularly for the unmatched Soviet initiatives for anti-satellite weaponry threatening to unravel the core concept of "freedom of space." Politically, Eisenhower's "open skies" had not succeeded in 1955, but the sputniks and U.S. satellites led to *de facto* overflights. The principles were instituted in the Space Act, and subsequently the Soviet Union had become a party to the international space law agreements.[11]

It was also consistent with President Carter's open stance on "the welfare of mankind" so he was not dissuaded from disseminating such a detailed blueprint of U.S. space policy principles and program guidelines. Some concurrent footnotes might be noted here, although future historians will have more information. In June 1978, Carl Sagan persuaded President Carter to include a personal message in the Voyager spacecraft slated to leave our solar system--"a message to the Stars." President Carter's contribution said, in part: "We earthlings are attempting to survive our time so we may live with yours. We hope someday, having solved the problems we face, to join a galactic civilization...."[12] Congressional space committee members, hearing of it, insisted that their names also be included in Sagan's celestial mail bag. They were. And, interestingly, candidate Carter's UFO publicity turned up in a memorandum from the White House to NASA. It requested that a scientific study be conducted on unidentified space observations. NASA refused this request on the grounds that the absence of any hard data made such a study infeasible. As it turned out, NASA was given the task of answering the White House mail on the subject.[13]

President Carter's "National Space Policy" release, which violated his own personal preference for one-page briefs on policy issues, received little news media review. Most of that was related to the SALT II process or the threat of "space wars" or first public mention of DOD space missions and challenge acknowledged by the White House.[14] Space goal advocates on Capitol Hill, including Congressman Don Fuqua, and Senators Adlai Stevenson and Harrison Schmitt and others, were concerned. Mr. Carter had promised to issue soon his paper on "Civil Space Policy," which appeared in October. In the meantime, the Space Shuttle appeared the operational goal. It did demonstrate the long lead times required to develop such systems from conception to reality.

The twentieth anniversary of the official birth of the National Aeronautics and Space Administration on October 1, 1978, provided the opportunity for President Carter to identify with the Space Shuttle and some former Apollo astronauts. At the Kennedy Space Center, the President awarded the Congressional Space Medal of Honor to Neil A. Armstrong, Frank Borman, Charles Conrad, Jr., John H. Glenn, Virgil I. Grissom (posthumous), and Alan B. Shepard. In his address, he offered few insights on his chart for the future of the space program: "The first great era of the Space Age is over. The second is about to begin... with the Shuttle," He pointed out that $100 billion historically had been invested in the U.S. space programs. He said: "It is now time for us to capitalize on that major investment even more." After referring to communication and weather satellites, he made news media headings by saying: "Photo reconnaissance satellites have become an important stabilizing factor in world affairs in the monitoring of arms control agreements. They make an immense contribution to the security of all nations. We shall continue to develop them." He referred to the utility of earth resources satellites. On the future, Mr. Carter said:

> I am often asked about space factories, solar power satellites and such other large-scale engineering projects in space. In my judgment, it is too early to commit the Nation to such projects. But we will continue the evolving development of our technology, taking intermediate steps that will keep open possibilities for the future.
>
> During the period of the Saturn-Apollo missions, we were pilgrims in space, ranging far from home in search of knowledge. Now we will become shepherds tending our technological flocks, but like the shepherds of old, we will keep our eyes fixed on the heavens.
>
> We are committed to the practical use of space. But we are equally committed to the scientific exploration of the solar system and the universe....[15]

President Carter confessed that his participation in honoring the anniversary of NASA and honoring some of those who had made space achievements was "one of the most exciting events of my life." Celebrating his own 54th birthday, he said he had been assured that the first Shuttle launching would take place before his next birthday.[16] He was to be disappointed.

Mr. Carter was keeping his future space options open in his appearance at the Kennedy Space Center. The long-range space policy paper was being completed by the NSC Space Policy Review Committee. In anticipation, Senator Adlai Stevenson had already submitted a Space Policy Act of 1978

President Jimmy Carter at the 20th anniversary of NASA ceremony in the VAB at Kennedy Space Center, October 1, 1978. He awarded the first Congressional Space Medal of Honor to former astronauts, Neil Armstrong, Frank Borman, Charles Conrad, John H. Glenn, Virgil Grissom (posthumously), and Alan Shepard. (NASA Photo No. 78-H-611)

(S.3530) to the Senate. And Senator Harrison Schmitt, former *Apollo 17* astronaut scientist, had called for tripling the U.S. space effort for the next thirty years. Their consequence remained to be seen.[17] But President Carter's punch line at KSC raised questions: "The age of space can now be said to have reached the threshold of its maturity."[18] It conveyed less of the endless frontier of space science and technology challenge as well as the White House perspectives in dealing with the endless catalog of daily problems at home and abroad. Indeed, the coming of the Space Shuttle era soon would present a new dimension of space capabilities, which would, in turn, generate a new threshold for subsequent opportunities. His mention of "photo reconnaissance satellites" was acknowledged in the press as an element in verification for a new strategic arms limitation agreement with the Soviet Union.[19]

President Carter announced his long-awaited "U.S. Civil Space Policy" on October 11, 1978. It was the result of a four-month inter-agency study by the White House Policy Review Committee, one chaired by Dr. Frank Press.[2] It was a shred out of the "National Space Policy" issued on June 20, one which, it stated, would "set the direction of U.S. efforts over the next decade." It underwrote what Mr. Carter had said at the Kennedy Space Center almost two weeks earlier, emphasizing existing technology and by using the Shuttle "to reduce the cost of operating in space over the next two decades to meet national needs." Its capstone was: "It is neither feasible nor necessary at this time to commit the United States to a high-challenge space engineering initiative comparable to Apollo. As resources and manpower requirements for Shuttle development phase down, we will have the flexibility to give greater attention to new space applications and exploration, continue programs at present levels or contract them...."

It was not an unreasonable Civil Space Policy, but it insured that many watchers would reexamine the earlier National Space Policy paper covering defense, commercial, and civil space parameters. NASA was charged with getting on with full utilization of the Shuttle as well as space applications by encouraging investment by the private sector for technological exploitation. NASA should also continue a "vigorous program of space science including planetary exploration and international cooperation." It cataloged space applications, putting NASA back into R&D for communications satellites. Its strategy for the Shuttle included the creation of a task force for its operational enhancement. And, the Policy Review Committee (Space) would oversee technology transfer in the space sectors to maximize efficiency with necessary security and management. It was a remarkable open and comprehensive document if it stated little that was not already functional in the space program. And, what had been hammered out in the Federal forum in the Executive Branch could now be coherently discussed on Capitol Hill and elsewhere.

Two days later in a press interview, President Carter was asked if there were grounds for the rumors going around the Kennedy Space Center that as soon "as the Shuttle became operational, you will order even more cut backs in an austerity program?" He replied that the U.S. had "a very aggressive space policy. Anyone who reads the document that has been prepared very carefully, very thoroughly, by the Defense Department, CIA, NSC, all those that use them, including Agriculture, Commerce, and finally approved by me, would say it is a very sound program based on scientific need and actually capitalizing on the great exploratory efforts that have been made in space. We look upon the Space Shuttle as a way to change dramatic, very costly, initiatives into a sound progressive and innovative program to utilize the technology that we have available to us....

"So we are not going to minimize or decrease our commitment to space at all....

"It is not a matter of playing down the importance of space. It is a matter of using what we have already learned more practically."[21]

Harry Truman never had a space program. But he did not hesitate to include the NSC or the Joint Chiefs of Staff or others to buttress his comments to the press on pointy questions. General Eisenhower never used his reliance on his NSC to justify in public his space policy, but he always pointed with pride to his science advisers. Mr. Carter apparently would like to compare himself to President Truman. And if he appeared to have a space policy setup somewhat comparable to Eisenhower's, even if not a practical space program, likewise comparable in a latter time of technology, such would make an interesting analysis. But not here.

At the end of 1978, the Washington weekly aerospace journal of note, read as avidly by the Soviet Embassy as by the aerospace community, *Aviation Week and Space Technology*, listed President Carter in its list of "Laurels for 1978"--"for recognizing the need for a new national space policy and dragooning government bureaucrats into developing and publishing one, and also prodding the Department of Defense into an aggressive mode for applying modern technology to growing military space needs."[22] There was little question that the development of the Space Shuttle during the 1970's, initially a mere $10 billion commitment when approved by Nixon, had to demonstrate utility and returns for both NASA and DOD missions before another large-scale program would be feasible in the White House. It was also evident that NASA's budget for FY 1981 had shown the least growth of any Federal program, without counting the escalating inflation.[23]

On space initiatives by the Carter administration, the increased visibility given to renewed interest in military technology had been evidenced. This occurred under the management of Secretary of Defense Harold Brown, a physicist with a long track record as Secretary of the Air Force and other managerial offices. The Air Force looked to the Shuttle as the few pioneers looked to the flyer of the Wright brothers.[24] Reentry of NASA into communications satellite R&D was a plus in an area already a huge commercial success with quantum growth projections. But it was President Carter's decision to normalize relations with mainland China that erected symbolic new dimensions regarding the world balance of power and the role of the United States in the Far East.

An American science and technology delegation headed by Science Adviser Press, including NASA administrator Frosch, visited Peking in November 1978. Dr. Frosch has said: "Vice Premier Ding Xia Peng looked right at me when he asked if he could buy a television satellite. They wanted it intensely and right away."[25] The return visit of the Chinese delegation to Washington and Houston in February 1979, confirmed modest availability of non-military technology for the most populated nation on earth. Problems of the Federal Republic of China on Taiwan or cost-cutting reductions of U.S. troop commitments in South Korea would have to be ameliorated for the tri-polar super-power world. Evangelical overtones of the Carter White House in its relations with the Kremlin as well as the Middle East were apparently to be tested, about which future historians will have their say. May they not ignore the space vectors.

POSTSCRIPT*

President Carter's hopes that *Columbia* would make the first test flight of the Space Shuttle before his next birthday in October 1979, or even before his possible nomination and reelection in November 1980, were to be dashed by technological problems not uncommon in space history. Eisenhower had his *Sputnik* and his *Vanguard*, then the Redstone and the Atlas problems for *Mercury*, never forgotten despite *Echo*, *Tiros*, and *Discoverer 13*, plus *Pioneer V*. Kennedy never saw anything fly he began. Johnson had the Apollo fire while Nixon inherited the Apollo triumphs. In fact Carter was the first President to see a major space launch before he was elected, *Apollo 17*. One of its astronauts was to become one of the President's most severe critics--Senator Jack Schmitt of New Mexico. *Columbia* cannot take off until it is ready, and it is not yet in a Soviet race.

Mr. Carter's unfamiliarity with much of the space program when he came to Washington accounted for his understandably slow involvement. But he appears to have gotten atop the space policy matrix, with its multifaceted elements, implications and potentials, from basic science to national security and practical benefits on earth. About his vital "sense of history" to anticipate well the initiation time for the greatest future benefits in space science and technology, President Carter's antennae seem to be turned on and sensitive to practical propositions as they may generate economic benefits in the near-term future. More than that seems a little too much to ask of any incumbent of the White House in any year, unless once again, as in John F. Kennedy's early months as President, national prestige and political warfare mesh with space destinies. The year of 1979, was truly a triumphant one generating breathtaking closeup images of the distant planets, their moons, and rings. Even the fall of the Skylab station about which Mr. Carter was personally concerned, was successful in its atmospheric reentry and breakup without damaging the prospects expected from the Shuttle. History has a trajectory of its own, and that events unforeseen happen should not prove surprising. Making things happen, rather than reacting, is the ultimate test of leadership which the American people expect of their Presidents. President Carter had at least kept the options open for future space endeavors.

---

* *This paper was delivered on March 28, 1979, but a few items need brief mention here.*

President Carter makes presentation of the Robert H. Goddard Memorial Trophy on March 24, 1980 to the NASA/JPL Voyager Team for execution of two historic flybys of the planet Jupiter and its moons. In the Oval Office left to right President John Lent of the National Space Club, President Carter, and NASA Administrator Robert Frosch. (NASA Photo)

JAMES E. CARTER

REFERENCE NOTES

1. Cf. Jules Whitcover, *Marathon: The Pursuit of the Presidency* (NY: 1977), pp. 644-77.

2. Warner Brown, "Carter Laughs Off His 'UFO Sighting'," *Washington Post*, 5/12/76; Paris Flammonde, *UFO's Exist!* (NY: 1976); "Postscript," *Washington Post*, 5/9/77, p. 5; H. L. Helms, "President Carter and Georgia's UFO Wave," *UFO Report* (May 1979), pp. 16-19.

3. P. M. Boffey, "Frank Press, Long Shot Candidate, May Become Science Adviser," *Science* (AAAS), 195 (25 February 1977), pp. 763-66; "Science Adviser Press: First Hints of How He is Doing," *Science*, 196 (22 April 1977), pp. 412, 460; "The President's Scientist," *Time* (4/11/77), pp. 73f.

4. Interview of Robert A. Frosch, by the author, 2/21/79; cf. Thomas O'Toole, "Science Party: Enlightning," *Washington Post*, 4/25/77, p. B-1, 3. Biography of Frosch in *NASA Activities*, 9 (June 1977), p. 3. Frosch had been a consultant of NASA since April 20, 1977, and Carter announced his nomination on May 23.

5. Excepted interviews.

6. Willis H. Shapley, *Research and Development in the Federal Budget, FY 1978* (Washington: 1977) is basic reference; cf. "Press on Space," *Astronautics and Aeronautics* (AIAA, May 1977), p. 20; A. L. Hammond, "Carter Budget Tilts 'Back to Basics' for Research," *Science*, 199 (3 Feb. 1978), pp. 507-10.

7. Cf. Ken Hechler, *Towards the Endless Frontier: History of the Committee on Science and Technology, 1959-79* (House Committee Print: 1980), pp. 338-42.

8. *Aeronautics and Space Report of the President--1977 Activities* (Washington: NASA, 1978), p. 7.

9. Most widely known were G. Harry Stine, *The Third Industrial Revolution* (NY: 1975); Gerald K. O'Neill, "Colonies in Orbit," *New York Times Magazine*, 1/18/76, pp. 10-11, 25-29, and other articles on O'Neill's "High Frontier" in most aerospace magazines. T. A. Heppenheimer, *Colonies in Space* (Harrisburg, PA: Stackpole Books, 1977).

10. *Papers of the Presidents, J. Carter, 1978*, June 20, 1978, pp. 1135-37; also in Senate Committee on Commerce, Science, and Transportation, *Space Law: Select Basic Documents* (2nd Ed., December 1978), pp. 558-60.

11. Cf. earlier discussion of Eisenhower's "Open Skies."

12. Carl Sagan, *et al.*, *Murmurs of Earth: The Voyager Interstellar Record* (NY: 1977), pp. 27-35.

13. Excepted interviews; cf. "Carter Is Facing Rebuff on Reviving UFO Probe," AP, *Washington Post*, 11/26/77, p. B-5.

14. Edgar Ulsamer, "The New Space Policy," *Air Force and Space Digest* August 1978, pp. 12-14; Anatol Johannson, "USA-USSR - The Real 'Star War'," *Aerospace International* (August-September 1978), pp. 5-6.

15. White House Release, "Remarks of the President at Congressional Space Medals Award Ceremony, 10/1/78, 7pp.

16. Text, pp. 1-2.

17. Henry T. Simmons, "Carter Keeps His Space Options," *Astronautics and Aeronautics* (November 1978), pp. 4-6; Hechler, *Endless Frontier*, p. 1010; Harrison Schmitt, "No-Vision Space Policy is Really No Policy," *Insight* (NSI), 3 (December 1978), pp. 10-13.

18. This key sentence, not in the official text release, is found on the 5"x8" cards (xerox cy in NASA News Room) used by President Carter in giving his speech, and in the text published by NASA, the sentence reads: "The age of space, as Dr. Frosch said, can now be characterized as having reached the threshold of its maturity. It began 21 years ago this week with the launching of Sputnik I..." *NASA Activities*, 9 (November 1978), p. 3. This is not highly significant except that sometimes well-known authors have been known to cite speeches that were never given, while very few indeed were ever written by presidents. Ike and JFK edited theirs.

19. Cf. E. Walsh, "Carter Vows U.S. Will Continue Leadership in Space," *Washington Post*, 10/2/78, p. 8; L. Olson, "U.S. Says Officially for 1st Time, It Uses Satellites to Spy on Soviets," *The Sun* (Baltimore), 10/2/78, p. 1; R. Toth, "Costly Spectaculars are Out: New Space Policy Outlined," *Washington Post*, 10/12/78, p. A-5.

20. White House Release, "Fact Sheet: U.S. Civil Space Policy," 10/11/78, 4 pp.

21. White House, "Interview with the President for Non-Washington Editors and News Directors, Cabinet Room, 10/13/78, Excerpt, 2 pp.

22. Harry S. Dawson, Jr., "Other Federal Programs Dwarf NASA," *Insight* (NSI), 3 (November 1979), p. 4. Health, Education, and income security (social security and welfare) programs amount to $260 billion in FY 1980, which is more than twice the size of the defense budget, while the three program areas spend an amount equal to NASA's total annual budget every 7 1/4 days.

23. White House Release (Immediate), "Fact Sheet: U.S. Civil Space Policy," October 11, 1978, 4 pp.

24. On October 13, 1978, while interviewed by Non-Washington Editors and News Directors in the White House, President Carter explained his civil space policy without mentioning NASA, as follows: "I think [it] a very aggressive space policy. Anyone who reads the documents that have been prepared very carefully, very thoroughly by the Defense Department, CIA, NSC, all those who use it, including Agriculture, Commerce, and finally approved by me, would say it is a very sound program based on scientific need and actually capitalizing on the great exploratory efforts that have been made in space. We look upon the space shuttle as a way to change dramatic, very costly initiatives into a sound progressive and innovative program to utilize the technology that we have available."

He went on to explain the continuation of interplanetary, astronomy, assessments of the earth, weather, and communications, and expanded cooperation with foreign nations and "private firms in our nation." It was, he said, "not a matter of playing down the importance of space. It is a matter of using what we have already learned more practically." Office of White House Press Secretary, interview with the President on October 13, 1978, excerpts from the interview for general release, 2 pp.

25. Interview of NASA Administrator Robert Frosch, by the author, February 23, 1979.

III

U.S. CONGRESS AND OUTER SPACE

Eilene Galloway *

Congress moved rapidly in 1957 to legislate for space policy, program, and budget so that the U.S. would become preeminent in using and exploring outer space. Scientists and engineers generated ideas for government organization and management. High priority was given to space activities by establishing NASA, distinguishing between NASA and the Department of Defense, thus creating an overall institution at the highest governmental level and emphasizing international cooperation. Congress created permanent committees and provided for annual authorization of NASA funds to ensure in-depth analysis and continuing expert attention to developing space programs.

Influences which made foresighted prudent decision-making possible are analyzed.

Recent space policy statements by the Executive Branch have led to legislative responses and initiatives now pending in Congress.

The preeminent accomplishments by the United States in exploring and using outer space were made possible by the confluence of several streams of thought and action toward the end of 1957 and the beginning of 1958. The occasion for the formation of basic policy decisions, immediately implemented by well-funded ventures into this new frontier, was the dramatic orbiting of the first sputnik by the Soviet Union on October 4, 1957. The worldwide psychological impact was instantaneous, arising from wonder over the phenomenon of flying through space far beyond the Earth, and surprise that the Soviet Union was first to demonstrate advanced space science and technology. There was also immediate recognition that the capability of orbiting satellites was evidence of the power to launch intercontinental ballistic missiles. The possibility of a new dimension of warfare spurred decisions to develop activities in outer space for peaceful purposes.

---

* *Vice President, International Institute of Space Law, International Astronautical Federation; President, Theodore Von Karman Memorial Foundation, Inc.; member, International Academy of Astronautics; special consultant to Congressional committees on international space activities.*

At this crucial juncture in our history, all the elements essential for swift, foresighted decisions and action were present: spearheading leadership, knowledgeable scientists and engineers, an industrial base, resources, harmony between the Executive and Legislative Branches of the government on objectives, and an institutional structure on which to build.

Although this paper is focused primarily on the role of Congress in outer space activities, it must be remembered that the dual Congressional role of passing legislation and exercising the investigative function is performed in conjunction with the Executive Branch. Still, there are unique features in the legislative process, particularly as related to the the U.S. space program and its support by the elected representatives of the people.

GENESIS OF THE U.S. SPACE PROGRAM

A remarkable assemblage of factors, already in existence in 1957, could be organized and channeled toward goals for the future success of the United States in conducting outer space activities.

First, the International Geophysical Year (IGY) was in progress. Worldwide studies of man's environment--the earth, oceans, atmosphere, and outer space--were being made by the international scientific community between July 1, 1957 and December 31, 1958. Sixty-seven nations were represented on committees which were simultaneously collecting data on research projects and laying a firm foundation of international scientific cooperation.[1] Both the United States and the Soviet Union undertook outer space projects as part of their participation in the IGY. Subsequent events revealed the broad scope of Soviet plans as a second sputnik was sent into orbit on November 3, 1957 and a third on May 15, 1958. Although the U.S. had planned only a small satellite assigned to the Navy--Project Vanguard--nevertheless there were scientists and engineers who were expert in rocketry and missiles and the Department of Defense had taken official cognizance of the possibility of outer space activities since World War II. The main values that had been established by the IGY were international cooperation and the conduct of space activities for peaceful purposes, values which were later to be enshrined in U.S. national legislation and in international law.

Second, the United States had an industrial base in operation by aviation firms with experienced personnel and resources ready to expand and become the aerospace industry. The relations between the government and the aviation industry had been worked out satisfactorily in regard to the difference between research and development as distinct from operations.

Third, there was a government institution already engaged in research and development on aerospace problems--the National Advisory Committee on Aeronautics (NACA) which had been created as an independent government agency on March 3, 1915 with authority to "supervise and direct the scientific study of the problems of flight, with a view to their practical solution... and to direct and conduct research and experiment in aeronautics."[2] Since 1952, NACA had been studying "problems associated with unmanned and manned flight at altitudes from 50 miles up, and at speeds from

Mach 10 to the velocity of escape from the earth's gravity." Furthermore, "NACA's major aeronautical research centers and field stations... concentrated increasingly upon solution of problems of ballistic missiles, hypersonic aircraft, and space vehicles."[3]

Fourth, the motivation for a U.S. space program was compelling. We wanted to preserve outer space for peaceful purposes and prevent warfare from being directed to the Earth. The Space Age did not begin with hostility but with benign intentions and consequences. Furthermore, the peaceful applications of space technology were quickly made known by scientists and engineers so that planning by government could be directed toward arrangements designed to produce benefits.

Fifth, effective political leadership arose as suddenly as a rocket to undertake responsibility for organizing the government for the conduct of space activities. Senator Lyndon B. Johnson, then majority leader of the Senate, was the first to take up the challenge and carry forward. Later, he was joined by the majority leader of the House, Congressman John W. McCormack. In 1957, Senator Johnson was Chairman of the Senate Preparedness Investigating Subcommittee of the Senate Armed Services Committee and on November 25, 1957, the month following the orbiting of the first sputnik, he began the Inquiry into Satellite and Missile Programs, hearings which continued through the remainder of that year and into 1958. By January 23, 1958, the Subcommittee was in a position to review testimony given by witnesses representing government and industry, science and technology, education, military and civilian fields of knowledge. Concluding that the satellite was not yet a weapon, the Subcommittee reported, however, that the Soviet Union had led the world into outer space and that it would be necessary for the United States to exert a tremendous effort to ensure its own preeminence in defense and space. The Subcommittee was well aware that its jurisdictional responsibilities were limited to defense, but pointed out the necessity of others exploring whatever had to be done to involve the total effort of the nation. The Subcommittee agreed that "... the same forces, the same knowledge, and the same technology which are producing ballistic missiles can also produce instruments of peace and universal cooperation."[4]

The spearheading leadership of Senator Lyndon B. Johnson in conducting these hearings and interviewing experts had the immediate effect of accelerating activity in the Department of Defense to strengthen the U.S. space program. On January 31, 1958 the U.S. orbited *Explorer I*, the satellite whose data led to the significant discovery of the Van Allen radiation belts. Pending decisions on how the government should be organized to conduct a total U.S. space program, and because of the urgency of the situation, Congress passed interim laws whereby the United States could speed ahead. On February 11, 1958, a law was enacted to provide the Department of Defense with $10 million in supplemental funds for space activities during the remainder of the fiscal year (Public Law 85-322). A second law, enacted on February 12, 1958 authorized the Secretary of Defense to engage in advanced research, and for one year to develop space projects which were designated by the President (Public Law 85-325). With the establishment of the Advanced Research Projects Agency in the Department of Defense, our space efforts could proceed while Congress considered legislation for permanent future arrangements.

Sixth, there was harmony between the Executive and Legislative Branches of the government as well as between the Senate and the House of Representatives concerning the necessity for effective action to achieve a leading U.S. space program. The most significant and dramatic instance was President Kennedy's Special Message to the Congress on Urgent National Needs, a message which he delivered before a joint session on May 25, 1961.*

The President said the United States has the resources and talents necessary to take a clearly leading role in outer space but that a national decision had not been made to marshal the resources for long-range goals to be pursued urgently on a time schedule. Asking the Congress for a firm commitment to appropriate essential funds to accomplish four national goals, the President said, "First, I believe that this nation should commit itself to achieving the goal, before this decade is out, of landing a man on the moon and returning him safely to the earth." The second goal was to accelerate development of the Rover nuclear rocket. Third, was acceleration of space satellites for worldwide communications. Fourth, was establishment of a satellite system for worldwide weather observation. The President was frank in pointing out that this new course of action would be very expensive and would require careful consideration by the Congress. Finally, he said, "This decision demands a major national commitment of scientific and technical manpower, material and facilities.... It means a degree of dedication, organization and discipline which have not always characterized our research and development efforts."

Congress responded affirmatively to this appeal and spectacular results were achieved. The manned landing on the moon overshadowed other national objectives because of its amazing nature, but outstanding achievements in global space communications and meteorology, contributing to new patterns of international cooperation and thereby to peace, were produced by the harmonious interaction between the President and the Congress.

EARLY CONGRESSIONAL ORGANIZATION

When the comprehensive nature of space activities was revealed in the hearings held by the Senate Preparedness Investigating Subcommittee toward the end of 1957, it became evident that the subject matter of major component parts of a United States space program cut across the jurisdiction of several standing committees of the House and Senate. A different combination of the substantive committees could be involved with each new piece of space legislation, in addition to the regular processes of the Committees on Appropriations. For example, bills dealing with meteorological and reconnaissance satellites could conceivably come within the jurisdiction of the Committees on Armed Services, Appropriations, Interstate and Foreign Commerce, Foreign Relations, Government Operations and the Joint Committee on Atomic Energy. The problem facing the Senate and House was how to combine related subjects so that the parliamentary situation involving responsibility and authority would be clear to all concerned. There was discussion of having only one committee to handle NASA legislation, the method used for the Atomic Energy Act of 1946 when Senator Brian McMahon was chairman, but this plan was not followed. The idea of a joint committee was also discarded.

* *Public Papers, President Kennedy, 1961*, pp. 403-5.

The Senate established the Special Committee on Space and Astronautics on February 6, 1958.[5] For the most part, the 13-member Special Committee was composed of the chairmen and ranking minority members of the standing committees which had a logical interest in space exploration: Appropriations, Foreign Relations, Armed Services, Commerce, Government Operations, and the Joint Committee on Atomic Energy. The importance attached to space exploration by the Congress was indicated by the election as committee chairman of the Majority Leader of the Senate, Lyndon B. Johnson.

The House of Representatives created the Select Committee on Astronautics and Space Exploration on March 5, 1958 with 13 members whose permanent committee assignments reflected the possible extension of their interests to space and space-related subjects.[6] Paralleling the Senate's "blue ribbon" committee, the Majority Leader of the House, Congressman John W. McCormack, became chairman.

WHERE DID IDEAS FOR SPACE LEGISLATION ORIGINATE?

The problem was how to organize the government to carry on a high priority space program. Although this could be identified as a political science problem, specifically one in public administration, this point was not perceived by that profession. Initially there was some speculation that outer space legislation could be identical with that of atomic energy, probably because the sudden psychological impact of the atomic bomb was remembered, an event that also resulted from the work of scientists and engineers. When it was pointed out that outer space was an environment in which a variety of activities could take place whereas nuclear fission was a source of energy, it was concluded that the two subjects required different approaches.[7]

The basic ideas which were eventually incorporated in the National Aeronautics and Space Act of 1958 originated with the scientists and engineers who had access both to the Executive and Legislative Branches of the government.

The scientific community was prepared to contribute to the legislative process in 1958 because their members had built up an understanding of government during the dozen years that had elapsed since the 1946 Senate hearings which resulted in the Atomic Energy Act of 1946. They had started various publications, notably the *Bulletin of Atomic Scientists*, and they were motivated to improve their knowledge and participate as citizens in the government. A contributing factor was that the "community" included scientists and engineers who had already devoted at least ten years to "pioneering the Nation's efforts in the research exploration of the threshold of space."

On November 21, 1957, the Rocket and Satellite Research Panel proposed A National Mission to Explore Outer Space. Among the 27 members of the panel were James A. Van Allen (chairman), Wernher von Braun, Krafft Ehricke, Homer E. Newell, W. H. Pickering and F. L. Whipple. On January 4, 1958 this panel joined with the American Rocket Society in issuing the National Space Establishment which summarized the basic premises common to both proposals.[8] The National Society of Professional Engineers analyzed

HOUSE SELECT COMMITTEE ON ASTRONAUTICS AND SPACE

April 25, 1958

Beginning hearings on space legislation, Hugh L. Dryden, Director of the National Advisory Committee for Aeronautics, as witness (center) with NACA models of inflatible satellites. Seated with Dryden (on left) is George Feldman, Chief Counsel, and (on right) is Chairman John W. McCormack, Majority Leader. Other members of the Committee standing (left to right): W. H. Natcher (D.-Ky.), Jas. M. Fulton (R.-Pa.), Lee Metcalf (D.-Mont.), B. F. Sisk (D.-Cal.), F. McDonough (R.-Cal.), Leo O'Brien (D.-N.Y.), Kenneth Keating (R.-N.Y.), and, Gerald R. Ford (R.-Michigan). (From Dryden Papers, Johns Hopkins University - III-1, NASA Photo No. 76-H-550)

America's Role in the Exploration of Outer Space in its statement of February 13, 1958. And on February 14, 1958, the National Academy of Sciences - National Committee for the International Geophysical Year 1957-58 - reported on Basic Objectives of a Continuing Program of Scientific Research in Outer Space.[9]

The major elements of these proposals were: (1) to create a national civilian establishment to carry on space research and activities; (2) to give this establishment independent statutory status so that it was separate from the Department of Defense and need not depend upon the appropriation of military funds; (3) to recognize the unique interest of the Department of Defense in space activities and provide for cooperation with the new agency; (4) to provide broad cultural, scientific and commercial objectives; (5) to ensure U.S. leadership and a continuing adequately funded program. All these proposals emphasized the practical benefits which would contribute to the U.S. and the people of the world. The statement of January 4, 1958 explained that--

> There will be a rich and continuing harvest of important practical applications as the work proceeds. Some of these can already be foreseen - reliable short-term and long-term meteorological forecasts, with all the agricultural and commercial advantages that these imply; rapid, long-range radio communications of great capacity and reliability; aids to navigation and to long-range surveying; television relays; new medical and biological knowledge,... and these will be only the beginning.[10]

It is interesting to note in this early history that all scientific groups looked forward to manned satellite ventures, including landing a man on the moon and returning him safely to Earth, as well as flights to Mars and Venus. The experts were well versed in the potential benefits of using outer space and they were able to explain to political decision-makers the significance of this multidisciplinary effort.

PASSAGE OF THE NATIONAL AERONAUTICS AND SPACE ACT OF 1958

On April 2, 1958, President Eisenhower sent a message to Congress proposing that the National Aeronautics and Space Agency be established by using the National Advisory Committee for Aeronautics as a nucleus and expanding its functions. On April 14th, Senator Lyndon B. Johnson and Senator Styles Bridges introduced S. 3609, the National Aeronautics and Space Act of 1958 to provide the research into problems of flight within and outside the earth's atmosphere. The provisions of this bill incorporated the proposals of the Executive Branch which benefited from the advice of the President's Science Advisory Committee of which Dr. James R. Killian was chairman. The explanatory statement prepared by the President's Committee, Introduction to Outer Space, included "the moon as a goal" explaining that "to land a man on the moon and get him home safely again will require a very big rocket engine...." Exploration of Mars and Venus was listed among scientific objectives. There was enthusiasm for attaining such practical benefits as global satellite communications, television via outer space, the improvement of weather forecasting and the opportunity of studying solar energy.

The Administration bill was introduced in the House (H. R. 11881 later changed to H. R. 12575) on April 14, 1958 by Congressman John W. McCormack whose Select Committee on Astronautics and Space Exploration began hearings on April 15th. The legislative process brought about many refinements and improvements in the bill as finally enacted. The modifications made by the House Committee in the original bill as proposed by the Executive Branch included changing the name of the agency to the National Aeronautics and Space Administration, giving the Director greater authority, broadening the scope of NASA, creating liaison committees between NASA and the Department of Defense and the Atomic Energy Commission, emphasizing the need for more information on space activities, and authorizing more international cooperation.[11]

The House bill accepted the same assumption as that of the Administration regarding the nature of the governmental organization problem to be solved. Their proposal was that an internal board of advisers within the new space agency, meeting only occasionally during the year, would give the new director adequate authority for a total U.S. space effort which involved relationships with numerous other departments such as the Department of Defense, the Department of State, the Department of Commerce, and the Department of Agriculture. This concept was carried over from the old NACA organization where it had worked successfully because NACA did not have to exercise authority over other federal agencies with a variety of programs and requirements. The Senate Special Committee on Space and Astronautics came to the conclusion, however, that an overall body was needed to plan for a complete coordinated U.S. space program. The Senate bill, S. 3609, therefore provided for centralized guidance and interagency coordination at a high level by establishing the National Aeronautics and Space Board. In conference, the House and Senate agreed on Title II - Coordination of Aeronautical and Space Activities, providing for the National Aeronautics and Space Council in the Executive Office of the President. The original version provided that the President preside over the Council meetings. The Council was composed of the Secretary of State, Secretary of Defense, NASA Administrator, Chairman of the Atomic Energy Commission, one additional member from government and not more than three other members distinguished for their achievements in the private sector. The Council's function was to advise the President concerning his functions to survey significant aeronautical and space activities, develop a comprehensive program for all agencies, designate and fix responsibility for the direction of major undertakings, provide for effective cooperation between NASA and the Department of Defense and resolve differences arising among departments and agencies.[12]

The Conference Report was adopted in both the House and Senate by voice vote on July 16, 1958 and approved by the President on July 29, 1958. Many changes were made in legislative proposals as they underwent scrutiny of the legislative process. As a staff participant in both House and Senate special space committees, this author was impressed with the wisdom resulting from the bicameral system. Between the time the House completed its April hearings on the NASA Act and the beginning of the Senate hearings on May 6, 1958, all participants--governmental and non-governmental--had an opportunity to study, use foresight, and assess probable consequences of legislative decisions.

Another result was Congressional emphasis on international space cooperation. Although the Administration bill contained a declaration that the United States was to cooperate with "other nations and groups of nations", there was no subsequent section to implement this policy. Section 205 was written into the Senate bill, providing that "The Administration under the foreign policy guidance of the President, may engage in a program of international cooperation in work done pursuant to this Act, and in the peaceful application of the results thereof, pursuant to agreements made by the President with the advice and consent of the Senate." When President Eisenhower signed the bill on July 29, 1958, he stated that this section was regarded as not precluding less formal arrangements, an interpretation which was never objected to by the Senate because it permitted many forms of beneficial international agreement between NASA and qualified foreign entities. Through the years the relevant Congressional committees kept close track of NASA's international program which they supported wholeheartedly.

International cooperation was strengthened by the opening policy declaration, also added by Congress, "that it is the policy of the United States that activities in space should be devoted to peaceful purposes for the benefit of all mankind." The peaceful exploration of outer space was further emphasized by Chairman John W. McCormack of the House Select Committee on Astronautics and Space Exploration in House Concurrent Resolution 332, 85th Congress, providing for peaceful purposes of space exploration and use of the United Nations.[13]

The Declaration of Policy was revised to clarify the relationship between the Department of Defense and NASA, and cooperation among all agencies was directed with no "unnecessary duplication."

CONGRESSIONAL ORGANIZATION FOR SPACE ACTIVITIES

Congress established a high priority for the U.S. space program at the beginning of the Space Age and expressed its continuing concern by organizing its committee structure to deal with space legislation and review of executive implementation. As we have seen, established committees were assigned outer space matters until the Select and Special Committees were created in the House and Senate. After completing work on the NASA legislation, four options were examined for future Congressional organization: (1) a joint space committee; (2) dividing jurisdiction among existing committees; (3) referring space matters to the Joint Committee on Atomic Energy; and (4) creating new separate standing committees in the House and Senate. Ultimately the fourth alternative was adopted.

The House Committee on Science and Astronautics was established on July 21, 1958, its jurisdiction being defined in the House Rules as--

a. Astronautical research and development, including resources, personnel, equipment, and facilities
b. Bureau of Standards, standardization of weights and measures and the metric system
c. National Aeronautics and Space Administration
d. National Aeronautics and Space Council

HOUSE SPACE COMMITTEE CHAIRMAN AND NASA WITNESSES

March 13, 1961

First appearance of NASA Administrator James E. Webb, with Deputy Administrator Hugh L. Dryden, before the House Committee on Science and Astronautics Chairman, Overton Brooks (D.-La.). (III-2; NASA Photo No. 61ADM-9)

SENATORS ROBERT KERR AND MARGARET CHASE SMITH
AND NASA ADMINISTRATORS JAMES WEBB AND HUGH DRYDEN

Mariner II press conference in NASA headquarters on December 15, 1962, covered the first successful Venus probe. Senator Kerr (D.-Okla.) was Chairman of the Senate Committee on Aeronautical and Space Sciences, and Senator Smith (R.-Me.) a member of the Committee. (III-3; NASA Photo No. 62ADM-36)

PRESIDENT JOHNSON WITH CONGRESSIONAL SPACE COMMITTEE LEADERS

In the White House, President Lyndon B. Johnson examines the first space photography of the surface of the planet Mars taken by *Mariner IV*, July 28, 1965. Shown (left to right) are: Congressman James Fulton (R.-Pa.); Senator Clinton B. Anderson (D.-N.M.), chairman of the Senate Committee on Aeronautical and Space Sciences; William Pickering, Director of the Jet Propulsion Laboratory; and, far right, Congressman George Miller (D.-Cal.) chairman of the House Committee on Science and Astronautics. (III-4; NASA Photo No. 65-H-1308)

HOUSE HEARINGS ON THE APOLLO 204 FIRE TRAGEDY

Rep. Olin E. Teague (D.-Tex.), NASA Administrator James E. Webb, Deputy Administrator Robert C. Seamans, and Rep. George P. Miller, chairman of the House Committee on Science and Astronautics, at opening session of Subcommittee hearings on the Apollo 204 Accident Report. April 11, 1967. (III-5; NASA Photo No. 67-H-395)

e. National Science Foundation
f. Outer space, including exploration and control thereof
g. Science scholarships
h. Scientific research and development

This jurisdiction tended to expand as aeronautical subjects were added as well as an increasing number of scientific and technological subjects. By House Resolution 988 of the 93rd Congress, the committee was renamed the Committee on Science and Technology, to be effective January 3, 1975.[14] The new jurisdiction added civil aviation research and development, environmental research and development, all energy research and development and the National Weather Service. Thus outer space became one subject placed in the total context of science and technology.

Meanwhile in the Senate, Resolution 327 was passed on July 24, 1958 to create the standing Committee on Aeronautical and Space Sciences with the following jurisdiction:

(A) Aeronautical and space activities, as that term is defined in the National Aeronautics and Space Act of 1958, except those which are peculiar to or primarily associated with the development of weapons systems or military operations.

(B) Matters relating generally to the scientific aspects of such aeronautical and space activities, except those which are peculiar to or primarily associated with the development of weapons systems or military operations.

(C) National Aeronautics and Space Administration.

Such committee also shall have jurisdiction to survey and review, and to prepare studies and reports upon, aeronautical and space activities of all agencies of the United States, including such activities which are peculiar to or primarily assoicated with the development of weapons systems or military operations.

It was clear that the Senate Armed Services Committee was to continue its legislative jurisdiction over military matters, although the new space committee was authorized to include defense matters in studies and reports.

The Senate Committee on Aeronautical and Space Sciences was abolished and its functions transferred to the Committee on Commerce, Science and Transportation on February 4, 1977 when the Senate passed Committee System Reorganization Amendments of 1977, S. Res. 4, 95th Congress, first session. Jurisdiction over outer space matters then became a function of several subcommittees with authority over communications, oceans, the weather, and "science, engineering, and technology research and development and policy."[15]

SIGNIFICANCE OF THE AUTHORIZATION PROCESS

Authorization required for the appropriation of funds to federal agencies may be on a continuing basis so that the Appropriation Committees can proceed without more frequent specific reports from other committees. The use of annual authorization bills enacted into law was unusual at the time this method was adopted for NASA, although the procedure has been generally expanded since 1958 (particularly by the Armed Services Committees) as a means of improving the relations between policy, program and budget and ensuring that the Executive Branch follows Congressional intent. Absence of an annual authorization might mean that an agency's program was not given detailed, in-depth consideration by the Appropriations Committees which necessarily have a workload covering all programs requiring funds.

On August 21, 1958, Congress took special action which resulted in the space committees making an annual review of NASA's activities. Two appropriation bills, one on military construction and the other on supplemental appropriations, were amended to provide for an annual authorization. The requirement was at first temporary but became a permanent feature of the legislative process for outer space:

> Notwithstanding the provisions of any other law, no appropriation may be made to the National Aeronautics and Space Administration unless previously authorized by legislation hereafter enacted by the Congress.[16]

The in-depth history of the space program can be studied in the annual hearings on the authorization of NASA funds and the reports. The hearings are illustrated and indexed and each year constitute up-to-date basic information and analysis on aeronautical and space activities of the United States from 1958 to the present time. Combined with the annual reports required by the NASA Act from the President to the Congress, one can easily comprehend the support by Congress of U.S. space activities. Not the least element in this sustained span of attention was the foundation provided by expert Congressional staffing.

CONTRIBUTIONS OF ADDITIONAL COMMITTEES

It is not possible in this conference paper to trace all the space and space-related legislation handled by committees other than those established specifically for aeronautical and space activities. Such an undertaking would result in several volumes. Space communications must be singled out, however, for special mention. At one time there were 14 committees and subcommittees with an interest in satellite communications. The most important legislation was the Communications Satellite Act of 1962 which was of concern to the Committees on Foreign Relations and Foreign Affairs, Commerce, Armed Services, and the space committees. Establishment of the Communications Satellite Corporation (COMSAT), leading to the creation of the commercially successful International Telecommunications Satellite Organization (INTELSAT) was an outstanding achievement involving the executive and Legislative Branches in creating unique relationships between government and industry, science and technology, and the conduct of U.S. foreign policy.[17]

ROLE OF THE SENATE FOREIGN RELATIONS COMMITTEE

The climate of international space cooperation was facilitated not only by the NASA Act but also by use of the national space program in establishing relations between the United States and other nations. The first dramatic episode occurred on November 17, 1958, when Senator Lyndon B. Johnson, at the Request of President Eisenhower, addressed the United Nations on The Peaceful Uses of Outer Space, explaining that the Executive and Legislative Branches of our government were united in dedicating outer space to peaceful purposes for the benefit of all mankind and that "Today outer space is free. It is unscarred by conflict. No nation holds a concession there. It must remain this way."[18]

The U.S. gave its support to establishing in 1958 the United Nations Committee on the Peaceful Uses of Outer Space which later formulated four space treaties which were referred by the President to the Senate for advice and consent. These treaties were under the jurisdiction of the Senate Committee on Aeronautical and Space Sciences which had produced analytical staff studies on these international agreements.

Hearings were held by the Senate Foreign Relations Committee which reported the treaties favorably with the result that the Senate gave its advice and consent to the following:[19]

1. Treaty on Principles Governing the Activities of States in the Exploration and Use of Outer Space, including the moon and other celestial bodies. Ratification was advised by the Senate on April 25, 1967 and the President deposited ratification on October 10, 1967 when the Treaty went into force.

2. Agreement on the Rescue of Astronauts, the Return of Astronauts and the Return of Objects Launched into Outer Space. The Senate advised ratification on October 8, 1968 and ratification was deposited on December 3, 1968 when the Treaty entered into force.

3. Convention on International Liability for Damage Caused by Space Objects. Ratification was advised by the Senate on October 6, 1972 and the Treaty entered into force with respect to the U.S. on October 9, 1973 when ratification was deposited by the President.

4. Convention on the Registration of Objects Launched into Outer Space. Senate advice and consent were given on June 21, 1976 and the instrument of ratification signed by the President on July 24, 1976. This convention went into force on September 15, 1976.

OVERALL GUIDANCE OF THE UNITED STATES SPACE PROGRAM

The National Aeronautics and Space Council was created by Congress in the NASA Act of 1958 to ensure a high-level focal point in government to analyze and plan for the overall space and space-related programs of all federal departments and agencies. Although the Council was not a strong institution, being advisory to the President in his Executive Office, nevertheless it gave outer space a priority and identified the necessity for interdepartmental cooperative coordination and the effective performance of its functions required an expert professional staff. President Eisenhower did not wish to make use of the Council, but when President Kennedy began his administration, the NASA Act was amended to make the Vice President chairman, the other members being the Secretary of State, Secretary of Defense, Secretary of Transportation, the NASA Administrator, and the Chairman of the Atomic Energy Commission. The report which the NASA Act required the President to make annually to the Congress on aeronautical and space activities was prepared by the Council staff.[20]

President Nixon did not wish to retain the Council, however, and sent to Congress Reorganization Plan No. 1 of 1973 to abolish the Council and its functions to be effective July 1, 1973.[21] The main motive was to reduce the size of the White House staff. All reorganization plans come under the jurisdiction of the Committees on Government Operations, and there is a general feeling in Congress that any President has the right to make his own staff arrangements. Congress did not pass a bill objecting to the reorganization plan and the Council was abolished. Some of its functions were transferred to the National Science Foundation which was for a time responsible for the overall annual report to the Congress on aeronautical and space activities. As might have been expected, this arrangement proved to be unsatisfactory because the National Science Foundation had no authority that it could exert over other government departments. In political science terms, a form of centralization at a high level gave way to a concept of decentralization at a level which could not effectively operate over the U.S. space activities of all agencies.

Meanwhile the scientists and engineers, who had lost their unique position in relation to the White House when the President's Science Advisory Committee was abolished, sought new arrangements for contributing advice to the government on science and technology and its impact on public policy. There then followed a period when Congressional committees, the White House staff, scientists and engineers worked closely together to define national policy and establish an organizational structure with specific functions.

The result was the passage of the National Science and Technology Policy, Organization, and Priorities Act of 1976 (P. L. 94-282) approved by President Ford on May 11, 1976.[22] The national policy for science and technology was set forth in Title I and connected with priority goals, one of which was "advancing the exploration and peaceful uses of outer space." Although outer space activities thus became only one among 13 objectives, it is significant that many of the other goals cannot be adequately achieved without the use of space technology; for example, contributions to national security, food supply, energy, protection of the environment, strengthening the economy, educational opportunities, improvement of transportation and communications and eliminating pollution. Space technology is a tool that can be used to improve numerous functions and it is necessary to study all the "spinoff" production of NASA to measure the impact of space developments on society.

A number of organizations provided for in the National Science and Technology Policy, Organization, and Priorities Act of 1976 were deemed unnecessary to President Carter who sent to the Congress Reorganization Plan No. 1 of 1977 which did not encounter Congressional opposition.[23] This plan abolished a number of organizational units and transferred their functions to the President: the Intergovernmental Science, Engineering, and Technology Advisory Panel, the President's Committee on Science and

Technology, and the Federal Coordinating Council for Science, Engineering, and Technology, all of which had been established in Titles II, III, and IV of Public Law 94-282. Two functions were transferred by President Carter to the National Science Foundation: the annual science and technology report and the responsibility for a five-year forecast of current and emerging problems. NSF has a contract with the National Academy of Sciences to assist with this work. Some functions were transferred to the Office of Management and Budget (OMB): reorganization and liaison between federal and state governments.

The Office of Science and Technology Policy (OSTP) was retained in the Executive Office because it was one of ten units that President Carter decided to work with directly because he needed "their constant advice and counsel, almost on a daily basis." Dr. Frank Press, a geophysicist, was appointed Director of the Office of Science and Technology and sworn in on June 1, 1977.[24]

Thus in a period of six years, many moves were made between centralization and decentralization, and different combinations seem to emerge for different programs, some of them on an *ad hoc* basis.

When we examine what is happening organizationally on outer space problems, we find a number of units mentioned in the President's Statement on Space Policy of October 11, 1978. President Carter directed a review of policy on outer space by the National Security Council, and as a result a Policy Review Committee (Space) was established in May 1978 with the Director of the Office of Science and Technology Policy, Dr. Frank Press, as chairman. The Committee's assessment resulted in the White House Statement on U.S. Civil Space Policy on October 11, 1978.[25] The conclusions of this Committee were that (1) "space policy will reflect a balanced strategy of applications...;" (2) future activities will be chosen to accord with national objectives; (3) "it is neither feasible nor necessary at this time to commit the United States to a high-challenge space engineering initiative comparable to Apollo."

A number of organizational units have been activated by these policy statements. NASA will chair an interagency task force to examine options for current and future institutional arrangemnts for remote sensing of the earth by satellites, looking toward an "integrated national system". The desirability of consolidating meteorological satellite programs will be reviewed by the Department of Defense, NASA, and the National Oceanic and Atmospheric Administration (NOAA) of the Department of Commerce. Proposals for increased participation by the private sector will be analyzed by NASA and the Department of Commerce. Domestic and international public satellite services will be reviewed by the Department of Commerce's National Telecommunications and Information Administration (NTIA). The Agency for International Development and the Department of the Interior will work with NTIA on programs for less developed countries. An interagency task force will study uses of the Shuttle and report to the Policy Review Committee (Space) before preparation of the Fiscal Year 1981 budget. This Committee will also seek the transfer of technology between space sectors, which are not defined but presumably are areas of space activity within the U.S. government.

There seems to be an assumption that in the course of time an organizational structure will evolve and consequently the final outcome cannot be outlined at this time. This approach is the opposite of that adopted in 1958 when problems of organization and management were solved at the outset of U.S. expansion into outer space.

These developments have provoked some negative responses in the Congress. For many years the Executive and Legislative Branches worked hand in hand in developing space activities, but they began to differ over future policy and programs, particularly on the issue of establishing an operational remote sensing system. NASA's LANDSAT system, although experimental, proved to be so successful both nationally and internationally that there was great enthusiasm in the Congress for creating an operational system for remote sensing of the earth by satellites.[26] The Executive Branch has resisted such proposals so that in several space policy and program areas the Congressional displeasure has been expressed on the content and pace of accomplishing space goals.[27]

CONCLUSIONS

Congress gave the use and exploration of outer space high priority which has been sustained from 1957 to the present. Permanent standing committees were given continuing responsibility for evaluating programs and ensuring adequate funds. Expert professional staffing was provided. Basic policies were adopted in 1958 on the organization of the government to conduct space activities by establishing the National Aeronautics and Space Administration, providing for cooperation between NASA, the Department of Defense and all other federal agencies, and by assigning responsibility for policy-making decisions at the highest level of government. A basic policy was that "activities in space should be devoted to peaceful purposes for the benefit of all mankind." The objectives emphasized that the United States should preserve its leadership in outer space, make effective use of scientific and engineering resources, avoid unnecessary duplication, and cooperate with other nations and groups of nations.

Although there have been organizational changes in Congressional committees, outer space has not lost its priority or its continuing budgets, even though there are differences of opinion on the size of any budget. Priority is maintained because Congress took the significant action of requiring annual authorization of funds for NASA, a practice which ensures the closest surveillance of programs and their cost. The tremendous number of analytical Congressional documents published between 1957 and 1979 constitute a foundation of knowledge to which contributions have been made by government and industry, scientists and engineers, and the academic community. Congressional committees have been strongly supportive of NASA's international program and a number of senators, representatives and professional staff have participated as delegates to the United Nations Committee on the Peaceful Uses of Outer Space and its subcommittees. The way in which the Congress has been organized and functions resulted in a number of long-term committee members becoming expert on outer space matters.

After two decades we can conclude that the NASA Act was drawn with remarkable foresight considering that its statutory provisions have facilitated the attainment of U.S. leadership in space ventures. Perhaps the greatest weakness of the Act, however, and one that has not yet been overcome, is the concept that NASA is only a research and development institution and not one to undertake operational programs. There is some legal fuzziness about this point as arguments have been made that the law does not confine NASA merely to research and development. Nevertheless, that stance has been strongly taken by some NASA personnel and by others in the government, e. g., the Office of Management and Budget, so that decisions have been long delayed and, for example, have not yet been made on an operational remote sensing system of the earth by satellites. Although it is well established that a large and varied program cannot be effectively managed by a number of committees in the Executive Branch, the trend has been in that direction and there is evident relunctance even to appoint NASA as the lead agency although the NASA Act provides that aeronautical and space activities "shall be directed by a civilian agency exercising control... except those peculiar to the military. (Underlining added. P. L. 85-568, Sec. 102 (b)).

In the opinion of this author, NASA's role should be clarified by statute so that it may operate space programs as designated by the Congress. NASA has the staff and experience to operate programs and has an excellent record of cooperating with other agencies and with other nations.

Admittedly it is difficult to coordinate and interrelate a number of agencies whose functions can be improved by using space science and technology, but the least effective method is to attempt the task by assignment to a number of committees representing different agencies and with no permanent professional staff to evaluate interacting forces. If different persons attend meetings from time to time there is no opportunity to build up an institutional memory. Delay in decisions can result if committees report at different times on a variety of subjects to a central committee. It is not sufficient to make building blocks if the blocks are left lying about. There should be more participation by experts in public administration so that practices which have proved ineffective in the past can be avoided in the future. There is something unsettling about each new Administration changing the system of organization and management at the top level for space activities and, indeed, for all science and technology. The most effective structural framework cannot be expected to come about by evolution. Organization must be determined in advance with as much foresight as possible exercised in estimating probable consequences. Too few of the experts in government administration and management have been willing to study science and technology so they can work with the subject in all its ramifications.

In looking toward the future it is necessary to examine assumptions often generated by attitudes leading to preordained conclusions. We must recognize that decisions based on getting a "balance" between programs can be decisions that destroy priority for a desired objective because balancing funds means leveling off among competing agencies. We may wish to establish a priority and stick to it. Once a decision has been made to undertake a program, responsibility must be given to a single administrator with the resources and funds necessary to accomplish the objective in a specified period of time.

ASTRONAUT JOHN GLENN BEFORE JOINT SESSION OF CONGRESS

First American to orbit the earth in Mercury spacecraft made a 20-minute address to a joint session of the Congress on February 26, 1962. He received a standing ovation. In November 1974, John H. Glenn was elected to the U.S. Senate (D.-Ohio). (II -6; NASA Photo No. 62-MA6-173)

THE UNITED STATES CONGRESS AND OUTER SPACE
FROM SPUTNIK TO THE SHUTTLE

REFERENCE NOTES

1. *Annals of the International Geophysical Year*, vol. IX (New York: Pergamon Press, 1959); the Membership and Programs of the Participating Committees; see also Report on the International Geophysical Year, February 1959; National Science Foundation, National Academy of Sciences; Hearings before the Subcommittee of the House Committee on Appropriations, 86th Congress, 1st session, Testimony of Dr. Joseph Kaplan, February 18, 1959, p. 6

2. 50 U.S.C. 151.

3. Dr. Hugh L. Dryden, National Advisory Committee on Aeronautics (NACA), in testimony before the House Select Committee on Astronautics and Space Exploration, 85th Congress, 2nd session, Hearings on H.R. 11881, April 22, 1958, pp. 404-410.

4. Inquiry into Satellite and Missile Programs, Hearings before the Preparedness Investigating Subcommittee of the Senate Committee on Armed Services, 85th Congress, 1st and 2nd sessions, pt. 3, 1958, p. 2429.

5. Senate Resolution 256, 85th Congress, February 6, 1958.

6. House Resolution 496, 85th Congress, March 5, 1958.

7. Eilene Galloway, "The Problems of Congress in Formulating Outer Space Legislation", Hearings before the House Select Committee on Astronautics and Space Exploration, 85th Congress, 2nd session on H.R. 11881, April 15, 1958, pp. 5-10.

8. "A National Mission to Explore Outer Space", a proposal by the Rocket and Satellite Research Panel, November 21, 1957. National Space Establishment proposal signed by James Van Allen, Chairman, Rocket and Satellite Research Panel and George P. Sutton, President of the American Rocket Society, January 4, 1958; Compilation of Materials on Space and Astronautics, No. 1, Senate Special Committee on Space and Astronautics, Committee print, 85th Congress, 2nd session, March 27, 1958, pp. 14-19.

9. *Ibid.*, pp. 20-44.

10. *Ibid.*, p. 19.

11. Astronautics and Space Exploration, Hearings before the House Select Committee on Astronautics and Space Exploration, 85th Congress, 2nd session on H.R. 11881, April-May 1958, 1542 p., H.R. 12575 (in lieu of H.R. 11881) passed the House of Representatives by voice vote on June 2, 1958.

12. Final Report of the Senate Special Committee on Space and Astronautics, pursuant to S. Res. 256, 85th Congress; 86th Congress, 1st session, Senate Report No. 100, March 11, 1959, 76 p., see particularly pages 2-12.

13. Hearing before the Subcommittee on National Security and Scientific Developments Affecting Foreign Policy of the Committee on Foreign Affairs, House of Representatives, 85th Congress, 2nd session on H. Con. Res. 326 (later H. Con. Res. 332) a concurrent resolution relative to plans for the peaceful exploration of outer space, May 20, 1959, pp. 4, 6, 24, 26, 29, 32-33. See also Senate Foreign Relations Committee Report 1728, July 23, 1958 which was agreed to by the Senate on July 23, 1958.

14. Monographs on the Committees of the House of Representatives, House Select Committee on Committees, 93rd Congress, 2nd session, Committee print, December 13, 1974, pp. 133-138.

Committee Reform Amendments of 1974, Explanation of H. Res. 988 adopted by the House of Representatives, October 8, 1974; Staff report of the House Select Committee on Committees, 93rd Congress, 2nd session, Committee print, pp. 49, 215.

Constitution, Jefferson's Manual and Rules of the House of Representatives; 94th Congress, 2nd session, House Document No. 94-663, pp. 390-391, Wm. Holmes Brown, parliamentarian (Washington, D.C., U.S. Govt. Printing Office, 1977).

15. Committee on Aeronautical and Space Sciences, United States Senate, 10th Anniversary, 1958-1968; 90th Congress, 2nd session, Senate Document No. 116, July 19, 1968, 109 p.; see also Standing Rules of the United States Senate, Committee on Rules and Administration, U.S. Senate, April 30, 1977, pp. 30-31.

16. Public Law 86-45, section 4, June 15, 1959 (73 Stat. 75, 42 U.S.C. 2460; see also National Aeronautics and Space Act of 1958, as amended, and Related Legislation; Senate Committee on Commerce, Science, and Transportation, 95th Congress, 2nd session, Committee print, December 1978, 185 p.; see also Eilene Galloway, "The United States Congress and Space Law", *Annals of the Air and Space Law Institute*, vol. III (Montreal: McGill University, 1978), pp. 395-407.

17. International Cooperation in Outer Space: A Symposium, Senate Committee on Aeronautical and Space Sciences, 92nd Congress, 1st session, Senate Document No. 92-57, December 9, 1971, pp. 609-651.

18. Final Report of the Senate Special Committee on Space and Astronautics, *op. cit.*, pp. 58-62.

19. Space Law, selected basic documents, 2nd ed., Senate Committee on Commerce, Science, and Transportation, 95th Congress, 2nd session, Committee print, December 1978, 600 p.

20. National Aeronautics and Space Act of 1958, as amended and Related Legislation, *op. cit.*

21. Reorganization Plan No. 1 of 1973, *38 Federal Register 9579* April 18, 1973, 87 Stat. 1089.

22. 42 U.S.C. 6601.

23. U.S. Congress, House, Committee on Government Operations, Reorganization Plan No. 1 of 1977, Report to accompany H. Res. 688, Report No. 95-661, 95th Congress, 1st session, 68 p.

Senate Committee on Governmental Affairs, Reorganization Plan No. 1 of 1977 relating to the Executive Office of the President, Report to accompanying S. Res. 222, Report No. 95-465, 95th Congress, 1st session, 19 p.

Reorganization Plan No. 1 of 1977, as amended Sept. 15, 1977, *Federal Register*, vol. 42, October 21, 1977, pp. 56101-56103.

24. Dorothy M. Bates, Science and Technology in Policy Formulation at the Presidential Level: Recent Developments, Issue Brief No. 1B78027, Congressional Research Service, Major Issues System, The Library of Congress, January 30, 1979, 20 p.

25. The White House Fact Sheet, U.S. Civil Space Policy, Office of the White House Press Secretary, October 11, 1978, 4 p.

26. Earth Resources and Environmental Information System Act of 1977, Hearings before the Subcommittee on Science, Technology, and Space of the Senate Committee on Commerce, Science, and Transportation, on S. 657, 95th Congress, 1st session, Serial No. 95-30, May, June 1977, 352 p. See also Senator Adlai Stevenson's remarks on the Shuttle and space policy before the National Space Club, Washington, D.C., January 16, 1979.

Examples of bills are S. 3530, a bill to establish national space policy and program direction, Sept. 27, 1978; S. 3589, a bill to establish an operational Earth Data and Information Service in NASA, October 12, 1978; S. 3625, a bill to provide for the establishment, ownership, operation, and regulation of a commercial earth resources information service, utilizing satellites and other technologies. 95th Congress, 2nd session, H.R. 14297, a bill to establish a Space Industrialization Corporation... 95th Congress. Senate Subcommittee on Science, Technology, and Space of the Senate Committee on Commerce, Science, and Transportation, Hearings on U.S. Civilian Space Policy, beginning on January 25, 1979 (not yet published).

27. Senator Adlai Stevenson, U.S. Civil Space Policy, *Congressional Record*, vol. 124, no. 168, October 14, 1978, 95th Congress, 2nd session. The text of the White House statement on U.S. Civil Space Policy appears at the end of Senator Stevenson's remarks.

IV

THE POLITICAL ECONOMY OF AMERICAN ASTRONAUTICS

Mary A. Holman *
and
Theodore Suranyi-Unger, Jr. **

This paper presents the status of American astronautics in the context of political economics, which emphasizes the close interrelationship between the viability of a strong technical capability in space, as well as continued progress in science and technology more generally, and support by the federal government. It is contended that federal support should be on a long-term sustained basis rather than by means of the annual federal budget process that allocates government funds in a short-term manner, usually in response to public fads or perceived national crises.

The title given to this paper may cause some wonderment, partly because of the vastness that it suggests and partly because of the choice of an uncommon phrase, political economy. But there are explanations. First, *The Political Economy of the Space Program* was the title of a 1974 book by M. A. Holman, in which the subject matter was laid out in some detail[1] A condensed sequel a decade later could have carried the same title with a "... Revisited" attached to it, but that would have been trite, and besides, many of us are not certain that the United States has a space program today that is either qualitatively or quantitatively comparable to what we had a decade or two ago.

Secondly, the phrase "political economy" was deliberately retained because its connotations are today as valid as they were a decade ago. During the infancy of our discipline, about a century ago, what is known as "economics" today was commonly called "political economy". This does not mean that we are necessarily a century behind the times; rather, it means that the phrase "political economy" is still used, although in a more restricted sense. Looking at it cynically, the phrase is often used to describe "soft" or inexact or nonanalytical treatments of a subject. But

* *Professor of Economics, The George Washington University, Washington, D.C. 20052*

** *Research Professor of Economics, The George Washington University, Washington, D.C. 20052* in a more positive vein, political economy may be considered a subdiscipline of economics in which the contents are so intimately interwoven with political or social trends and events of a given period, or point in time, that the reduction of all variables in the examination to rigid abstractions would deprive the subject of its timely and lively interest. Hopefully, it is in this second sense that we use the term.

Having classified the subject matter as one of political economy, we have also established its close connection with the government. In fact, as this audience is well aware, the initiation and the promotion of the space effort have been just about exclusive functions of the federal government, with only a minimal contribution from the private sector. In the lingo of the economist, the fruits of the space effort fall into the category of social goods, but they are social goods with some noteworthy characteristics. First, they are very expensive; second, because of their obvious military application potential, they are unpopular in certain circles. In fact, in the mind of the careless or the uninformed, they are simply lumped into the category of "mass destruction hardware". And thirdly, even though the bounty of peaceful, civilian-oriented spinoffs, fallouts, and secondary benefits are so well known to this audience, much of the country's population is not aware of them, simply because the connection is not readily apparent. After all, how many consumers know that they owe their freeze-dried camping foods and their pressure spray cleaners to the U.S. space program? The gist of all of this is simply another very well known fact; the space effort is not among the most popular public undertakings in this day and age.

Putting questions of popularity aside for a moment, we might become somewhat more exacting and ask whether we can justify and verify the widespread claims that the U.S. space effort has been neglected and has been barely limping along during the course of the past decade. What is the economist's role in answering that question? What specific economic phenomena, if any, may be identified in arriving at the answers, and if called for, in what way can the economist contribute to solutions, if such solutions exist, other than to say that we need to spend more money?

Here, the economist has two separate bags of tricks; one is filled with concepts and principles, the other with real world facts. The first of these bags is much larger because it contains economic theories and principles that pertain to all real world phenomena, whether astronautics, air pollution, or population control. There exists no theory of astronautical economics. For that reason, the conceptual portions of the ensuing must necessarily be more general than what the title of this paper suggests. They must be broad enough to cover the entire arena of scientific and technological progress and its promotion by the federal government.

The second bag contains empirically oriented subject matter, related to specific phenomena in the real world such as agriculture, health, or astronautics. This is the world of hyphenated economics where it becomes possible to develop an empirically focused body of knowledge. Accordingly, our specific treatment of the space program will be within that realm of economic inquiry.

THE BUDGET AS AN INDICATOR OF THE POLITICAL ECONOMY OF ASTRONAUTICS

It is not possible to examine the political economy of astronautics without looking at alternative and competing allocations for federal funds. If the U.S. had a unique National Goals Policy, such a task would be fairly simple. Although a number of statements have been made about establishing a National Goals Policy since the end of World War II, this country remains without such a structure.[2] In effect, it is only the annual federal budgeting process that establishes goals and these annually determined priorities are typically short-run in nature.[3]

Table 1 presents information on the allocation of federal funds to various national programs during the last 40 years. The pre-World War II year, 1940, was included to show that the U.S. is currently spending approximately the same relative share of the federal budget on income security as it did in 1940. Of course, the purposes of income security programs today are much different from those in 1940, when the rate of unemployment of the civilian labor force was about 15 percent. Since the early 1960's, the largest shifts in federal spending have been sharp declines in spending for national defense and large increases of funds for education, health, and income security. It must be recognized that many of the programs shown in the dozen major kinds of expenditure programs are "noncontrollable" forms of federal spending. The term "noncontrollable" means that funds cannot be diverted to other kinds of programs without changes in the law. One reason for the recent creation of the new Department of Education was that the single agency for Health, Education, and Welfare did not have the discretionary authority to shift funds from income maintenance to education.

Generally, the trust funds (e.g., social insurance and federal highway programs) are financed through permanent appropriations. Similarly, the law that relates to the national debt mandates Congress to appropriate funds to pay interest on the national debt as the funds are needed. Levels of spending for veterans benefits, as well as military retirement pay, are determined by statute rather than by annual budget review.[4]

Because national defense is a "controllable" kind of federal expenditure, it is ultimately the responsibility of Congress to evaluate external risk and to make suitable appropriations. Annual expenditures for national defense can be extremely volatile; the annual changes, no doubt, reflect changes in international crises. This is exemplified in the recent (1980) willingness of the President and most of the presidential candidates to insist that Congress increase spending for defense after the Soviet invasion of Afghanistan in December 1979. On one hand, given its size, spending for defense is the most "controllable" item in the federal budget. This is seen in the sharp relative decrease in defense spending between 1970 and 1980. But during this period, the post-Vietnam War era, there were only a few relatively minor defense crises, such as the Mayaguez incident, and détente with the Soviets was the main concern. On the other hand, if what the Executive or the Congress perceive as an emergency vital to national security exists, the defense budget cannot be used as a buffer to cushion spending for other "priorities." In the context of the dozen

Table 1

FEDERAL BUDGET EXPENDITURES FOR "NATIONAL GOALS." FISCAL YEARS PERCENTAGE DISTRIBUTION

National Objective	1980	1970	1960	1950	1940
Total (Billions of $)	563.6	196.6	92.2	39.5	9.1
		Percentage Distribution			
Defense	23.1	40.0	49.8	34.2	16.5
Space & Science	1.1	2.3	0.4	0.3	--
Energy	1.4	2.3	--	--	--
Agriculture	0.8	2.6	3.6	7.1	16.5
Natural Resources and the Environment	2.3	1.6	1.1	3.0	5.5
Transportation & Commerce	3.5	3.6	5.2	4.6	--
Community Development	1.5	1.2	1.1	0.8	--
Education	5.5	4.4	1.4	0.3	--
Health	10.0	6.6	0.8	4.6	33.0
Income Security	33.9	21.9	19.5		
Veterans Benefits	3.7	4.4	5.9	16.7	6.6
Interest	11.2	9.3	9.0	14.7	12.1
Total Above	98.0%	98.4%	97.8%	86.3%	90.2%

Sources: *Economic Reports of the President*, Various Years

major programs listed in Table 1, there are only a few that can be changed without major changes in the law--those that can be amended most easily relate to space and science, energy, and natural resources and the environment. It seems quite likely, that if the budget is to be substantially cut during fiscal 1981, a large part of that cut will come from proposed spending on R & D and on astronautics.

The data in Table 2 separate federal expenditures for R & D from government spending in general. The message is approximately the same. There has been a marked change in national priorities since the mid-1960's. That change, of course, has been away from the continued development of a strong scientific and technological capability in space and into the more tangible, practical or "real world" directed research endeavors in energy, health, economic development and community services.

THE RESEARCH REVOLUTION OF THE 1950'S AND THE 1960'S

Behind the numbers shown in Table 1, is the fact that between 1950 and 1965, this country witnessed a Research Revolution, with a sizeable share of total federal spending going for R & D. The amount reached a peak during the mid-1960's, with about 10 percent of federal outlays allocated to R & D, including defense R & D. In 1964, federal R & D spending amounted to about $12.6 billion and total federal outlays were about $118.6 billion. Another measure of R & D spending is in relation to the Gross National Product. As a percentage of GNP, federal R & D spending was 1.97 percent in 1964.

Table 2

FEDERAL R & D SPENDING IN THE UNITED STATES FOR "NATIONAL COALS"

	National Defense	Space	Energy	Health	Economic Development & Community Services
			(Billions)		
1961-62	$ 7.3	$ 1.2	$ 0.8	$ 0.5	$ 0.4
1966-67	8.3	5.3	0.9	1.0	1.1
1976-77	12.0	2.9	2.1	2.4	3.2
			Percentage Distribution		
1961-62	71	12	7	5	4
1966-67	49	32	5	6	7
1976-77	51	13	9	10	14

Source: *National Science Board, National Science Foundation, Science Indicators 1978, (Washington, D.C.: U.S. Government Printing Office, 1979), p. 147.*

The roots of the Research Revolution came from the World War II period when the practical potential of science-based technology became impressively visible. These potential results of R & D were perceived by the layman as well as by the specialist. World War II saw the successful smashing of the atom and birth of the Nuclear Age, in addition to the many accomplishments that were of a purely military-oriented nature.

In the immediate post-World War II period, there was a continued national desire for maintaining the pace of technological progress. With the end of the war also came the realization that the strongest single moving force of R & D--the national defense objective--was no longer present, and with the resumption of a privately operating peacetime economy, R & D activity was likely to decline. This perception was best expressed by Vannevar Bush in his classic, *Science, the Endless Frontier.*[5] In that work, the former Director of the Office of Scientific Research and Development delivered two messages to the American public. First and self-evident in the title of the work, that the potentials of new knowledge to benefit humanity are virtually endless. The second message, however, was that if the rate of progress is to be maintained and increased, the amount of R & D spending must also be maintained and increased. Furthermore, the necessary volume of R & D spending, especially in basic research, had to become a special concern of the federal government for two cogent reasons. Vannevar Bush first observed:

> ...we can no longer count on ravaged Europe as a source of fundamental knowledge. In the past we have devoted much of our best efforts to the application of such knowledge which has been discovered abroad. In the future we must pay increased attention to discovering this knowledge for ourselves particularly since the scientific applications of the future will be more than ever dependent upon such basic knowledge.

New impetus must be given to research in our country. Such new impetus can come promptly only from the Government. Expenditures for research in colleges, universities, and research institutes will otherwise not be able to meet the additional demands of increased public need for reseach.

Second, Bannevar Bush believed:

...we cannot expect industry adequately to fill the gap. Industry will fully rise to the challenge of applying new knowledge to new products. The commercial incentive can be relied upon for that. But basic research is essentially noncommercial in nature. It will not receive the attention it requires if left to industry.[6]

This philosophy, as expressed by Bush, resulted in the establishment of the National Science Foundation in 1950. With the creation of that agency, the formal assumption of responsibility for fundamental research was accepted by the federal government. During the earliest years after the advent of the National Science Foundation, the federal R & D effort came to be spearheaded by the DOD-AEC-NSF trilogy. However, in the late 1950's, a fourth federal R & D effort objective began to unfold rapidly: the objective of space.

The astronautics objective, or later, space race, also finds its roots in the post-World War II era when Wernher von Braun and a group of scientists from Germany began their research at White Sands Proving Grounds.[7] The beginnings were small, but as is well-known today, the space program, following the establishment of the National Aeronautics and Space Administration in 1958, proved to be a most powerful prime mover of the national R & D objective.

This was the institutional environment and the national mood immediately following World War II. Research and development emerged as a national objective, but the Research Revolution did not materialize until the late 1950's, as a strong and intense reaction to the Soviet launching of *Sputnik I* in late 1957.

Leonard Silk coined the term "Research Revolution," and he used it to designate the period from the late 1950's until the mid-1960's. The term "Research Revolution" can be justified on two grounds, First, it was triggered by the Russian Sputnik; hence, similar to all revolutions, it was reactive. Secondly, the intensity of sentiment and the ensuing action (i.e., the sharp rise in the R & D effort) are also deserving of the label.

The promises of R & D were widely publicized, and NASA made a manned lunar landing by the end of the decade of the sixties, the prime objective of its mission. This objective was inspiring to the nation, even though there were many who believed that the goal was symbolic rather than real because they did not really think that such an achievement was actually within the realm of possibility in such a short time, if indeed, ever. The promise of astronautical research was such that between 1960 and 1966, about 85 percent of the increase in federal R & D spending was for space research.[8]

During the Research Revolution, the results of research, in space or in any other technology, were not the prime considerations in making the R & D investment. So intense was the enthusiasm during the research revolution that the very performance of R & D seemed to fulfill the national goal. In the perception of many, what should have been the means became the end.

Government agencies attempted to outdo each other in R & D spending, and in many academic institutions, research displaced teaching as a main endeavor. A large number of research institutions were born, both in the industrial and in the nonprofit sector. The total ranks of "R & D" personnel grew from 237,000 to 565,000 between 1958 and 1968. Research and development began to be used as an advertising device, and even some comic book heroes became research scientists. Given that national mood, many activities that had but a remote connection with real research and development came to be classified as R & D and listed under R & D budgets. This was true in both the private and the public sectors. Every year, with great pride, the National Science Foundation reported ever increasing expenditures for R & D.

During this time, members of the scientific community and government science policy officials, although well pleased, did not rest on their laurels. Rather, they worked hard to maintain or to increase the pace of R & D activity and to keep the Research Revolution going, as if they were certain that once the momentum lagged, all would be lost. They did so mainly by attempting to forecast the results of the R & D effort and by attempting to educate the population on the large and tangible benefits that could be expected from research. Despite their substantial efforts, by the mid-1960's the Research Revolution began to lose momentum. By the beginning of the 1970's, the country found itself in a deep "R & D Depression."[9]

THE R & D DEPRESSION

There was no single event that triggered the R & D decline; rather, the depressive reaction and changing attitudes came gradually. One of the earliest harbingers of the new mood was a refusal by the Congress to approve the customary and expected increase in the annual budget requested by the National Science Foundation for fiscal 1965. Clearly, another setback for research came with the assassination of President Kennedy. As is well known, President Kennedy was a staunch advocate of technological progress and of science; only a few weeks before his death he made strong pleas to the Congress for the restoration of the requested funds for the National Science Foundation.

Yet another blow to the prestige of R & D was dealt by the tragic Apollo 204 accident in January 1967. In that accident, Astronauts Chaffee, Grissom, and White, symbols of the conquest of space, were killed on the ground. During the ensuing months, if not years, a large part of NASA's professional talent was compelled to spend significant amounts of time and money in the almost endless discussions with Congress on the subject of the accident and how that accident might have been prevented. After some reasonable rectification of the circumstances that had caused the accident, many of the resources that had been diverted to Congressional hearings might have been spent much more effectively in the pursuit of the objectives of the Space Agency.

Perhaps the most important setback for R & D activities was--and probably is--the fear of runaway inflation that accompanied the huge increases in federal spending for the Vietnam War, and the "need" to reduce spending in the non-defense "controllable" programs in the budget.

The position of R & D as a preferred item on the national agenda began to deteriorate even more rapidly in the final years of the decade of the 1960's and continued that deterioration into the seventies. To Jerome B. Wiesner, former presidential science adviser, that period was viewed as one in which there had been no time in the post-World War II period as bleak for R & D nor one in which scientists were more discouraged.[10]

Hunter Dupree attributed the deterioration of the R & D climate to "four shocks" suffered by the R & D community, especially the community of fundamental research. According to Dupree, the first shock was the

> rise of the existentialist university and those affected by the whole outlook of the New Left. ... Their concern over the military-industrial complex led to a wide questioning of the role of government in American society.

This new group within the academic world, according to Dupree

> can support an idealized version of research tightly controlled for short-run social and political ends determined by charismatic processes within dissident movements, and at the same time hope to enjoy the level of support for basic research attained by the old system.[11]

According to Dupree, a second shock resulted from the growing internationalization of science that disrupted the established and domestically controlled administration and support of R & D. The third shock was a perturbation in the area of the social sciences. In Dupree's view:

> In the 1960's, the scientific horizons of the social sciences expanded considerably, not necessarily in step, however, with the explosive demands made under the heading of external criteria.[12]

By this Dupree must have meant that society began to demand the same precisness and predictability from the social scientist as it had historically demanded of the physical and natural scientists.

Lastly, the fourth shock arose from total confusion about the role of the military in American life. As a result of the Vietnam War, an image was revived that made warfare appear as it had before World War II--incompatible with the high tradition of science.[13]

Certainly, these four shocks, as suggested by Dupree, appear correct. However, there is another cluster of three closely related phenomena that also helped to bring about the R & D Depression of the 1970's. The first of these can be called the increasing demand for R & D results rather than merely the R & D activity; the second was the rapidly growing interest in technology assessment; and the third was the increasingly popular concern about the environmental and ecological effects of scientific and technological progress. In some parts of society, the last factor of the trilogy has reached hysterical proportions and will probably continue to present a major challenge to the progress of R & D in the eighties.

The close interrelationship of the three factors is clear because the growing demand for R & D results focuses on the unanticipated and the sometimes undersirable by-products of progress (i.e., the concern of technology assessment) and on the environmental and ecological impacts.

Examining the increasing demand for R & D results, it was indicated above that in the early stages of the Research Revolution, the results of research findings were relatively unimportant. As a national objective, what mattered was that the performance of R & D be increased and fostered. Although there had always been some skeptics, the voices of skepticism became much stronger during the mid-1960's when politicians and students of the public scene began to ask many more questions about the usefulness of R & D and when they began demanding to know what the country was getting in return for the billions of dollars invested in R & D annually.[14] To be sure, the scientific community had always tried its hardest to convince the public that R & D, including fundamental research, has many practical benefits. This particular effort was concentrated within the National Academy of Sciences and resulted in such publications as *Basic Research and National Goals* plus a number of "special reports," to include, *Ground-based Astronomy--A 10 year Program; Chemistry, Opportunities and Needs; Physics--Survey and Outlook*. As the questioning continued, some members of Congress even began to dispute the mission of the National Science Foundation.

For example, during the course of Congressional hearings in the spring of 1970,[15] Congressman Joe L. Evans addressed an unmistakable remark to William McElroy, then Director of the National Science Foundation:

> We want the National Science Foundation to tell us how to solve the problems of the environment. If we solve these problems, Doctor, we are going to give the credit to the NSF and your board of directors. If they are not solved, we will point the finger of scorn at you.

McElroy replied to the statement, "We accept the challenge."

Of course, the acceptance of that challenge by McElroy did not mean that either he or his fellow spokesman for the scientific community abandoned their interest in pure science and joined the bandwagon of the skeptics. To the contrary, in their small numbers, the scientists of that day were as vociferous in their defense of science as were the skeptics in their attacks against science. Perhaps the culmination of the outcry for a restitution of R & D as a top national priority came in the form of a series of hearing and in the publication of a Congressional report entitled, *Toward a Science Policy for the United States*.[16] The findings of the report highlight the following factors that were believed to be most detrimental to the process of science and technology.

> Congressional action moves to curb science support along two lines: (1) a general tendency to consider research as a partially expendable item with regard to overall budget reductions, and (2) specific efforts to move the mission-oriented agencies away from basic research unless some sort of "relevancy" can be demonstrated.
>
> A public disenchantment with technology of uncertain dimension, induced by environmental, social, and educational factors, among others.
>
> A movement away from science as the glamorized activity to which government, scientists, businessmen alike had responded favorably during the nuclear and space-engendered excitement of the past quarter century.
>
> Preoccupation of Government with seeking solutions to immediate demanding crises--such as unemployment, crime, environment, welfare, urban decay, military, and foreign exigencies and the like using "off-the-shelf technology."[17]

To reverse, or at least to counteract the trends described, the report recommended that "the Federal Government should formally recognize its debt to and dependence on science and technology, and establish herewith a national policy for their support and furtherance.[18] In addition to the recommendation for higher-levels and more stable funding rates, there were recommendations for the establishment of a National Institute of Research and Advanced Studies (NIRAS) with a budget of over $2 billion each year, and for the establishment of an Office of Technology Assessment (OTA) with an appropriation of $5 million for the first year of its operation. Of these two proposed science organizations, only the latter was actually established.

SOCIO-ECONOMIC-POLITICAL ASPECTS OF THE R & D REVOLUTION

The preceding argued that spending for science and technology in the United States has largely been a function of the federal government and the annual budget process. However, from that it does not necessarily follow that the architects of government R & D policy and those responsible for the main decision process are located within the federal government--either in the Executive Branch bureaucracy or in the Congress,

In the pluralistic system that pertains to the promotion of science and technology are the President, the Congress, and the various R & D oriented agencies in the Executive Branch. However, other equally important parts of the system are located in industry, in the universities, and in the arena of nonprofit research institutions, and last but certainly not the least, with the media. The way that this institutional framework functions can be illustrated by the impressions of foreigners to whom the American brand of pluralism remains somewhat of a mystery.

Shortly before the onset of the R & D Depression, a sizeable study was undertaken by the Organization for Economic Cooperation and Development (OECD) on the mechanism of formulating science and technology policy in the United States. The report of the findings of the team provides an informative illustration of the insight gained by foreigners on U.S. science policy formulation. The London *Economist* gave the following facetious, but insightful summary:

> Who runs American science and shapes its policy? Not the Administration and not Congress either, although they put up two-thirds of the funds, but an underground of eggheads whose agents infiltrate everywhere ... a brief that was given to four experts from Europe: an Englishman, a Frenchman, a Belgian, and a Dutchman--was on the face of it routine: to visit the United States and report to the Organisation for Economic Co-operation and Development in Paris on American science policy. But when they got there, they found there wasn't any. There were specialised agencies and advisers by the score. There was the evidence of their eyes that the country was spending in hard cash something like $24 billion a year on science and in real terms a steady 3 per cent of its gross national product. But no one would admit responsibility for deciding overall how this money was going to be spent, not even the $15 billion of it that came directly out of federal funds. Science was burgeoning all around them, but of science policy they could find not a trace.[19]

What the Europeans found, of course, was that there existed no single American group that formulated science policy. Further, it found that the most important influence over what was to happen on the scene of R & D promotion came from the "underground of eggheads," whom the London *Economist* labeled the "Scientific Mafia." The editors of the *Economist* had reference to two kinds of phenomena when they used the colorful phrase, "Scientific Mafia." First, they referred to the formal and informal involvement of scientists and engineers in the governmental decision-making process as members of various consulting entities. The second reference was to the influence on public opinion that scientists exert by means of their writings, speeches, and media appearances.

During the era of the Research Revolution, the core of the scientific structure was located in three closely cooperating entities: the President's Science Advisory Committee (PSAC), the Federal Council for Science and Technology, and the President's Science Adviser. A single individual, the Science Adviser, also served as Chairman of the PSAC and the Federal Council for Science and Technology.

In the next tier there was an elaborate network of lower-level and much more specialized advisory bodies, and all had very close ties with the research and academic communities. The research community, of course, was spread over every sector of American society, not just in the universities.

The brief description about the structure of science policy formulation above was deliberately put in the past tense. With the onset of the R & D Depression, President Nixon unceremoniously abolished the Office of the President's Science Adviser, the Federal Council for Science and Technology, and the PSAC. President Ford, in the first days of his administration, reestablished the Office of the President's Scientific Adviser. In a number of statements, President Carter has mentioned the need for federal R & D spending to maintain the technological lead of the American people. In constant 1972 dollars, spending for applied research only rose from $3,433 million in 1976 to $3,479 million in 1978 and federal funds for basic research, also in 1972 constant dollars rose from $2,508 million to $2,758 million during the same period.[20] This increase in R & D spending cannot be viewed as strong support for science and technology by the Carter Administration.

MEASURES OF OUTPUTS AND BENEFITS

Since the late 1960's, there have been few new studies on the socio-economic effects of R & D spending in general, and of astronautical research in particular. The explanation for this must be at least three-fold. First, the R & D depression necessarily limited the amount of money for social research as well as for scientific research. Secondly, the change in national priorities away from R & D and space goals shifted socio-economic research into the areas of environmental concern, poverty, inflation impact studies, and the effects of discrimination on labor participation. In 1967 and at the peak of NASA's interest in social science research, about six percent of grants and contracts let by NASA to nonprofit organizations were for research studies in the social sciences.[21]

Thirdly, it seems apparent that the state of the art in measuring the outputs, the benefits, and the relationship between R & D inputs and economic growth has not progressed vary far, if at all, during the past decade.

PUBLICATION, PATENT, AND PRODUCT COUNTS

The best quantifiable surrogates for output measures, i.e., measures of new knowledge produced are available in three forms: (1) numbers of publications, (2) numbers of patent applications and patents granted, and (3) a count of new products or processes resulting from R & D. In this trio, the latter is extremely difficult to obtain and is usually compiled for given purposes on an *ad hoc* basis. Data on the first two kinds of measures appear in Table 3.

The acceptance of publications as a proxy for the measure of R & D output is based on two reasons. First, publications do result from the substance of R & D and they can be counted easily. Secondly, almost all research findings eventually get published. This is especially true of research findings that relate to fundamental research. However, because the quality and the acceptance standards of scientific and technical journals vary widely and because the standards tend to be fluid over time, the apparent high validity of these two considerations quickly is impugned. In addition, it is the practice of many authors to publish the same, or almost the same, findings in several places in the guise of different titles or in different contexts. What is true of publications counts is also true of counts of citations to the literature, another form of R & D output measurement.

Table 3

MEASURES OF GROWTH IN SCIENTIFIC AND TECHNICAL INFORMATION FOR THE UNITED STATES IN SELECTED PERIODS

Information Media	Period	Growth	Average Annual Rate of Change
Scientific and Technical Book	1965 to 1974	8,800 to 14,440	5.6
Scientific and Technical Journals	1965 to 1975	1,670 to 2,010	1.8
Scientific and Technical Articles	1965 to 1975	120,461 to 158,860	2.8
University (Ann Arbor) Microfilm Dissertations	1965 to 1974	8,865 to 15,610	6.4
Published Conference Proceedings	1965 to 1972	1,730 to 2,290	4.1
Processed NTIS Reports	1965 to 1975	14,000 to 61,100	15.2
Processed GPO Technical Reports*	1965 to 1975	1,870 to 2,770	11.2
Patent Applications	1965 to 1974	100,400 to 108,000	0.8
Issued Patents	1965 to 1974	66,600 to 79,900	1.9

SOURCE: *Market Facts, Inc., Statistical Indicators of Scientific and Technical Communication (1960-1980), Volume II: A Research Report. For the National Science Foundation under Contract No. NSF-C878, D. W. King, Principal Investigator, pp. 97, 129-30, 176, 185, 193, 194.*

* *Included about 1,120 reports in 1965 and 810 reports in 1975 that were also available from NTIS.*

Patents are another kind of surrogate measure for R & D output. Patent counts are justified as indicative of R & D output because each patent application must pass the criteria established by the United States Patent and Trademark Office before it becomes an issued patent. Yet, there is widespread belief in industry and among students of science and technology that business corporations patent only those inventions that will prove to be untenable as trade secrets. *Science Indicators,* published biannually by the National Science Board of the National Science Foundation, provides information about the numbers of patents and patent applications.[22]

At first glance, the most appealing method of measuring the output of R & D seems to be counting new products and new processes. However, with this technique there are at least two insurmountable problems. First, it is exceedingly difficult to distinguish between significant and insignificant inventions. One example of a study that looked at new products and processes is the oft-quoted "Project Hindsight," sponsored by the Department of Defense.[23] That study concluded that the most important defense system consisted of dozens of "minor or insignificant inventions." The second difficulty in measurement is that of identifying the time lag between R & D and the innovation or the new product. For example, a study that investigated only those products that were commonly acclaimed as important found that the time lag between R & D and innovation for the jet engine was 14 years, the lag for radar was 13 years, that for the zipper was 27 years, and the lag for television was 22 years.[24]

Given the difficulties, the data displayed in Table 3 reveal some interesting facts that, in turn, may lead to interesting conclusions about the relationship between inputs into and outputs of the R & D process during the R & D depression. Despite the fact that R & D inputs were about the same in the mid-1970's as in the mid-1960's, the results of R & D, as measured by the most realistic surrogates, had grown at considerable rates. For example, dissertations in the sciences and in engineering, as well as scientific and technical books, had been growing in numbers at an average annual rate of about six percent.

As the data in Table 3 show, the highest growth rates of 15 and 11 percent, respectively, are for the two media that service primarily the public sector, namely, the National Technical Information Service of the U.S. Department of Commerce and the U.S. Government Printing Office. This is true despite the fact that government-financed R & D expenditures stabilized at about the same level as those in the private sector during the R & D depression.

This finding might lead one to speculate that what the government is now accomplishing in terms of promoting science and technology is widening, as well as deepening, the existing base of knowledge; that is, the knowledge produced during the R & D Revolution. This is a positive view. The same phenomenon might also be viewed negatively; that is, by claiming that the latter-day federal efforts at the promotion of science and technology have been directed toward quantity at the expense of quality. The paucity of valid measures of R & D output, and for that matter input data, make this dichotomy in interpretations impossible to resolve analytically.

ECONOMIC IMPACT STUDIES

During the late 1970's, NASA contracted for a few studies that related to the economic impact of R & D spending in general and NASA R & D spending in particular. Broadly, the studies can be grouped as mainly micro versus mainly macro studies. In economics, the notion micro relates to the particular, a particular innovation, firm, industry, etc. Whereas macro analysis covers aggregates, such as total spending, total employment, and total output. One of the micro studies was analytical and conceptual in nature; it was prepared by Zvi Griliches. A macro study and a sizeable research project was prepared by Chase Econometrics; it provided numbers for alternative levels of NASA spending. The different approaches of the two studies highlight both the political and the apolitical aspects of socio-economic research.

The Griliches Study. It is almost certain that the research results of Griliches will be used, modified, expanded and cited by academicians interested in the transfer of technology. It is equally certain that the results of the research will be ignored by federal bureaucrats without academic interests and by politicians in their quests to vie for federal R & D funds.

After making the important point that one of the major problems in measuring the contribution of R & D to economic growth arises from the fact that the values of products incorporating the R & D (e.g., defense, health, and space) are poorly measured, if measured at all, Griliches shows how possible assumptions about markets can result in vastly different measures of productivity for the same invention. At one extreme, under competition and free licensing of the invention, there would be no increase in measured productivity because total costs and total revenues would rise by the same amount and quality improvements would not be picked up as real decreases in price. The productivity gain of a monopolist is at the other end of the spectrum used by Griliches. That producer, for the same invention, would show a measured gain in productivity of 250 percent.[25]

The Chase Econometrics Study: A Macro View. NASA commissioned Chase Econometrics Associates, Inc., to evaluate the economic impact of NASA R & D spending on the U.S. economy. The Chase Econometrics report was completed in April 1976, and undoubtedly because the results of the research showed substantial aggregate economic benefits arising from spending on space, they were submitted in support of NASA's fiscal 1977 budget request in hearings before the Subcommittee on Housing and Urban Development and Independent Agencies of the Senate Committee on Appropriations. That Subcommittee was chaired by William Proxmire, who can, by no means be proclaimed as a champion of R & D spending and who is known by many people for his "Golden Fleece Award."

Senator Proxmire requested the General Accounting Office to make a careful assessment of the Chase study, to include assumptions, analytical techniques, and the validity of conclusions.[26]

With Senator Proxmire's request, the GAO's research on research began, and *ex ante*, there could be no doubt that that study, as well as virtually any other study, could be disputed on the basis of either assumptions, technique, or conclusions. On a larger, and certainly more publicly known scale, the same kind of research on research is currently being conducted on the initial findings of several studies and the proposed ban on open sales of saccharin and products containing saccharin by the Food and Drug Administration.

With the one comment that the Chase Report was probably packaged wrong,[27] it is not the purpose here to discuss the academic quality of the research findings of the Chase Group. Rather, the purpose is to comment on the reactions of the General Accounting Office and the reactions of officials at NASA in response to the Congressional request to have the Chase study reviewed. It is the actions and reactions of these kinds of groups that constitute part of the political economy of astronautics.

Using sophisticated macro econometric simulation techniques that were not possible before the advent of the computer, the principal finding of the Chase report was that the historical return from NASA spending had been about 43 percent.[28]

The research report of the General Accounting Office was academic in tone, but the GAO report clearly stated that as a matter of consideration by the Congress "... Technical studies are frequently presented to the Congress in support of agency budgets and as evidence for or against proposed legislation and that when important questions are at stake, such studies should be subjected to independent examination and appraisal." In addition, the GAO report stated, but outside the purpose and intent of the Chase study, that that work"...did not try to evaluate how effectively NASA carried out its primary objectives, such as space exploration and satellite communication."[29]

The thrust of the analysis of the GAO study was to make changes in the assumptions used by Chase, and thus obtain different results while using the same kind of statistical analysis. Using the time frame between 1956 and 1974, the GAO research found that the rate of return on NASA's R & D spending was between 25 and 28 percent rather than the 43 percent that the Chase results had produced, using data between 1960 and 1974.[30]

Without going into greater detail about either the research conducted by GAO or that by Chase Econometrics, the main recommendation of the GAO report to the Administrator of NASA was to limit the Space Agency's research to less aggregate analyses, such as studying particular innovations rather than the agency's total R & D budget and to examine the effects of individual innovations in particular industries rather than on an economy-wide basis.

These recommendations, in effect, specified the exact kinds of socio-economic research that NASA had been sponsoring for the past fifteen years or more. The GAO encouraged NASA not to engage in new approaches in its desire to ascertain the effects of space R & D spending. Even if the Chase results are subject to question, or for that matter, completely incorrect, it is unfortunate not to be able to explore all avenues of academic research in attempts to find answers to important questions.

The reply of officials in NASA to the GAO report showed acumen and immediate grasp of the possible political ramifications of that report. Without abandoning the Chase group, as many federal agencies are willing to do, NASA correctly stated that a rate of return on R & D in the range of 25 to 28 percent can be considered a very good investment of federal funds. Intelligently, NASA did not take issue with the GAO results, rather NASA simply stated that the GAO results showed that because empirical measurement in economics is an inexact science, ranges rather than absolute magnitudes are important. With this reaction, NASA could, and did indeed, say that the GAO findings, in fact, reinforced the results of the Chase Study.[31]

Other Micro Studies: Reports about Innovations. About the same time that NASA let the contract with Chase Econometrics, the Agency also contracted for several studies on the economic effects of particular technologies. These are the kinds of studies that the GAO recommends that the Space Agency support. If the results of research that examine particular technologies are used in support of federal funds for R & D, the results can also be easily axed in some of the following ways: (1) The studies are anecdotal evidence because of the case study nature of the research; (2) the wrong technologies were selected for examination or the selection was not based on random sampling procedures; and (3) the wrong economic method was used in the analysis, including inappropriate discount rates and time intervals.

One of the research groups made a cost-benefit analysis of selected programs in NASA's Technology Utilization Office. For biomedical projects spanning a ten-year period, the researchers found benefit-cost ratios ranging from a low of 4.1 for pace-maker projects to a high of 41.0 for applications relating to the cataract tool.[32] For applications in engineering, the low benefit-cost ratio for Track-Train dynamics was 2.6 and the high was 340.0 for zinc-rich coatings projects.[33]

Similar micro-oriented research was undertaken by Mathematica, Inc. for the following four technologies: cryogenic multilayer insulation materials, integrated circuits, gas turbines in electric power generation, and NASTRAN. The group at Mathematica believed that total net benefits attributable to NASA were "probably" of the magnitude of $7.0 billion for the four technologies combined. The highest benefit was estimated to accrue from technology covering integrated circuits, which alone counted for $5.1 billion of the total.[34]

TECHNOLOGICAL PROGRESS AND PRODUCTIVITY GROWTH

Since the mid-1970's, much has been said about the sharp decrease in U.S. productivity, especially when compared with the gains in productivity in foreign countries. Table 4 presents changes in productivity in five major industrialized countries between 1960 and 1977. The measure of productivity used in the data is output per worker-hour. The information contained in the data clearly shows that productivity gains in the U.S. have sharply lagged behind those in Japan, France, and West Germany. and are even below the productivity growth in Great Britain. The National Science Board, which is extremely conservative in attributing changes in economic variables to changes in scientific activities, states:

> The current slowdown in U.S. productivity growth is due to a variety of factors; one of them may be the fact that national R & D expenditures in constant dollars experienced little or no growth from 1968 to 1975. If the United States had continued to devote at least the same fraction of national resources to R & D as it did in the 1960's, the U.S. productivity gains might have been greater. It is likely that increases in R & D investments in Japan and West Germany have been positive factors contributing to the large productivity gains in those countries.[35]

For perspective, Table 5 provides information on the performance of R & D as a percentage of the GNP's of selected foreign countries. The same information is also provided for the U.S. The data show that as a percentage of GNP, resources allocated to R & D have sagged more in the U.S. than any other country since the mid-1960's. For West Germany, Japan, and the Soviet Union, the percentage of R & D spending increased quite sharply during the decade of the mid-1970's.

In a 1979 study for the New York Stock Exchange, John W. Kendrick does not hedge about the relationship between a slowdown in technological progress and the decrease in U.S. productivity during the 1970's. Kendrick attributes almost one-fourth of the decline in productivity between 1973 and 1979 to sharp decreases in private and federal outlays in R & D projects after the mid-1960's. Table 6 presents the relative contributions

Table 4

CHANGES IN PRODUCTIVITY IN THE MANUFACTURING SECTOR: SELECTED COUNTRIES

1960 = 100

	France	West Germany	Japan	United Kingdom	United States
1960	100	100	100	100	100
1965	129	134	150	120	125
1970	176	171	279	141	132
1975	219	218	332	162	150
1977	251	250	379	164	160

SOURCE: *National Science Board, National Science Foundation, Science Indicators 1978, (Washington, D.C.: U.S. Government Printing Office, 1979), p. 156.*

Table 5

THE PERFORMANCE OF R & D AS A PERCENTAGE OF GNP: SELECTED COUNTRIES

	France	West Germany	Japan	United Kingdom	United States	Soviet Union
1961 or 1962	1.4	1.3	1.5	2.4	2.7	2.6
1966	2.0	1.8	1.5	2.3	2.9	2.9
1972	1.9	2.3	1.9	2.1	2.4	3.6
1976 or 1977	1.8	2.3	1.9	-	2.3	3.5

SOURCE: *National Science Board, National Science Foundation, Science Indicators 1978, (Washington, D.C.: U.S. Government Printing Office, 1979), pp. 143-145.*

Table 6

THE DECLINE IN THE RATE OF PRODUCTIVITY GROWTH: SOURCES OF DECLINE BETWEEN 1973 and 1977

	Percent of Decline
Slower Output Growth (low rate of capacity of utilization)	32
Less Mobile Captial and Labor	23
Slowdown in Technological Progress	23
Increased Negative Impact of Government	13
Lower Land Quality	9
Total Decline	100

SOURCE: *John W. Kendrick, Reaching a Higher Standard of Living, The New York Stock Exchange, 1979.*

of five forces to the recent decline in productivity. In addition to decline in R & D spending *per se*, Kendrick believes that part of the slowdown in technological progress results from little or no change in the average age of the U.S. stock of capital between 1973 and 1979. This compares with an estimated average decrease in the average age of capital stock of three years between 1948 and 1966. The relationship between the age of capital stock and a slowdown in technological progress rests on his finding that capital goods produced in a current year embody the newest technological advances.[36]

HAS THE SPACE PROGRAM SAGGED?

The answer to that question must be yes, based on numbers, programs, and the short-run prospects for astronautical research.

There is little doubt that lower levels of funding result in lower levels of product or outputs. The simplest way to look at the lagging space program is in terms of constant dollar spending. Federal funds for the space program peaked at $5.9 billion, or roughly 0.8 percent of the GNP, in fiscal 1966. In fiscal 1979, federal spending for space (including general science and technology) was $5.0 billion, or 0.2 percent of the GNP.[37] Using equivalent purchasing power of 1972 constant dollars, the fiscal 1966 peak would have been $7.8 billion, whereas the 1979 level was only $3.1 billion. By this measure, real spending for space and astronautics dropped by more than 60 percent during the 13-year period. Although there is no reason to argue that space spending should have remained some constant percentage of the GNP, if such spending had remained at roughly 0.8 percent of the GNP, it would have amounted to about $19.0 billion, or almost four times the actual fiscal 1979 level.

Another way to look at whether the space program has sagged is by comparing the current space effort with benchmark projections made a decade ago by scientists and science policy makers looking into the future. In 1969, the President's Space Task Group developed a set of options for space exploration projects in the late 1970's and beyond. Here, we will ignore, except to mention that one option, what the Group called a "maximum pace - limited not by funds but by technology," the highest option that the group included in its report.[38]

Within reasonable funding levels, the Space Task Group outlined three policy options. The high-option policy might have resulted in the establishment of an earth orbiting space station and an earth-to-orbit space-shuttle transportation system by the mid-1970's. In the low-option program, these systems would have been deferred until the late 1970's, but presumably would have been operational by 1980. Using 1969 constant dollars, the President's Space Task Group suggested 1980 funding levels of $9.4 billion for the high-option program and $5.5 billion for the low-option system.[39] In 1979 current dollars, that would have amounted to $17.9 billion for the high-option program and $10.5 billion for the low-option package.

A recent report (1979) of the National Research Council, National Academy of Sciences, that presents a five-year outlook for Science and Technology virtually ignores space and astronautics as separate components of general scientific endeavors. In the 500-plus page report, the largest and the most recent missions of the Space Agency, namely, the Spacelab and the Space Shuttle, are discussed in about six pages. Space technology in general is given another two pages.[40]

A number of political leaders who are protagonists of R & D in general, and astronautics in particular, have expressed strong views that the U.S. is lagging behind in science and technology. Here we select two short quotes to sum up the current status of the space effort--especially its sagging state.

In his opening remarks at a symposium on the future of space science and space applications, Senator Adlai E. Stevenson said:

> The opportunity is real. But the harsh truth is that our present efforts to plan the Nation's future in space lack the foresight and dedication that characterized the decisions made 20 years ago.... Today the planning process for the U.S. space program is little more than the product of an annual battle between Congress and the Office of Management and Budget over specific line items in the Federal Budget. This process is almost immune to establishing longer run purposes and directions. In addition, no concerted effort is being made to devise the new institutional arrangements which will be necessary to manage a maturing and operational space program as distinguished from the experimental R & D projects of the first 20 years....[41]

Speaking about the current space program and the benefits of technology, Senator Harrison H. Schmitt believes:

> We must do something to repair the damage that has been done over the last decade. If we do not repair that damage, then there is no other major deflationary force that we can establish within this economy. We must fight the budget.... But if we do not establish a technology base from which we can build the economies of the future, then many of these, if not most of these, efforts will be for naught.[42]

CONCLUSIONS AND RECOMMENDATIONS

Following any discourse of problems of shortcomings, something positive, or better yet, something constructive is in order. We will be happy to oblige on condition of three caveats. First, that we are not expected to offer instant, complete, or permanent cure-alls. Secondly, that we are not expected to come up with suggestions so novel that no one has ever heard of or thought of anything like them. Finally, that the suggestions may be proffered in a very general form rather than as prescriptions for immediate implementation, point by point.

Prior to making such suggestions, it will be well to describe the problems and shortcomings in a form in which they can become responsive to the suggestions to be developed. This can be most easily done by gleaning some generalizations from the foregoing discussions as follows:

A number of discrete and to a large part fortuitous events in the late 1940's and the early 1950's brought about what Leonard Silk dubbed a "Research Revolution," followed by what we might dub a Golden Age of Science and Technology. This was not a golden age of tranquility; on the contrary, it was restless and dynamic, where progress and discovery became pervasive prime national goals. It is true that the mainstay of the Golden Age was growing federal support of what William Fellner calls progress-generating activities: R & D, education, and innovation. But the part played by the private sector of society was just as important. People in industry as well as in academe were interested and enthusiastic about progress, and scientists and engineers came to enjoy unprecedented prestige. Indeed, scientific and technological progress assumed the dimensions of a national fad, one that in the manner of all fads, came to an abrupt halt with the R & D Depression.

We may now return to the need for suggestions for improvement. What we need to know, is: How do we go about creating a national fad, i.e., how do we go about recreating the Golden Age of Science and Technology? This is essentially the question that plagued President Carter when, not too long ago, he called for a program to attack the energy crisis, one that would amount to "the moral equivalent of war."

Unfortunately, we do not know how to do that and we are not certain that it has ever been done before. Some claim that the New Deal became a national fad, but in the light of too many accounts to the contrary, we are not ready to accept that as a case in point. This leaves us with the conclusion that the goal of synthesizing a national fad, by government action short of war, may well be impossible to realize. That, in turn, leaves us with but one rational conclusion and that is that we must wean scientific and technological progress from its close dependence on national fads.

The only plausible way to accomplish such weaning is to establish technological progress as a long range, government-sponsored national goal. This is not a recommendation of collective planning. Rather it is a suggestion that the federal government, in its entirety, begin to look upon long-range progress in the same frame of mind in which the National Science Foundation was mandated to look upon the promotion of basic research. That is to say, there must be room for development orientation rather than immediate-accomplishment orientation. It must be realized that the major problem of progress cannot be solved by the instant use of "off-the-shelf technology."

It is well known that the scientific community accurately predicted our energy crisis decades ago. They called for the development of energy sources not dependent on fossil fuels or nuclear fission or fusion. They made concrete suggestions as to how such objectives could be tackled, by exploiting the sun, the winds, and the tides, but no one in the government listened. Actually, it is more correct to say that no one in the government was in a position to listen even if he wanted to listen. There were no provisions in the days of 16¢-a-gallon gasoline to embark on a long-range program of energy development that would pay off twenty or thirty years later. And there are no such provisions in the government today, or certainly not on a scale that could be considered annually adequate in the fact of future needs and opportunities.

This concept, should anyone ever consider it, will be more difficult to ram down the throats of public officials than was the Planning-Programming-Budgeting System (PPBS) in the mid-1960's. And it is not even as painful as PPBS. But there are good reasons for its expected unpopularity. After all, how many politicians or Executive officials can be expected to show much enthusiasm for something that offers little prospect of any pay-off during the period of their tenure? This is true especially when there are things to be done that will--or at least it is hoped that they will--show results in the next budget year. What incentive is there for abandoning those in favor of projects whose benefits will only accrue to future generations of public officials--not to mention the American public, of course?

A somewhat similar suggestion, in much more detail and precision has recently been advanced by our good friend Ellis Mottur.[43] Although we may not agree with everything that Mottur is suggesting--in particular we are leery of the desirability of creating new government agencies to look at the problem--the spirit of the plan is very similar to what is motivating us here. We commend it to your attention.

In recapitulation then, what we have said is that the Golden Age of American Astronautics was also a Golden Age of American Science and Technology. That golden age was dynamic and it brought about unprecedented progress; it was born out of a number of fortuitous circumstances and events. And inasmuch as we do not believe that it is possible to synthesize a second golden age by evoking a moral equivalent of war, we suggest that the progress of science and technology be weaned from its vital dependence on fads and fashions, by causing the government to assume a long-range responsibility for scientific and technological development in an undisturbed and safe atmosphere.

THE POLITICAL ECONOMY OF
AMERICAN ASTRONAUTICS

REFERENCE NOTES

1. Mary A. Holman, *The Political Economy of the Space Program* (Palo Alto: Pacific Books, 1974).

2. The National Science and Technology Policy Organization and the Priorities Act of 1976 directed the Office of Science and Technology Policy to prepare a Five-Year Outlook on science and technology on a periodic basis. The National Academy of Sciences was asked in 1978 to help with the preparation of the first such five year outlook. The report of the National Academy of Sciences explored the current state of the art in such areas as the structure of matter, computers and communications, energy, health, and demography. The report, however, did not discuss alternative science policies nor did it make policy recommendations about levels of financing and priorities. See, the National Research Council/The National Academy of Sciences, *Science and Technology: A Five Year Outlook* (San Francisco: W. H. Freeman and Company, 1979).

3. A representative of the Office of Management and Budget has a detailed explanation of how the annual budget review process affects levels of R & D spending. That discussion includes a description of "cross-agency-zero-base" budgeting. W. Bowman Cutter, "R & D in the Federal Budget," *R & D, Industry, and the Economy* (Washington, D.C.: American Association for the Advancement of Science, June 20-21, 1978), pp. 23-29.

4. Although somewhat dated, Murray L. Wiedenbaum has an excellent discussion of controllable versus noncontrollable kinds of government spending. See, U.S. Congress, Joint Economic Committee, *The Analysis and Evaluation of Public Expenditures: The PPB System,* vol. I, (Washington, D.C.: U.S. Government Printing Office, 1969), pp. 357-369.

5. Vannevar Bush, *Science, the Endless Frontier, A Report to the President on a Program for Postwar Scientific Research* (Washington, D.C.: U.S. Government Printing Office, 1945).

6. *Ibid.,* as reprinted by the National Science Foundation, NSF 60-40 (Washington, D.C.: National Science Foundation, 1960), p. 22.

7. Sir Bernard Lovell gives an excellent history of the foreign, scientific, and military origins of space R & D in his delightful and informing book, *The Origins and International Economics of Space Exploration* (Edinburgh, England: University Press of Edinburg, 1973).

8. Leonard A. Lecht, *Dollars for National Goals, Looking Ahead to the 1980's* (New York: Wiley, 1974), p. 202.

9. The term, "R & D Depression" was originated by Stephan Dedijer in "The R & D Depression in the United States," *Science,* 168 (April 17, 1970), p. 345

10. Jerome B. Wiesner, "Rethinking Our Scientific Objectives," *Technology Review,* vol. 71 (January 1969), pp. 15-17.

11. Hunter Dupree, "A New Rationale for Science," *Saturday Review,* February 7, 1970, p. 57.

12. *Ibid.*

13. *Ibid.*

14. Based on survey results, the National Science Board presents an excellent discussion of the loss of confidence by the public in the research effort, especially basic research. The surveys showed that the two main reasons for the decline in public confidence was a growing "anti-intellectualism" in society and a more general growth of a "disdain" for the establishment during the mid to late 1960's. National Science Board, National Science Foundation, *Science at the Bi-Centennial: A Report from the Research Community* (Washington, D.C.: U.S. Government Printing Office, 1976), p. 73 ff.

15. These Congressional proceedings were reported in several issues of *Business Week* in mid-1970. For example see, *Business Week,* "Mission: Relate Science to Society." May 23, 1970.

16. U.S. Congress, House Subcommittee on Science, Research and Development of the Committee on Science and Astronautics, *Toward a Science Policy for the United States,* 91st Congress, 2d Session, October 15, 1970.

17. *Ibid.,* p. 7.

18. *Ibid.,* p. 10.

19. "The Scientific Mafia," *Economist,* January 13, 1968.

20. National Science Board, National Science Foundation, *Science Indicators 1978* (Washington, D.C.: U.S. Government Printing Office, 1979), p. 180 and p. 187.

21. Mary A. Holman, *The Political Economy of the Space Program, op. cit.,* p. 174.

22. *Science Indicators 1978* furnishes a great deal of information about patent counts as a measure of R & D output. National Science Board, National Science Foundation, *Science Indicators 1978* (Washington, D.C.: U.S. Government Printing Office, 1979), pp. 99-105 and pp. 218-219.

23. Raymond S. Isenson, "Technological Forecasting: Lesson from Project Hindsight." in *Technological Forecasting for Industry and Government,* edited by James R. Bright (Englewood, New Jersey: Prentice Hall, 1969).

24. *Ibid.*, pp. 35-54.

25. Zvi Griliches, "Issues in Assessing the Contribution of Research and Development to Productivity Growth," *The Bell Journal of Economics*, vol. 10, no. 1 (Spring, 1979), pp. 92-116.

26. Letter from Senator William Proxmire, Chairman, HUD-Independent Agencies Subcommittee, to The Honorable Elmer Staats, Comptroller General of the United States General Accounting Office, dated September 15, 1976.

27. For example, with almost no discussion, a summary table in the Executive Summary of the Chase Report shows the startling statistics that NASA spending of a successive $1 billion annually in the mid-1970's will result in a $23 billion increase in constant dollar GNP, a 5.8 percent <u>decrease</u> in the Consumer Price Index, and a 2.0 percent increase in productivity by the mid-1980's. See Table 6, page 14, Executive Summary, *Final Report: The Economic Impact of NASA Spending*, prepared for NASA, Contract No. NASW - 2741, April 1976.

28. A rate of return of 43 percent is not out of line with the findings of other researchers. For example, in his study of 17 important innovations, Mansfield found that the median social rate of return on the investments was about 56 percent and that the private rate of return was about 25 percent. Edwin Mansfield, "Returns from Industrial Innovation, International Technology Transfer, and Overseas Research and Development," Papers for a Colloquium on the Relationship between R & D and Returns from Technological Innovation, sponsored by the National Science Foundation, May 21, 1977. Published in U.S. Congress, Hearings before the Senate Subcommittee on Science, Technology, and Space of the Committee on Commerce, Science, and Transportation, *Oversight of Science and Technology Policy*, pt. 2, February 10 and April 26, 1978 (Serial No. 95-77), pp. 302-303).

29. Report of the Comptroller General of the United States: *NASA Report May Overstate the Economic Benefits of Research and Development Spending*, October 18, 1977, pp. i-iii.

30. *Ibid.*, p. 13.

31. Letter from Nathaniel B. Cohen, Director, Office of Policy Analysis, NASA, through Kenneth R. Chapman, Assistant Administrator for DOD and Interagency Affairs, NASA, to R. W. Cutman, Director, Procurement and Systems Acquisition Division, General Accounting Office, dated July 7, 1977 and July 11, 1977.

32. Some economists believe that federal projects with a benefit-cost ratio of 1.0 should be undertaken. In that context, a ratio of 4.1 can be viewed as being "high". Typically, the benefit-cost technique is used to rank projects and more frequently than not, the benefit-cost ratios are used to find the cut-off point for various projects, given the level of funds.

33. Robert J. Anderson, *et al.* with MATHTECH, Inc., P.O. Box 2392, Princeton, New Jersey, *A Cost Benefit Analysis of Selected Technology Utilization Office Programs, NASA Contract no. NASW 2731, November 7, 1976.*

34. Mathematica, Inc., *Quantifying the Benefits to the National Economy from Secondary Applications of NASA Technology*, NASA Contract No. NASA CR - 2674, March 1976.

35. National Science Board, National Science Foundation, *Science Indicators 1978, op. cit.*, p. 23.

36. Two similar studies of the recent decline in productivity do not attribute much, if any, of that drop in productivity to investment in R & D. Robbin Siegel cites rising energy prices, a decrease in the capital to labor ratio, and low rates of capacity utilizations as the major causes of the decline in productivity between 1965-1973 and 1973-1979. See, Robin Siegel, "Why has Productivity Slowed Down?" *Data Resources U.S. Review*, March 1979.

Lester C. Thurow attributes the recent decline in productivity to: an increase in employment in the service sector; a rise in energy prices; negative effects of health, safety, and environmental regulations; and employment stagnation and excess capacity of capital equipment. See, Lester C. Thurow, "The U.S. Productivity Problem," *Data Resources U.S. Review*, August 1979.

37. Data on spending for space are from *Economic Reports of the President*.

38. The technology limited policy included the possibility of an initial expedition to Mars in 1981 and a 50-man (earth-orbiting) space base in 1980.

39. U.S. President, Space Task Group Report to the President, *The Post-Apollo Space Program: Directions for the Future*, September 1969, p. 20.

40. National Research Council/National Academy of Sciences, *Science and Technology: A Five Year Outlook, op. cit.*, pp. 40, 45-47, 52, 162, 160, and 493.

41. Statement of Adlai E. Stevenson, *Symposium on the Future of Space Science and Space Applications*, Hearings before the Subcommittee on Science, Technology, and Space of the Senate Committee on Commerce, Science and Transportation, (Serial No. 95-58), February 7, 1978, pp. iii-iv.

42. Statement of Harrison H. Schmitt, *U.S. Civilian Space Policy*, Hearings before the Subcommittee on Science, Technology, and Space of the Senate Committee on Commerce, Science, and Transportation, (Serial No. 96-10), January 25, 31 and February 1, 1979, p. 3.

43. Ellis R. Mottur, *National Strategy for Technological Innovation*, a Committee Print, prepared at the request of The Hon. Howard W. Cannon, Chairman, U.S. Senate Committee on Commerce, Science, and Transportation, 96th Congress, 1st Session, October 1979.

V

EVOLUTION AND PROBLEMS OF SPACE LAW ON PLANET EARTH

Stephen E. Doyle *

This paper characterizes four main stages in the first half century of space law development (1930-1980), highlights main constituents of space law developed in the first quarter century of spaceflight (1957-1982), identifies some of the prominent and less prominent personal and institutional contributors to that development, and extrapolates from this brief history some modest predictions of near-future space-law growth.

This rapidly developing subset of international law, now widely recognized as space law, has been growing in stages for almost half a century. Since about 1930, jurists, legal scholars, and pundits have elaborated with increasing detail and expanding scope the principles, rules, and regulations under which spaceflight activities progress. This evolution has proceeded in at least four phases:

1. The initial phase involved exploration and speculation in principles and concepts of space law, dating from the early 1930's to the first successful spaceflight in October 1957.

2. The second phase involved consolidation and definition of fundamental principles and precepts dating from the successful orbiting of a man-made object in 1957 to the promulgation by the United Nations of the 1967 Space Treaty (limited supplementation may occur).

3. The third phase involves elaboration and regulation. Having begun in the late 1950's, it will continue indefinitely.

4. The fourth phase, involving regulation of extraterrestrial activities, has just begun.

This paper focuses on an historical period that spans two of these phases--extending from the first successful orbiting of a man-made object in 1957 to the operational certification of the United States' manned, reusable orbiting spacecraft called the Shuttle, expected by 1982.

---

* *Program Manager, Telecommunication, Information and Space Studies, Office of Technology Assessment, U.S. Congress.*

Legal literature concerning the impacts and implications of spaceflight dates from almost exactly a quarter of a century before the U.S.S.R.'s launch of *Sputnik I* in 1957. The first milestone in space law literature was a modest but perspicacious and prescient monograph written by a Czechoslovakian lawyer, published in Germany in 1932. Vladimir Mandl's *Das Weltraum-Recht, Ein Problem der Raumfahrt* (The Law of Outer Space, a Problem of Spaceflight) is a masterful compilation of legal research, scientific and engineering knowledge, joined with practical judgments in a rare and unique work.[1]

Not only was Germany the site of publication of the first serious work onspace law by a practicing lawyer, it was also the country in which the first doctoral dissertation on the topic of space law was written. Welf Heinrich, Prince of Hannover, submitted his thesis entitled "Air Law and Space" to the Georg August University of Goettingen in 1953.[2] Also in Germany, Alex Meyer, at the University of Cologne, had begun exploring the legal aspects of spaceflight from the early 1950's;[3] and in the orient an occasional, general speculative article appeared.[4]

Another of the early commentators on space law was Eugene Korovine of the U.S.S.R. Korovine appeared very early in European literature with an article published in France in 1934. Although he had done earlier legal work on sovereignty in airspace,[5] after the 1934 article, entitled "The Conquest of the Stratosphere and International Law,"[6] few other articles by him appeared until 1958;[7] and thereafter Korovine was a prolific contributor to legal literature in the early years of the Space Era, especially from 1958 to 1960.[8]

In the United States, there were a number of pioneering writers on space law before Sputnik: John Cobb Cooper presented his first paper on the question of upper limits on national sovereignty in 1948;[9] Oscar Schachter, a legal officer at the United Nations, was asking "Who Owns the Universe?" in 1951;[10] Andrew G. Haley presented a survey of legal issues involved in 1955;[11] and Myres McDougal of Yale University addressed such issues in 1956 at a meeting participated in by Cooper, Meyer, Schachter and others.[12]

Few were prepared for the flood of legal literature that was released by the launch of *Sputnik I*. But it was not only commentary, questioning, proposals and analysis that appeared. Given the reality of man's artifacts in space, a new branch of international law was established and its substance began to take shape.

The first essential and pragmatic international steps were taken out of operational necessity, in the form of bilateral agreements. Little notice is given to that aspect of space law today, but the early foundations were laid in numerous bilateral arrangements. Between 1959 and 1965, the United States had established more than 40 bilateral agreements involving manned-flight support communications, earth-orbiting satellite tracking and telemetry facilities, reimbursable launch arrangements, unmanned scientific and application satellite programs, and deep-space network support for tracking, telemetry and control of space objects on trajectories away from the earth.[13]

These early international agreements took several forms:

- o Some were agency-to-agency agreements, usually of limited scope, of a technical nature and finite duration.

- o Some were executive agreements, usually involving diplomatic-level exchanges based on existing legislative or treaty authority, or which may have exceeded agency authority but were subject to Congressional approval or implementation by funding, or which were entered into under the President's constitutional powers.

- o Some were formal treaties, usually involving matters of national scope and interest, generally of longer duration and appropriately cast in a more formal mode.

In this early period of new lawmaking without clear precedents, the principal actors were not always able to take much time to write about and describe their roles. But early senior legal officials at NASA, including John Johnson, Paul Dembling, Neil Hosenball, and their associates in the Office of the General Counsel, and Arnold Frutkin, Marvin Robinson, Oscar Anderson, Philip Thibideau, Richard Barnes, Lloyd Jones, and their associates in the Office of International Affairs, played important, demanding, creative roles with little visibility but high effectiveness.[14]

In the Department of State there were counterpart mid- and senior-level managers of whom little is seen in the historic literature of the period. Robert Packard, Wreatham Gathright, Leonard Meeker, Herbert Reis, Gerald Hellman and others were centrally involved but rarely credited with important contributions. At the United Nations we have often heard reference to the presidential statements and ambassador's speeches, but rarely is there a focus on the researchers and speech writers.

The achievements and personnel of Congressional committees were well documented in this forum, just one year ago, by an unpretentious and indefatigable science policy research specialist, now retired from the Library of Congress, Mrs. Eilene Galloway. And there is a recently retired colleague of Eilene's, who has labored with quiet and professional effectiveness for decades interpreting and reporting on, among other topics, Soviet space programs and Soviet space technology, law, policy and strategy, Dr. Charles Sheldon.

These and many other unfortunately nameless actors created a great deal of the foundation of what we call today "space law." We should not lose sight of the dedication, the professionalism, the vision and the courage of all these early laborers. As we meet annually to commemorate the contribution to spaceflight of Robert H. Goddard, it seems to me fitting also to recall that many, many others--doctors, lawyers, engineers, technicians, managers, accountants, secretaries and clerks--all had to contribute their individual efforts to make possible the accomplishments in space of this nation and of other nations.

We ought to mention also the historian. Historians provide a valuable but often ignored service. They record when, where, why, how and by whom things were done, not simply for the academic archiving of history, but to help us learn by studying our experience. Much that we will argue, very likely has been argued before. Much that we will do, very likely has been done before. We should learn from the past, not ignore it.

When *Sputnik I* went up, there was a truly global response. There is a well-chronicled history of the United Nations' creation of an *ad hoc* Committee on the Peaceful Uses of Outer Space. Among the first major undertakings of that body was the identification, analysis and prioritization of the legal problems and issues generated by spaceflight activity.[15] Some of the immediately identified issues were addressed and resolved in relatively rapid formulation of fundamental principles that won early unanimous support. The world was not only willing, it was eager to declare that outer space should be the province of all mankind, to be free of national appropriation, and to be used and explored for peaceful purposes and on the basis of equality and without discrimination.

Resolution 1721 of the Sixteenth General Assembly of the United Nations, in December 1961, articulated the rudimentary principles of emerging global space law, and laid out a program of work and study in areas of international cooperation, meteorology, and communications. That resolution expanded the membership of the new Space Committee, directing it to meet again and to continue its work.

The subsequent formulations of UNGA Resolutions 1884 (XVIII), of October 17, 1963, and 1962 (XVIII) of December 13, 1963, became explicit references in the Preamble of the 1967 <u>Treaty on Principles Governing the Activities of States in the Exploration and Use of Outer Space, Including the Moon and Other Celestial Bodies</u>. And with the entry into force of that treaty, the first decade of spaceflight was crowned with a charter of principles that would be decades more in their elaboration.[16] But that was all the work of lawyers, diplomats and politicians and that was a different environment from the pragmatic world of industrialists, businessmen, engineers, and physicists. It was the latter group who felt clearly that they were where the action was. That group wasn't talking, it was doing things.

Do you remember that in 1961 an *ad hoc* group of communication companies recommended to the FCC, and the FCC endorsed their recommendation to Congress, that satellite communications should be inaugurated in this country by a consortium of the then extant large vested interests in the communication field? Do you remember Senator Gore who responded by proposing a federal authority to own and operate the communication satellite system--a TVA in the sky?! And the Kennedy Administration took a hard look at those proposals and said there ought to be a people's dividend here somehow! And do you remember how all that came out with long hours of Congressional floor fights, hundreds of proposed amendments, filibuster and cloture, and all the tests of tempers and patience that finally produced the Communications Satellite Act of 1962? Senator John O. Pastore is now a banker in Rhode Island, but in 1962 he was a floor manager for the compromise Comsat bill, who learned and taught some clever parliamentary practices.

Well, some might say, "those were exciting years, those were challenging years! That's when the fights were fought, when big issues were at stake, and billions of dollars of investments were being decided. That was a generation ago!"

Do you know what is going on in the Congress today? Do you know what kinds of bills are pending right now that have to do with space? Anyone who thinks the golden era of space legislation is behind us has made a mistake.

We have bills pending in Congress that could structure this nation's space policy for another generation. We have bills addressing in detail how we should organize the next major operational industry in space applications--the remote sensing industry.[17] We have issues pending now, before this Administration, about where we want to go, how we want to get there and when.[18]

The United States and the world community have come a long way in space law since *Sputnik I*. From the 65-nation cooperative experiment called the International Geophysical Year (IGY), we have evolved a 103-nation Intelsat Organization, a ten-nation European Space Agency, a 22-nation cooperative Arab Communication Satellite Corporation and other, newer international organizations for space applications.

We are a long way away from a bilateral negotiation for a tracking station in Africa. In fact, there are now whole new international cooperative space programs and projects being formed, of which neither the United States nor the U.S.S.R. is a part.

The law that is being formulated today, in the United States, in Europe, in the United Nations, is law that is focused on the detailed regulation of the process of realizing the promises of space. We are setting up insurance provisions, we are covering liability risks, we are concerned with international interpersonal relations between citizens of different countries flying in space on the spacecraft of a third country. Suppose a tort or a criminal act is committed. Who takes jurisdiction? Under what law are claims filed or is prosecution brought?

We have quite clearly moved from the formulation of fundamental principles, embodied in several early treaties by the United Nations, through a period of consolidation and definition of the main precepts, into the elaborative, regulatory phase of lawmaking. In the United Nations, countries are focusing on guidelines for remote-sensing operations, guidelines for direct television broadcasting by satellite, and possible regulation of the use of nuclear power sources in space.

But there are other, more challenging, longer-term issues under debate as well. Should we define "outer space?" Should the geostationary orbit have a special status? Under what kind of governing regime might we one day exploit the resources of the moon and other celestial bodies?

These are questions now on the agenda, not only of the United Nations' Committee on the Peaceful Uses of Outer Space, but also of the Congress. And as Congress begins to move, for the first time to address issues of how to regulate, not only extraterritorial activity but literally extraterrestrial activity, what should be its guidelines, its guiding principles, its touchstone of validity?

In the plethora of commentary that emerged in the wake of *Sputnik I*, a great deal was written about the upper limit on national sovereignty. It is clear that over the next decade we are going to be busily engaged in perfecting our rules of behavior for space-related activities below and above that elusive limit, wherever it may be. Also in this coming decade, we must begin to strain our limits and to think in new and unearthly ways about how to regulate man beyond his home planet, in a hostile environment, seeking to explore, to build new homes and facilities, and to survive.

This is the beginning of the extraterrestrial law period. As we preoccupy ourselves with our anthropocentric concerns and our national priorities, we must also learn to think about all these new issues associated with extraterrestrial activity in revolutionary ways. I hope that the lawyers concentrating on what has to be done will take a little time to look ahead and also think about what ought to be done!

EVOLUTION AND PROBLEMS OF SPACE LAW:
FROM SPUTNIK TO THE SHUTTLE

REFERENCE NOTES

1. V. Mandl, *Das Weltraum-Recht, Ein Problem der Raumfahrt,* 48 pp. J. Bensheimer, Mannheim (1932). An excellent short biographical essay about Mandl, his life and work, by Vladimir Kopal, entitled "Vladimir Mandl: Founding Writer on Space Law," is contained in Durant and James (eds.), *First Steps Toward Space* 87, Proceedings of the First and Second History Symposia of the International Academy of Astronautics at Belgrade, Yugoslavia, September 26, 1967, and New York, U.S.A., October 16, 1968 (Smithsonian Institution Press, Washington, D.C., 1974).

2. W. Heinrich, "Air Law and Space," first published in English in 5:1 St Louis U.L.J. 11-69 (Spring 1958) and reprinted in Galloway (ed.), *Legal Problems of Space Exploration: A Symposium,* Document No. 26, Senate Committee on Aeronautical and Space Sciences, 87th Congress, 1st session, pp. 271-329 (GPO, Washington, D.C., 1961), hereinafter cited as "1961 Senate Symposium."

3. Meyer had been involved in research and writing on aspects of air law from as early as 1908. In *Beiträge zum Luft-und Weltraumrecht* (a special edition of the *Zeitschrift für Luft-und Weltraumrecht* issued as a "Festschrift" to honor Alex Meyer on the occasion of his 1975 retirement as Director of the Institute of Air Law and Space Law of the University of Cologne), there is a bibliography of more than 200 works by Meyer covering the period 1904 to 1975. The bibliography, compiled by Hans Pick, is at pp. 457-482. Meyer presented a paper, "Legal Problems of Flight into Outer Space," to the Third International Astronautical Congress in Stuttgart, Germany, September 5, 1952, reprinted in the 1961 Senate Symposium pp. 8-19. Meyer repeatedly cites and discusses the 1932 monograph by Mandl, thereby contributing to a heightened awareness of Mandl's amazing early study.

4. E.g., Fuimo Ikeda, "Jinko eisei to Kokusaiho" (Sputnik and International Law), *Juristo* (The Jurist), No. 132, pp. 20-25 (Tokyo, 6/15/57); and Ryoichi Taoka, "Kuiki no ryoyuken" (On the Right to Airspace), *Kuho* (Journal of Air Law), No. 2, pp. 1-30 (Tokyo, 10/31/56). Curiously, although Ikeda's 1957 article cites a number of European, Canadian, and American writers and their works, including J. C. Cooper, C. W. Jenks, N. Mateesco, O. Schachter, A. Meyer, E. Danier, J. Kroell, and the Chinese scholar Ming-Min Peng, and Soviet co-authors A. Kislov and S. Krylov, he cites no Japanese or other oriental sources in his work.

5. See Y. A. Korovin, "Problema Vozdushnoy Okkupatsii v Svyazi s Pravom na Polyarnyye Prostanstva" (Problems of Aerial Occupation Concerning Law in Polar Space): Published in Vol. 1, pp. 104-110 *Voprosy Vozdushnogo Prava, Spornik Trudov Sektsii Vozdushnogo Prava Soyuza Aviyakhim (Soyuz Obshchestv Druzhey Aviatsionnoy i Khimicheskoy Oborony i Promyshlennosti)* Problems of Air Law, A Symposium of Works by the Air Law Sections of the U.S.S.R. and RSFSR Unions of Societies for the Promotion of Industry and Chemical-Air Defense) (U.S.S.R. i Aviakhin RSFSR, Moscow, 1927), 300 pp., in Russian; cited from the 1961 Senate Symposium, *op. cit. supra,* note 2. That same 1927 Symposium includes an article entitled "Mezhdunarodnoye Publichnoye Vozdushnoye Pravo" (Public International Air Law), V. A. Zarzar, pp. 90-103.

6. E. Korovine (a variation in spelling used for the French journal) "La conquête de la stratosphère et le droit international (The Conquest of the Stratosphere and International Law), 41 *Rev. Générale de Droit Int'l Public*, pp. 675-686 (Paris, December 1934).

7. According to R. D. Crane, in the 1961 Senate Symposium, *op. cit. supra*, note 2, at p. 1025, Korovin edited a book on international law (Mezhdunarodnoye Pravo) published in Moscow in 1951. Crane traces chapters and sections of that book dealing with air law and the question of an upper limit on national sovereignty to earlier published and unpublished Russian works dating from 1947 and 1939. Crane also credits Korovin and Zarzar (see note 5, *supra*) with formulating principles in the 1920s and 1930s "which have served as a basis for subsequent writings of Communist legal scholars." R. D. Crane, "Guides to the Study of Communist Views on the Legal Problems of Space Exploration and a Bibliography," in the 1961 Senate Symposium, *op. cit. supra,* note 2, pp. 1011-1036, at 1011.

8. See bibliographic references in the 1961 Senate Symposium, *op. cit. supra,* note 2, and K. L. Li, *World Wide Space Law Bibliography,* 700 pp. with indexes (Institute and Center of Air and Space Law, McGill Univ., Montreal, 1978).

9. December 20, 1948, Cooper presented a lecture, titled "International Air Law," at the U.S. Naval War College, Newport, Rhode Island. This was the first public discussion of his views on the need for definition of an upper limit on national sovereignty. Extracts of this lecture are reproduced in Vlasic (ed.), *Explorations in Aerospace Law: Selected Essays by John Cobb Cooper* pp. 264-267 (1968). The earliest full elaboration of Cooper's views was presented on January 5, 1951, in an address at the Escuela Libre de Derecho in Mexico City, included in Vlasic's compilation of Cooper's essays, just cited, at pp. 257-264; also published in *International Law Quarterly* pp. 411-418, July 1951 (London); and reprinted in the 1961 Senate Symposium, *op. cit. cupra,* note 2, at 1.

10. Schachter's early commentary on space law took the form of a lecture delivered at the first Hayden Planetarium Symposium on Space Travel (October 12, 1951), which was later published in *Journal of the British Interplanetary Society,* pp. 14-16, entitled "Legal Aspects of Space Travel," January 1952; then somewhat expanded in the book, edited by Cornelius Ryan, *Across the Space Frontier,* pp. 118-131 (Viking Press, New York 1951),

as a chapter entitled "Who Owns the Universe?" The only previously published, contemporary referred to by Schachter is John Cobb Cooper's article on upper limits of sovereignty, in the 1951 *International Law Quarterly*, cited in note 9, above. Schachter's work proved to be very perceptive and prophetic. He could not have suspected, in 1952, the extent to which one of his proposals would generate debate and concern when it later appeared in a United Nations proposed treaty. He wrote:

> "Beyond the airspace,..., we would apply a system similar to that followed on the high seas; outer space and the celestial bodies would be the common property of all mankind, over which no nation would be permitted to exercise domination. A legal order would be developed on the principle of free and equal use, with the object of furthering scientific research and investigation. It seems to me that a development of this that would dramatically emphasize the common heritage of humanity and would serve, perhaps significantly, to strengthen the sense of international community which is so vital to the development of a peaceful and secure world order."

From Schachter, "Who Owns the Universe?" in Ryan (ed.) *Across the Space Frontier*, pp. 130-131 (1951).

The substance of Schachter's view on common heritage and the philosophy of common property for free and equal use have become important principles in two major contemporary treaty documents. First, the treaty being drafted in the United Nations on the Law of the Sea substantially incorporates the Schachter views in dealing with natural resources in the deep seabed. See, e.g., B. H. Oxman, "The Third United Nations Conference on the Law of the Sea," 74 *A.J.I.L.* 1 (1980). In addition, these same principles have been a subject of increasing public debate because of their inclusion in the Agreement Governing the Activities of States on the Moon and Other Celestial Bodies, approved by the UNGA on December 5, 1979; A/RES/34/68, December 14, 1979.

11. A lawyer and ardent internationalist in his astronautical activities, Haley's first published work on space law was "Basic Concepts of Space Law," a paper presented at the 25th Anniversary Annual Meeting of the American Rocket Society, Chicago, November 14-18, 1955. He had published previously on international cooperation in astronautics, on communications and immigration law, but after 1955 his space law articles multiplied rapidly. When the U.S. Congress, House Select Committee on Astronautics and Space Exploration compiled and published a staff report, *Survey of Space Law*, 86th Congress, 1st session, House Document No. 89, 1959, the bibliography included 35 articles, comments and a book by Haley dealing with aspects of astronautics and space law. Before his death in 1966, he wrote more than 300 articles and an award-winning book on space law, *Space Law and Government*, (Appleton-Century-Crofts, New York, 1963). In his November 1955 paper, the only contemporary commentators discussed were Cooper and Schachter.

12. American Society of International Law, Proceedings of the 50th Annual Meeting, Washington, D.C., (April 25-28, 1956) papers and comments included by Colclaser, Cooper, Knauth, Lissitzyn, McDougal, Meyer, Roberts, Roy, Rudzinski, Schachter and Wright.

13. See Staff Report prepared for the Senate Committee on Aeronautical and Space Sciences, 89th Congress, 1st session, Document No. 44, July 30, 1965 (GPO, Washington, D.C.) entitled *United States International Space Programs: Texts of Executive Agreements, Memoranda of Understanding, and Other International Arrangements, 1959-1965* and NASA pamphlet, "International Programs," prepared by the Office of International Affairs, NASA Headquarters, January 1973. See an excellent analytical introduction in the Senate staff report by Eilene Galloway.

14. Although NASA, as an institution, has not ever fully documented a history of its international programs, activities and contributions to space law, various General Counsels have done selected articles on particular aspects of legal practice, treaty negotiations and policy development. Arnold Frutkin, Assistant Administrator for International Affairs and Director of International Affairs at NASA for about 18 years (1960-1978), did write an analytical and interpretive history of the early years of experience that remains the best, firsthand account of the formative years--Frutkin, *International Cooperation in Space* (Prentice-Hall, New York, 1965).

15. Report of the United Nations *Ad Hoc* Committee on the Peaceful Uses of Outer Space, a report to the U.N. General Assembly, 14th session, New York, 1959, U.N. Document No. A/4141, July 14, 1959. Discussion of this report is contained in Official Records of the 14th General Assembly, First Committee, 1079th to 1081st meetings, and, *ibid.*, Plenary Meetings, 856th meeting. Text of the report is contained in the 1961 Senate Symposium, *op. cit., supra,* note 2, 1246-1272; also in Staff Report prepared for the Senate Committee on Aeronautical and Space Sciences, 88th Congress, 1st session, Document No. 18, May 9, 1963 (GPO, Washington, D.C.) entitled *Documents on International Aspects of the Exploration and Use of Outer Space, 1954-1962,* pp. 101-152.

16. For a compilation of the principal extant texts of major space law treaties and agreements as of December 1978, see *Space Law: Selected Basic Documents,* (2nd ed.), a committee print of the Senate Committee on Commerce, Science, and Transportation, 95th Congress, 2nd session, December 1978 (GPO, Washington, D.C.).

17. See Senate Bill 663, submitted by Senator Stevenson on March 14, 1979, co-sponsored by Senators Cannon, Ford and Riegle, being a bill to establish an Earth Data and Information Service in the National Aeronautics and Space Administration, and for other purposes; also Senate Bill 875, submitted by Senator Schmitt on April 4, 1979, co-sponsored by Senators Pressler, Tower and Young, being a bill to provide for the establishment, ownership, operation, and regulation of a commercial Earth Resources information Service, utilizing satellites and other technologies, and for other purposes. These two bills are reminiscent of the industry vs. government debate over Comsat in 1962!

18. An excellent survey of pending areas of legal and policy development was recently published by Mossinghoff, "New AIAA Task Force Wades Into the Legal Waters," *Astronautics and Aeronautics* 20-24 (October 1979). While this article focused on some of the real and pragmatic problems of governmental and industrial involvement in spaceflight activities, other developments, primarily in the Congress, are looking toward the formulation of new,

long-term national goals and policies: see Senate Bill 212, introduced by Senator Schmitt on January 24, 1979, co-sponsored by Senators Baker, Dole, Domenici, Goldwater, Hayakawa, Heinz, Lugar, Randolph, Thurmond and Wallop, being entitled "The National Space and Aeronautics Policy Act of 1979; also, Senate Bill 244, introduced by Senator Stevenson on January 29, 1979, co-sponsored by Senators Cannon, Cranston, Ford, Hayakawa, Hollings, Mathias, Steward and Zorinsky, being a bill to establish a national space policy and program direction; and for other purposes. In the House of Representatives, Congressman Brown (CA) introduced House Bill 6304 on January 28, 1980, being a bill to establish the national space policy of the United States, to declare the goals of the Nation's space program (both in terms of space and terrestrial applications and in space science), and to provide for the planning and implementation of such programs.

VI

SPACE TRANSPORTATION: REFLECTIONS AND PROJECTIONS

John H. Disher *

I am going to discuss space transportation from the perspective of seeing the past twenty years in Washington and one year in the Space Task Group at Langley. At Langley I was a project engineer on the Mercury Atlas program and I will have a few comments on that, but by and large my comments will relate to those things I have participated in or was a close observer of in Washington. So it will be a Washington perspective, with frequent forays into the field in Huntsville and Houston and Florida. I am going to attempt to confine my remarks to those things that I did participate in rather than those things that are hearsay. The programs that I will address are the (Fig. 1) Mercury, Gemini, Apollo, Skylab, Apollo Soyuz, and the Space Shuttle, in terms of history and the present and then I will describe briefly some of the upcoming programs that we see over the next ten to fifteen years.

In the 1959 period in Washington, the period in which I arrived, NASA had just been formed and Mercury was in its early stages of development. There had been a lot of studies at NASA, in industry, in the Air Force, JPL and various other places on manned landings on the moon. That project got a great amount of attention it seems, and one of my first jobs when I came to Washington was working with George Low on the formative plans for what turned out to be the Apollo program. This was a schedule (Fig. 2) that George and I prepared in February 1960 for the "Manned Spacecraft Program." At that time in preparing our plans we could talk guardedly of a circumlunar mission but we could not talk of a lunar landing as a proposed program and that lower bar on the right that says qualification and manned flight culminated in a circumlunar flight in 1968 and by coincidence we just barely made it with Apollo VIII at Christmas time in 1968. As I said there were many lunar studies that had gone on prior to the early 1960's. This was the first documented Office of Manned Space Flight plan. Interestingly the Office of Manned Space Flight at that time was a branch in Abe Silverstein's total Space Flight Development organization and had three people in it: George Low, Warren North and myself. And these were the kinds of charts we did - handwritten charts - in those days. This was a representative piece of work that led up to a Summer 1960

* *Director, Advanced Programs, Office of Space Transportation Systems, National Aeronautics and Space Administration.*

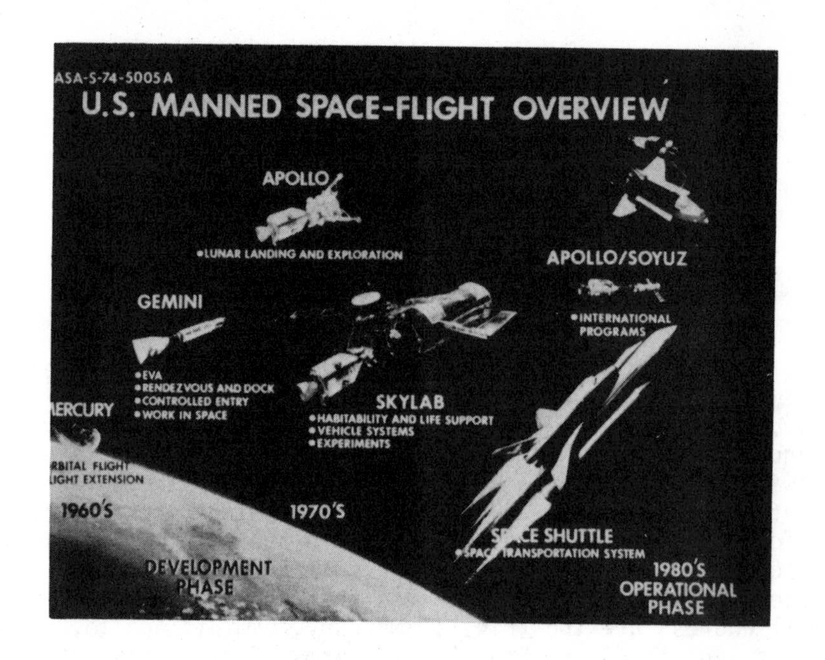

Fig. 1

MANNED SPACECRAFT PROGRAM

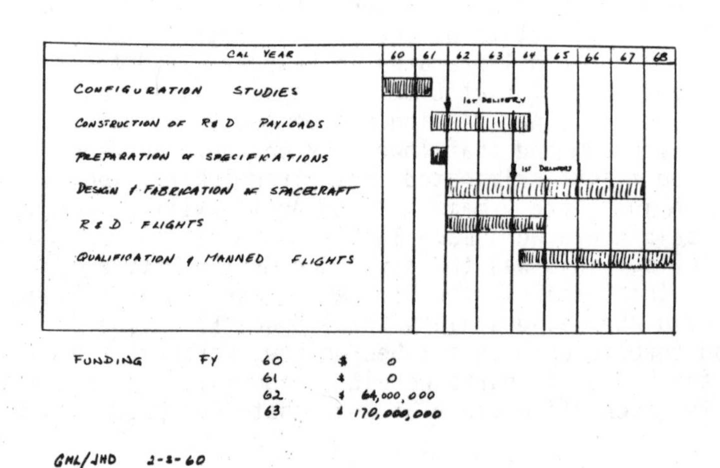

Fig. 2

conference on what had by then been named Apollo, a plan of study for a possible manned circumlunar mission in the late 1960's. The administrator at that time was Dr. Keith Glennan who administered the very conservative space policies of the Eisenhower Administration. Of course following the fall elections with the coming in of President Kennedy, things changed drastically and we quickly moved on to a manned lunar landing proposal under the same Apollo name.

Question from audience: "I see the funding was rather low the first couple of years."

Answer: Yes, we started very, very modestly.

Concurrent with that early Apollo planning of course the Mercury development activity was going on. The particular launch (Fig. 3) you will all remember - May 5, 1961 when Al Shepard flew, and it was preceded by a very, very critical series of developmental launches, on the Atlas. During 1960 the Atlas had a series of failures for one reason or another and one of those failures had been Mercury Atlas 1, in July 1960. The interface between the development capsule on Mercury and the Atlas failed structurally during launch, and it was a very, very crucial time in the program. John Yardley at McDonnell, and Bob Gilruth, Max Faget and Jim Chamberlain at the Space Task Group, devised a quick fix for the interface failure. The fix was simply an eight-inch wide stainless steel band in the form of a belly band or a corset. I think I will go to the Atlas illustration (Fig. 4) so that you will see it better. This of course is John Glenn's flight which was a year later. But take it back to February 1961 now, the Mercury Atlas had been fixed, the weak spot in it had been fixed in a very band-aid way, with the stainless steel reinforcement around

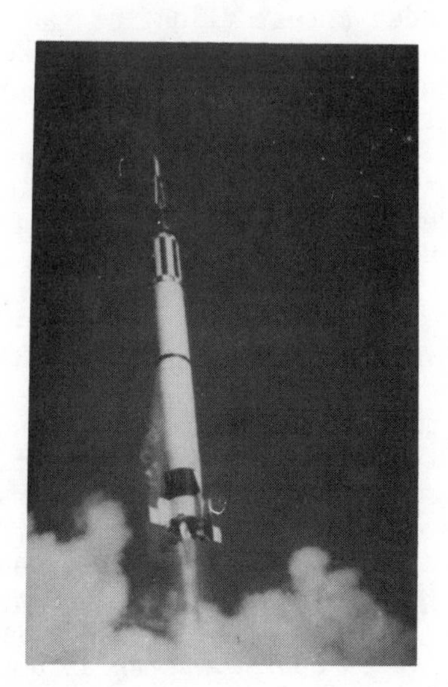

PROJECT MERCURY

1ST U.S. MANNED FLIGHT

DATE: MAY 5, 1961
PILOT: ALAN B. SHEPARD
SPACECRAFT: FREEDOM 7
LAUNCH VEHICLE: REDSTONE
MAX. ALTITUDE: 116 MILES
RANGE: 302 MILES
WEIGHTLESSNESS: 5 MIN.
MAX. SPEED: 5180 MPH

Fig. 3

PROJECT MERCURY

1ST U.S. MAN IN ORBIT

DATE: FEBRUARY 20, 1962
PILOT: JOHN H. GLENN, JR.
SPACECRAFT: FRIENDSHIP 7
LAUNCH VEHICLE: ATLAS
APOGEE: 163 MILES
PERIGEE: 100 MILES
PERIOD: 88 MINUTES
ORBITS: 3
WEIGHTLESSNESS: 4 1 2 HOURS
AVG. SPEED: 17,500 MPH

Fig. 4

the interface. The Air Force and General Dynamics naturally were very concerned for the reputation of their bird, they wanted in the worst way for NASA to wait for the thick-skinned Atlas which was the proper fix and was the fix which would be made before John Glenn flew, but at the same time NASA and McDonnell desperately needed a test flight of the capsule to verify its flight worthiness. In the end the decision was NASA's and the Air Force said we go along with your decision, but it is your decision and you will have to take responsibility for it, and a failure too, if one should occur. Dr. Seamans, NASA Associate Administrator and General Manager at the time, made the decision, but of course at that time Jim Webb had come in as administrator and Jim Webb was the man for whom the buck went no further, but Bob Seamans was the decision point and Jim Webb backed him. And the Mercury Atlas 2 then did fly successfully in February 1961 with that corset or belly-band on it. It was a pretty gutsy call because at that point the Apollo decision was in the offing; another failure at that point could have had very serious implications. Sometimes the rewards come to the venturesome, and it was a good gamble. With that success, Al Shepard's good flight on May 5 and the Russian Gagarin flight, events moved rapidly toward the Apollo decision on May 25, 1961. The manner in which Apollo would be carried out however was very much in question--direct to the moon, orbit rendezvous at the moon, earth orbit rendezvous or some combination of those were all in debate. The chart (Fig. 5) simply illustrates those several alternatives. The assumption initially was made that the mission would be direct but that of course was an assumption and the baselining did not stop the debate, nor was there any attempt to stop the technical arguments that went on regarding the pros and cons of the several methods. And it was not until the middle of 1962 in fact, a year

Fig. 5

after the initial Apollo decision had been made, that the lunar orbit rendezvous mode was selected as the basic route to the moon landing. Throughout 1961, development of Mercury was continuing, leading up to John Glenn's flight in February 1962. Early development of the Apollo spacecraft and of the Saturn launch vehicles was continuing. Initial contracts were being let most expeditiously. I remember in August 1961 the Massachusetts Institute of Technology Draper Laboratory had made a proposal to NASA that it do the guidance for Apollo, (the laboratory rightfully had an outstanding reputation) and Dr. Seamans was interested in taking up that proposal. I remember in the first week of August 1961 the procurement paper and the specifications and all had been put in a package and delivered to my management. I went on vacation thinking that it would take at least a month for those papers to get processed and through. I was up at the beach in Saugatuck, Michigan, a little town on Lake Michigan and I bought the Wall Street Journal that morning, August 8 or 9, and there on the top of the Wall Street Journal was, "MIT Selected for Apollo Guidance" so in those days things could move very rapidly. The sole source procurement was with a non-profit organization so I guess among other things that it made it a little simpler.

Well, throughout 1961 was a very hectic period with the simultaneous Mercury development, the Apollo contracting and the like and the mode decision going on. At the same time the spacecraft development was going on, the Saturn launch vehicle configuration was getting very great detailed concern, and I will discuss that in a moment. But I want first to use this single chart (Fig. 6) just to mark a quite interesting milestone which I remember well. This was October 1962, about four months after Jim Webb, Dr. Dryden and Dr. Seamans had made the lunar orbit rendezvous decision, and this was the first full scale review of Apollo planning for the administrator following that decision. The review was an all day review on October 10 of 1962, as I remember, and one of the illustrations we used showed the Apollo spacecraft (and this is literally the illustration that was used) which by that time, with the LOR decision having been made, the complete spacecraft configuration conceptually was known and it is shown on the left. The point of this chart was to convey to Mr. Webb and his high-level associates, whom he had invited in, the Department of Defense representatives and all, that we were talking about a spacecraft which in its total was larger in most respects than the whole Mercury Atlas launch vehicle and I remember that occasioned quite a bit of comment. As I indicated earlier, about the same time that the Apollo spacecraft was being contracted, Dr. Von Braun and his folks in Huntsville were considering the Saturn variations and this drawing (Fig. 7) represents the December 1961 decision that was made on Saturn V, with five F-1 engines on the first stage, five J-2 engines on the second and one J-2 on the third. All during 1961 there had been considerations of C2, C3, C4 configurations with respectively 2, 3, 4 and finally C5 with five F-1 engines in the first stage. And of course an important variable was the mode that would be used for lunar landing--would it be earth orbit rendezvous, would it be lunar orbit rendezvous, which was a very dark horse all during 1961, or would it be direct? The five engine configuration was selected as being easily capable of earth orbit rendezvous, and marginally with direct flight. Really it was not capable of the direct landing, but the Saturn V was easily capable of the earth orbit rendezvous and lunar orbit rendezvous modes.

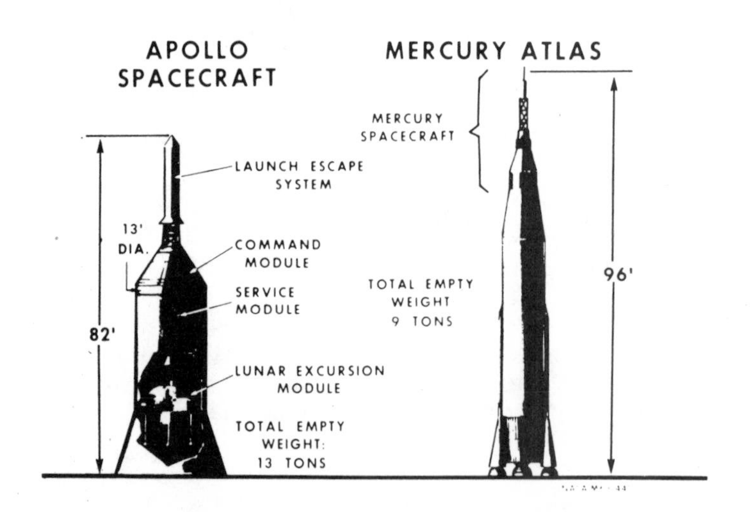

Fig. 6

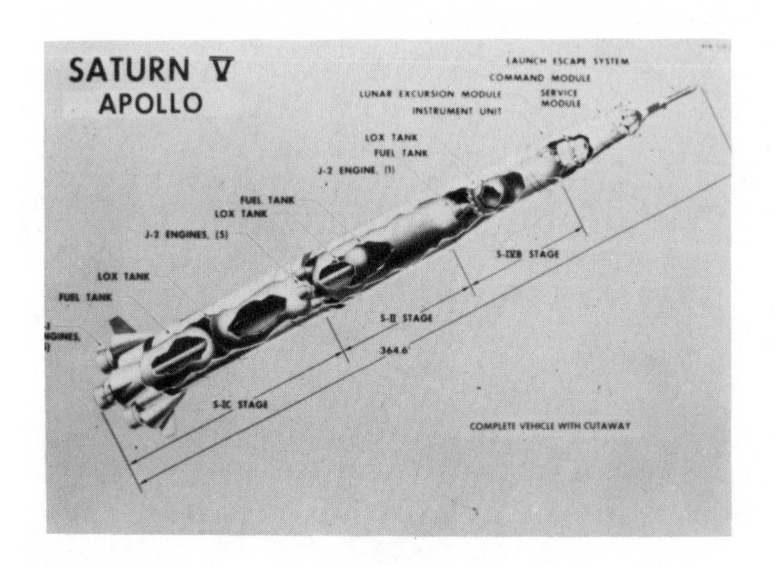

Fig. 7

At the same time, the Saturn I - Saturn IB series was undergoing development. This particular flight (Fig. 8) which was early in 1964 shows the first complete or two-stage orbital flight of the Saturn, and that relates to a point that I want to get to now that the Saturn I series went very conservatively step by step through the development process. The first stage only was flown first with inert payload (water). Not until the fifth flight in fact was a complete launch vehicle flown, and the fifth flight did not have a spacecraft on it, so it, in fact, did not reach what we call an all up configuration until much later. It was in this time period then in the Fall 1963 that George Mueller came on the scene. George had a very, very profound effect on the program in many ways. It is my conviction that clearly we could not have gotten to the moon in 1969 without George and his innovative approach and daring willingness to make decisions that were revolutionary at that time. In the Fall of 1963 George arrived and by early 1964 he had sold his point that all Saturn V testing would be "all-up"--from the very first Saturn V, we would have a complete launch vehicle with all live stages with a spacecraft on top. That was received with strong resistance from many elements of the program, but George always had the courage of his convictions and he carried the day. The test program as we laid it out in early 1964 is shown in Fig. 9.

Question from audience: "Was the resistance at Huntsville?"

Answer: Yes, initially; I cannot say how long it persisted, I have forgotten when they capitulated.

This then (Fig. 10) was just illustrative of the first Saturn V launch 501 in October 1967 which, of course, was several years later, but it was all-up as George Mueller had said it was going to be. As you will gather,

SATURN SA-5 FLIGHT

JANUARY 29,1964

Fig. 8

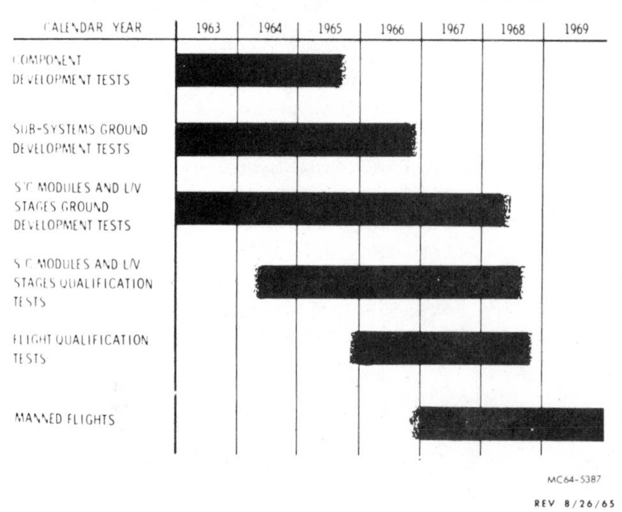

Fig. 9

Fig. 10 First Saturn V Launch, Oct. 1967

I am quite an admirer of George because I think he is the single individual most responsible for our lunar landing within the decade.

In the 1964 period there was a strong sentiment for cutting the fiscal 1966 Apollo budget. It was being alleged that Apollo was a crash program, that it should be slowed down, that the funding requests for 1966 could be reduced several hundred million dollars. In addition to his other attributes George was politically astute, and he set a team of us to show that Apollo was not in fact a crash program and to show what a catastrophic effect would accrue if the 1966 budget were reduced. I am going to show two charts (Figs. 11 and 12) to illustrate that.

COMPARISON OF DURATION OF MAJOR UNITED STATES R & D PROGRAMS

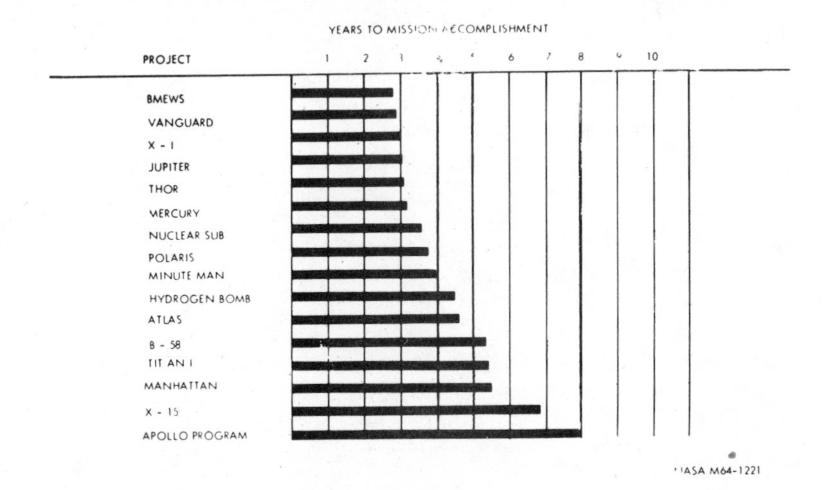

Fig. 11

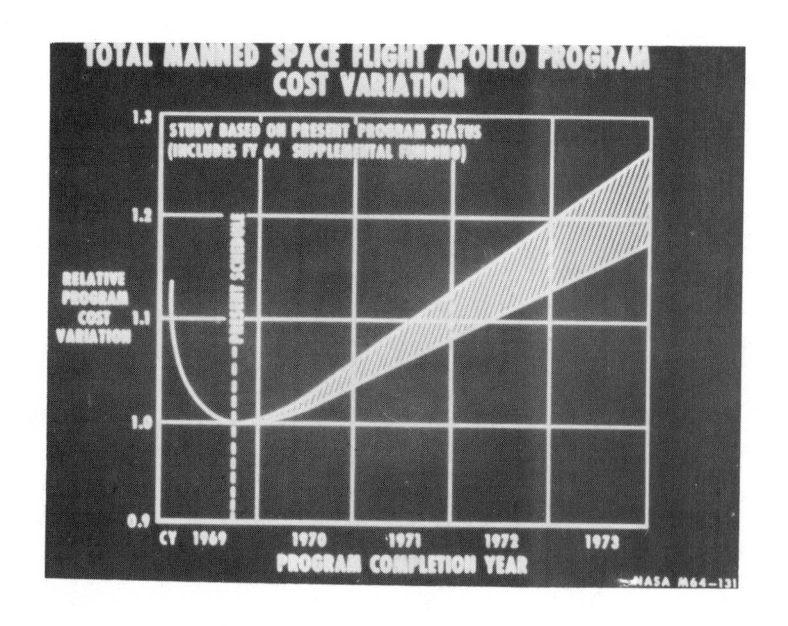

Fig. 12

This is a simple comparison of Apollo schedule R & D period with all of the other major programs that had occurred prior to that time. There were eight years of R & D in our plan for Apollo as compared with say three and a half for the nuclear sub, four years for Minuteman, four and a half for the hydrogen bomb, Apollo was not in fact a crash program. The other, the companion finding of the study, which was carried out in very substantial depth and detail, with adequate and appropriate documentation showed that at that time our target schedule, our planned landing within the decade by 1969 was optimum from a cost standpoint. We evaluated program alternatives, both attempting to speed up the program, as one would do in a crash program, and slowing it down as would happen with a lower FY '66 budget. This summary chart shows at that time, which was in 1964, we assessed there was some speed-up that could be done, perhaps as much as a year, but no more than that with even a crash program and the cost would rise very substantially. On the other hand, the same cost increases would accrue if the program were unnecessarily stretched out. Although I think that these conclusions were intuitively obvious and accepted by most engineers on the program, I found a surprising skepticism on that on the part of people outside the program until it was in fact shown in detail. In summary, George carried the day. The 1966 budget went through at substantially the levels that were needed to keep the program on its schedule.

Now, all during this period from 1961, before Mercury had flown, into the early Saturn flights the Gemini program had gone through with relatively little notice and was making very, very substantial contributions to the success of Apollo. The birth of Gemini was St. Patrick's Day, 1961, at a Wallops Island "hideaway meeting" when a "Mark II Mercury" spacecraft proposal was carried to Abe Silverstein by Bob Gilruth and Jim Chamberlain of the Space Task Group. At that point it was called Mercury or Mark II or Mercury II, but its mission was that which Gemini fulfilled; rendezvous development and demonstration (Fig. 13), EVA and enough duration in earth orbit to verify the ability to conduct a lunar mission. I remember that day 18 years ago at Wallops Island quite vividly. Jim Chamberlain, who is now an engineer with McDonnell Douglas, was the in-house "chief design engineer" at Space Task Group. He had an ambition that not only could the Mark II capsule provide this rendezvous, docking, and long-duration experience for Apollo, but it in fact should be a lunar spacecraft as well. Jim in his presentation to Dr. Silverstein had a fluorine stage on top of a Titan, and he had his Mark II spacecraft then on top of the Titan fluorine stage for a circumlunar mission. We were just sitting around a table looking at paper copies and when Jim came to those pages, Silverstein just looked at them and he tore them out of the book and threw them on the floor, and no mention was ever made of them and the lunar Gemini died there. But Jim Chamberlain later confirmed that that day was in fact the day that Gemini was born, although not until November/December 1961 was Bob Seamans in a position to formally give the go-ahead to the contract.

Question from audience: "The fluorine hydrogen stage? Flox?"

Answer: Not Flox, fluorine hydrogen.

Question from audience: "That was Bell Aircraft wasn't it?"

Answer: It probably was.

This is simply another illustration (Fig. 14) of one of the key Gemini missions that were in early 1966. The first docking in space was carried out on March 15, 1966 and similarly, one of the many other contributions of Gemini showing the first EVA activity (Fig. 15). It was interesting that the EVA experience with Gemini was quite sobering in its implications and I think that it was not until Skylab some years later that we learned to operate and to overcome the difficulties inherent in zero gravity. But this was quite a sobering experience in Gemini in that work outside the capsule was extremely difficult and exhausting.

Now I want to talk a moment about the beginnings of Skylab, which initially was called the Apollo Applications Program. You may remember the wet workshop--the conversion of the Apollo SIVB structure, after serving its propulsion function with a Saturn IB into living and working quarters for a three-man crew. Of course we ended up with the dry workshop using that same structure on a Saturn V--whereby it did not have to provide propulsion and could be outfitted on earth as living and working quarters. Figure 16 shows a comparison of the volume available for Mercury, Gemini and Apollo with the very, very large volume of Skylab--some 11,000 cubic feet. One of the fringe benefits that we have at NASA these days is the ability to walk across the street and see Skylab in the Smithsonian Institution. That is quite a thrill any noontime to be able to walk through a real duplicate of the Skylab that is still in orbit. Skylab is another feather in George Mueller's cap and this is a photograph of George's original sketch on a piece of 24" x 30" chart paper, in Summer 1966 (Fig. 17). He put down on paper things he had been telling us much of the year earlier, as to what he wanted in this bird, and we could not seem to get the message as clearly

FIRST RENDEZVOUS IN SPACE
GEMINI VII AS SEEN FROM GEMINI VI

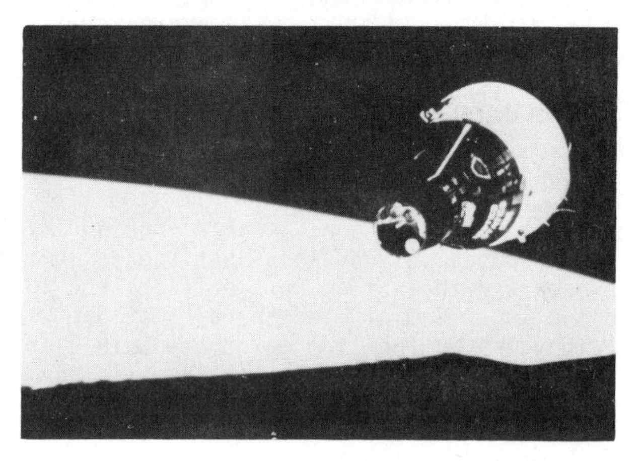

Fig. 13

THE FIRST DOCKING OF TWO SPACE VEHICLES

GEMINI VIII AND THE GEMINI AGENA TARGET VEHICLE MARCH 15, 1966

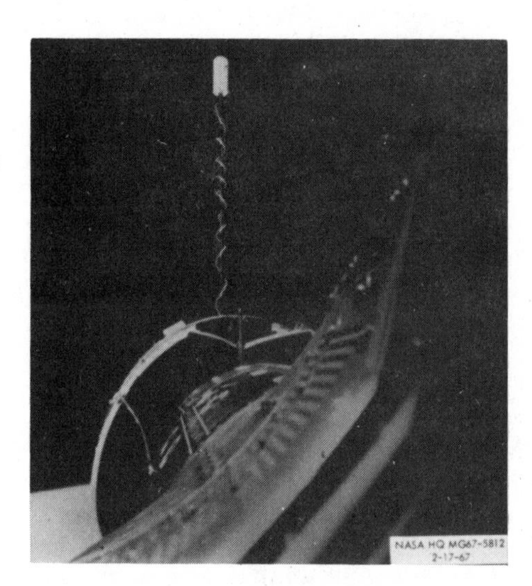

Fig. 14

as he wanted us to, so he wrote it down on paper. Down at the bottom of the drawing is what turned out to be the ATM, the Apollo Telescope Mount, or Astronomical Telescope Mount, and George's concept at this point included use of a free-flying module coupled by a soft tether and umbilical to the workshop. As it ended up we used hinges and what amounted to a hard tether, but all of the elements are there--the SIVB on the right, the airlock and docking adapter in the center and the logistics module (command service module) on the left.

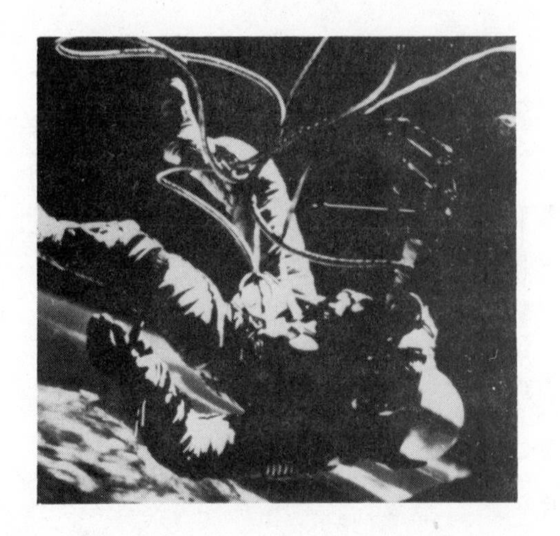

GEMINI IV EXTRAVEHICULAR ACTIVITY, ASTRONAUT EDWARD WHITE

Fig. 15

EXPERIMENT VOLUME IN EARTH ORBIT

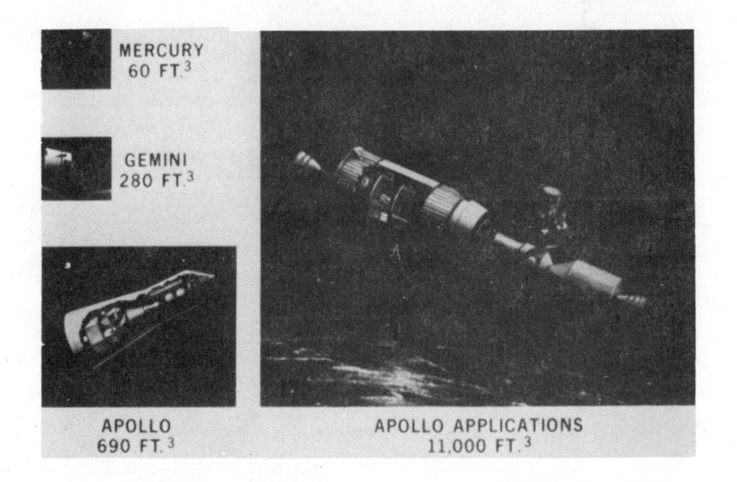

Fig. 16

Now going back across the intervening years between the initiation of Skylab and its culmination, Skylab was a victim of the budget crunches that George Mueller was able to stave off in Apollo. But Apollo applications which became Skylab in 1970, in fact had to idle along until Apollo XI was successfully accomplished. When Apollo XI flew, Skylab was able to take off. This (Fig. 18) shows the Skylab as it ended its mission in February 1974, just about five years ago. As is so often the case with experiments, the biggest contributions of Skylab were those that we did not

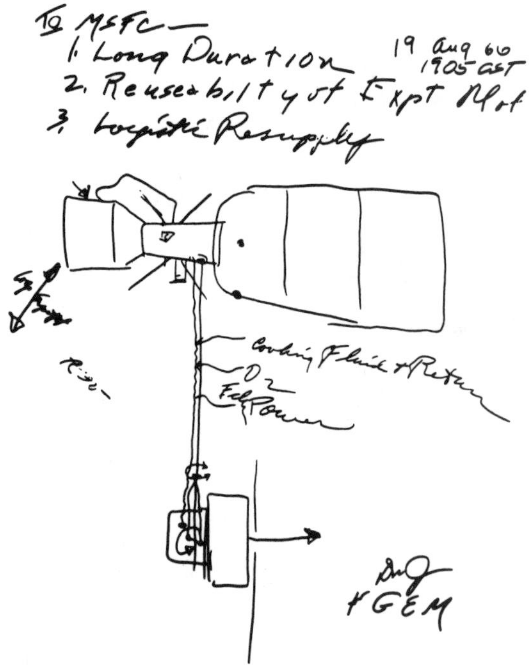

To MSFC —
1. Long Duration
2. Reuseability of Expt Mod
3. Logistic Resupply

19 Aug 66
1905 CST

Cooling Fluid & Return
O2
Fed Power

GEM

Fig. 17 George Mueller's Sketch of Skylab

NASA 74-HC-78

Fig. 18 Skylab

anticipate. The planned-for contributions were many and provided much new data; however, the biggest benefit was the experience that we had in unplanned maintenance and repair--first, the saving of the mission through Pete Conrad and his buddies' work in freeing the stuck solar array to provide power, then unfurling the umbrella-like sunshade to keep it cool. Then later the Skylab II crew assembling that sunshade over the living quarters of the workshop. As you will remember that literally was assembled stick by stick, tent-pole style, and then the big reflecting fabric (plastic) was deployed over it by a pulley-cable arrangement. The procedures and time-lines for those repairs were developed on the ground in the big neutral buoyancy water tank at Huntsville. To this day that is the only meaningful earth-space correlated time-line experience we have. The data that we got from the Skylab space assembly, the data that we now get from underwater simulations are the useful data that we have today in planning future space construction. We do not know how many years operational large structures in space are away, but they are clearly in our future. The point I am making is that Skylab gave us the confidence and factual basis for our planning. It also turned around many people who thought men in space were a hindrance rather than a help; a number of the scientists on the mission who had been skeptical early and looked at the crew as an impediment or a resource-user turned around completely and were lavish in their praise of the contribution of humans in orbit to the right kind of experiment.

Now those were some of the unplanned benefits of Skylab; the planned-for benefits in fact came too. The 84-day mission duration provides us the confidence that 90-day crew exchange is a reasonable planning figure for our future long-duration missions. The Skylab data were the first data that we had exceeding durations of fourteen days and they are extremely valuable in our future mission planning. More recently, of course, the Russian experience with their 140-day mission will give additional data.

Figure 19 is illustrative of the many solar experiments carried out on Skylab. With the special instrumentation that was carried it was possible

Fig. 19

to observe solar phenomona like the big sunburst on the left which was very significant to solar physicists. I cannot explain the physics of it, but people like Dr. Robert McQueen of Boulder, Colorado, say this was a revolutionary finding in solar physics. Figure 20 illustrates the use of the large volume of Skylab in evaluating space maneuvering in zero gravity but within the safety of a pressurized volume.

Fig. 20

Fig. 21 First Manned Lunar Landing

Returning now to Apollo, Fig. 21 illustrates the dramatic first moon landing in July 1969. This I believe is Buzz Aldrin, with the picture being taken by Neil Armstrong--a picture of the second man stepping on the moon. And the tremendous series of Apollo missions then culminating in Apollo XVII in 1972 (Fig. 22) and all of the success and cliff-hangers in between, an achievement that is still hard to believe to this day.

Fig. 22 Apollo 17

Now proceeding to the Apollo Soyuz Test Program (ASTP), Fig. 23 shows the Soyuz as the American craft was approaching it in 1975. This mission was not an advance in transportation, using existing technology for the most part as it did, but was more of a social and political experiment in an advanced technological area.

Let us talk now about the Shuttle. Figure 24 is an illustration that Max Faget used last month in his lecture at the annual AIAA meeting, I think it quite dramatically shows the more-than order-of-magnitude increase in reentry development associated with the Shuttle compared with Mercury, Gemini, and Apollo; these are all to scale. The weight and surface areas show a 20-to-60-fold increase respectively over prior vehicles that have reentered, and of course Apollo, the Apollo command module, is the largest device that has been returned to earth by man to date. I hoped to have a picture of the Shuttle arriving in Florida; it is there but the weather delay seems to have hindered our logistics a little so we will have to settle for this illustration (Fig. 25) of one of the earlier test flights of the Orbiter, but the Orbiter is in Florida and preparations are under way for a launch late this year.

Fig. 23 Apollo Soyuz Test Program (ASTP)

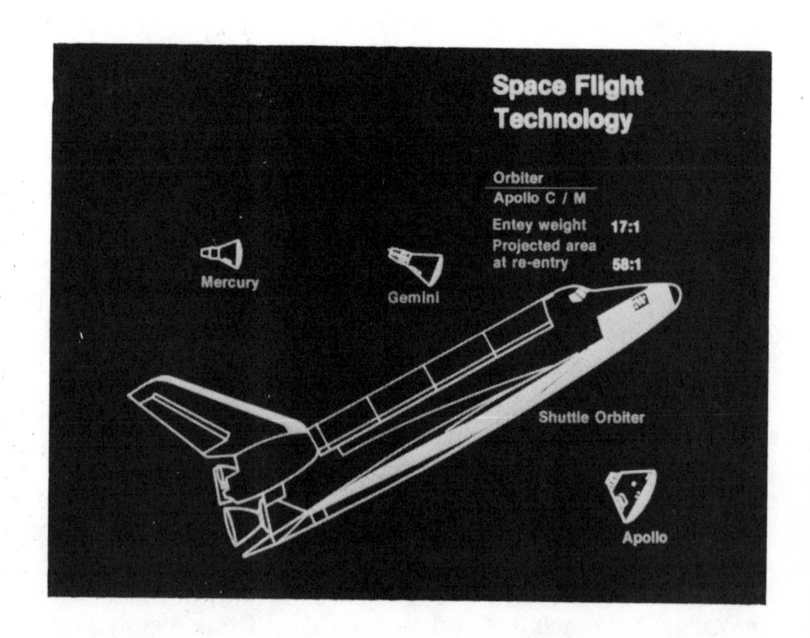

Fig. 24

Fig. 25 Space Shuttle Flight Test

Now from here on I want to address where we go from here. The immediate next step (Fig. 26) in transportation will be a rather modest thrust augmentation of the Shuttle for those missions that require a higher payload than the equivalent of 65,000 pounds due east out of Florida. Relatively small rockets strapped under the existing large solid rockets can increase the due east payload about 10,000 pounds from 65,000 to 75,000 pounds, but more importantly, for a polar mission out of the West Coast which is the more demanding mission that requires the augmentation, such strap-ons can increase the capability from about 22 to 32 thousand pounds, and that is extremely significant. The planning is to bring this capability on by 1984 to coincide with Vandenberg readiness. A further augmentation through addition of solids to the base of the external tank is also being examined. I have chosen not to address any of the trade-offs that went into selecting the basic Shuttle configuration. I had no affiliation with Shuttle definition although I was an interested observer at the time. Charlie Donlan has given several most interesting papers on that subject and I commend those to your reading as the definitive treatment of the origins of the Shuttle configuration, including the early trade-offs.

Now let us go on to the upper stages for the Shuttle. A reusable upper stage with round trip capability to geosynchronous orbit (GEO) will be the next big development in transporation. The initial upper stage of the space transportation system is an expendable solid propellant rocket which can take up to 5,000 pounds geosynchronous orbit one way, but does not have retrieval capabilities. Clearly, in the latter part of the 1980's and into the 1990's, we are not only going to have much larger payloads than 5,000 pounds in geosynchronous orbit, but we will need to retrieve and service them. We are going to need, I believe, manned access to and from geosynchronous orbit. We are going to need that manned access for

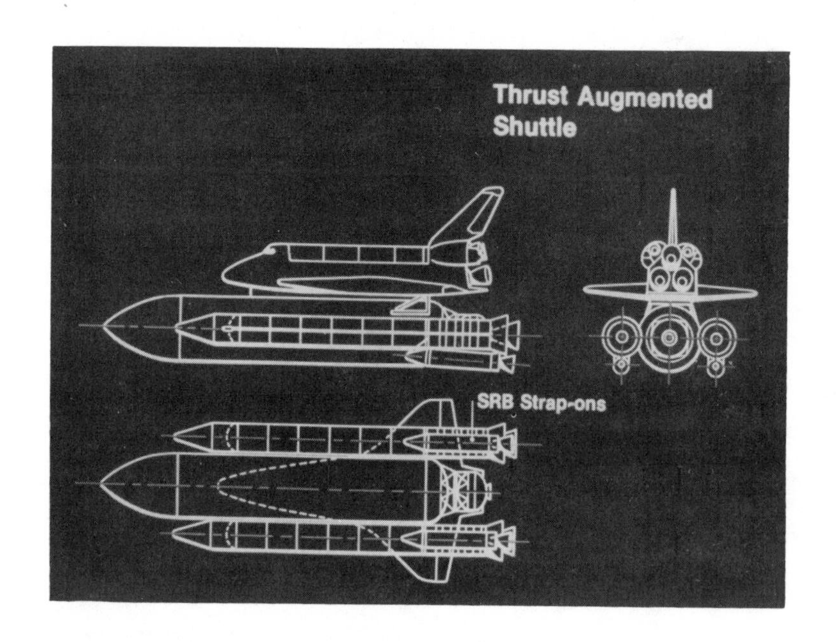

Fig. 26

maintenance, repair, refurbishing, updating of those very large sophisticated systems whose expense will justify an admittedly costly venture to synchronous orbit.

Figure 27 shows the range of upper stage payload sizes we see being needed over the next ten to fifteen years. Shown for comparison are the capabilities of the currently planned expendable upper stage (the IUS with 5,000 lb. one way capability) and an earlier planned reusable upper stage (reference tug) which is likely too small as now foreseen, unless used as an interim based on adapting a high-performance currently expendable stage such as the Centaur. The larger vehicles cited (Manned Orbital Transfer Vehicle--MOTV) and OTV for large space station support will require multiple Shuttle launches and orbital assembly at separately launched stages, or propellant transfer on orbit (or both). When one talks of a 50,000 lb. capability to geosynchronous orbit, that then is at least four Shuttles' worth of payload in low orbit for assembly to carry out the mission on the geosynchronous orbit. And that kind of a mission in particular is the one that will benefit strongly from Shuttle payload increase, to something like 100,000 pounds. Most Shuttle missions are volume-limited in their capability, however, upper-stage missions in the Shuttle are generally mass-limited, and it is these kinds of missions, upper-stage missions, that would benefit strongly from an upgrading of payload capabilities in the Shuttle. Figure 28 illustrates several different concepts for an orbital transfer vehicle that can move men or cargo from low to high orbit. The upper left is the tug concept which was under consideration prior to the selection of the solid, inertial upper stage now under development. The tug had about an eight-thousand-pound round-trip capability; we think that

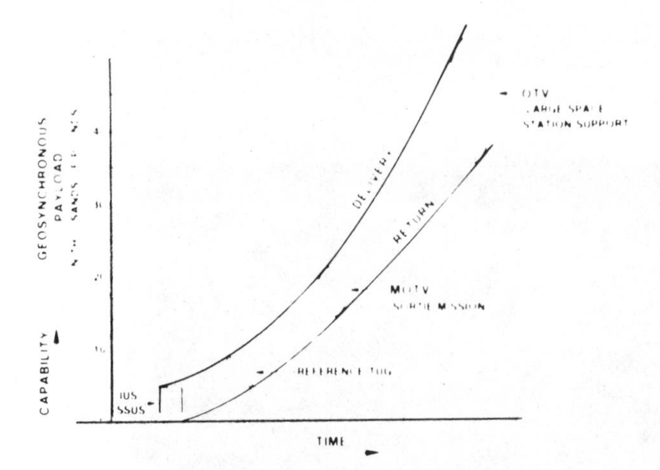

Fig. 27

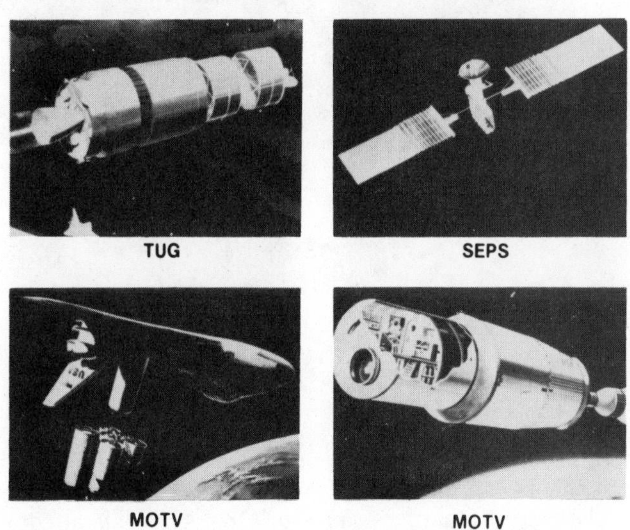

Fig. 28

the orbital transfer vehicle that will appear now in the late 1980's will be at least several times that size. Shown also on this chart in the upper right is the SEPS or Solar Electric Propulsion Stage, which is representative of a whole different class of upper stage--one that provides very low thrust for a very long time with very high efficiency. This is the second major area of upper-stage development we see in the 1980's. The first utilization of solar electric propulsion will be for high-energy kinds of comet or planet missions, but eventually, certainly, solar electric propulsion will be used to move large payloads from low orbit to high orbit as well--those payloads that can tolerate the long travel time.

The first use planned for the SEPS is a 1985 launch from the Shuttle for flyby of the Comet Halley and rendezvous with the Comet Tempel II. We propose to start development of the SEPS in FY 81 and an illustration of the current concept with its 25 kilowatts of solar power and 10 thrusters providing a total of 1 newton (about 1/5 pound) of thrust, is shown in Fig. 29.

Fig. 29 Solar Electric Propulsion Stage

Now looking beyond the basic Shuttle and its growth kind of capabilities, into larger payloads of the late 1980's, we have made a number of studies of possible configurations for such systems. Of course what we do will be dependent upon hard requirements which are not really known that well. Figure 30 is illustrative of several studies that have been made of vehicles that could carry large payloads on the order of a quarter million pounds to low orbit. This particular configuration would use Shuttle components and basic geometry to minimize new development. In effect, a large payload would replace the crew cabin and payload bay of the Orbiter and the propulsion, electronic parts of the launch system would be recovered for reuse.

Looking further ahead, Fig. 31 is illustrative of the Single Stage to Orbit, fully reusable cargo and passenger carrier concept. Such a vehicle may become attractive for its rapid turnaround and low cost per mission. It is not however likely to provide the lowest cost per unit mass to orbit. Substantial improvements over current state of the art in structure

Unmanned Heavy-Lift
Launch Vehicle
Derivative of Space Shuttle

Fig. 30

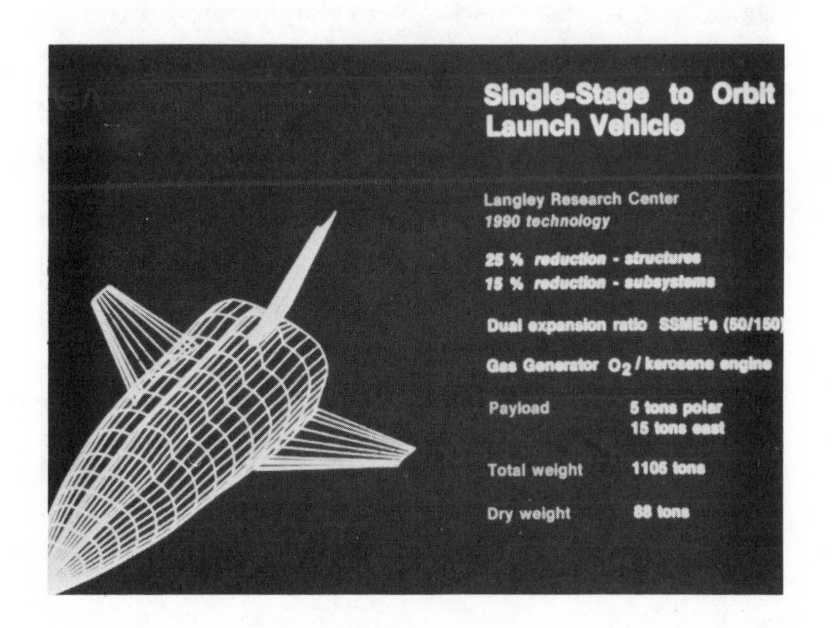
Single-Stage to Orbit
Launch Vehicle
Langley Research Center
1990 technology
25 % reduction - structures
15 % reduction - subsystems
Dual expansion ratio SSME's (50/150)
Gas Generator O2 / kerosene engine
Payload
5 tons polar
15 tons east
Total weight
1105 tons
Dry weight
88 tons

Fig. 31

and subsystems weights would be necessary to achieve the indicated performance and the concept is viewed as a research driver at present.

In the same time frame for future consideration are very large payload launches such as would be required for the solar power satellite. Figure 32 illustrates a conceptual design for such a vehicle--in this instance a two-stage fully reusable system with gross lift-off weight (GLOW) of 9,340 tons and payload of 450 tons. A vehicle such as this offers promise of payload costs to orbit of the order of $25 per pound, given sufficient traffic volume. Figure 33 shows a phasing summary of the developments in space transportation as we see them over the next decade. Specific timing of course will depend on budgetary decisions and the phasing of future needs that are beyond the present space transportation system capabilities.

In conclusion, I would like to speculate briefly about the commercial market that will exist for future space transportation services. Figure 34 projects commercial revenues from space-related operations over the next three decades.

The class of activity designated "information" dominates the near term, and is also the area certain to grow dramatically. These projections are for new information services, beyond the current space-communications business (which alone had revenues of about a billion dollars in calendar 1977).

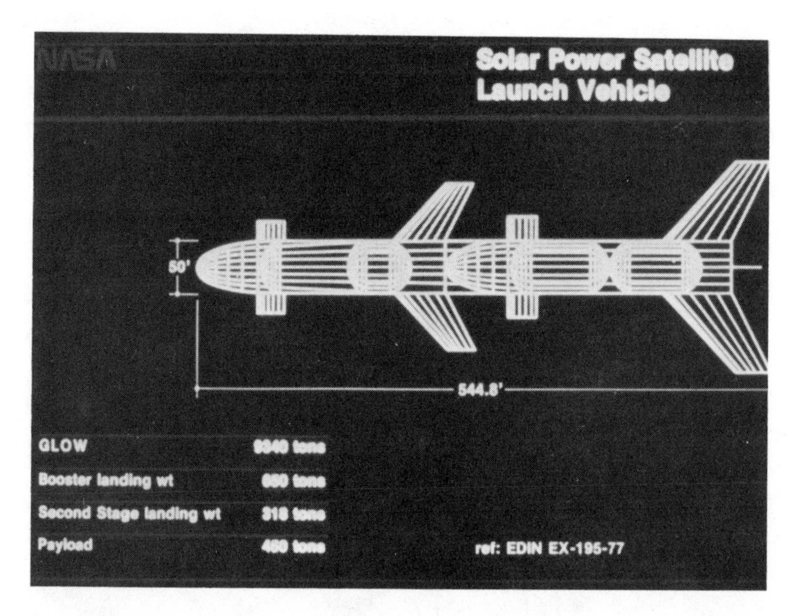

Fig. 32

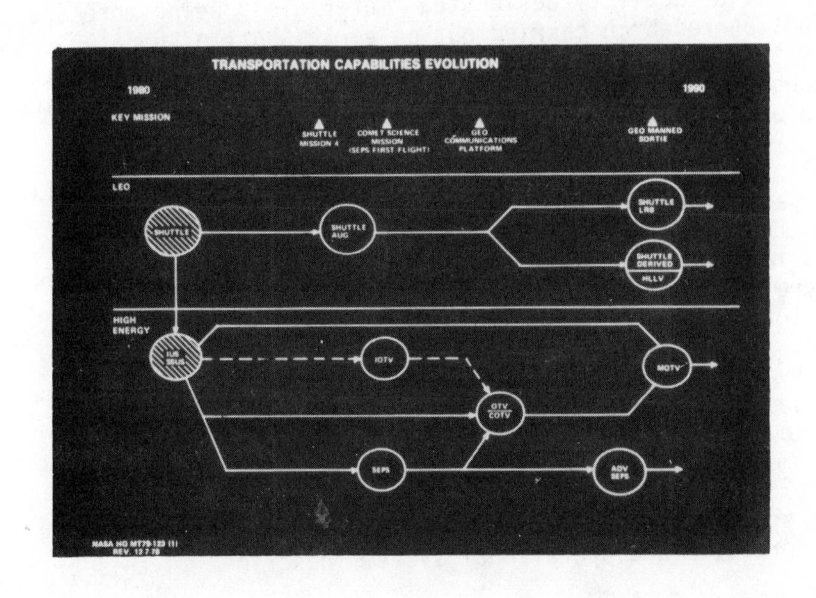
TRANSPORTATION CAPABILITIES EVOLUTION
1980
1990
KEY MISSION
SHUTTLE MISSION 4
COMET SCIENCE MISSION (SEPS FIRST FLIGHT)
GEO COMMUNICATIONS PLATFORM
GEO MANNED SORTIE
LEO
SHUTTLE
SHUTTLE AUG
SHUTTLE LRB
SHUTTLE DERIVED
HLLV
HIGH ENERGY
IOTV
MOTV
OTV
COTV
SEPS
ADV SEPS

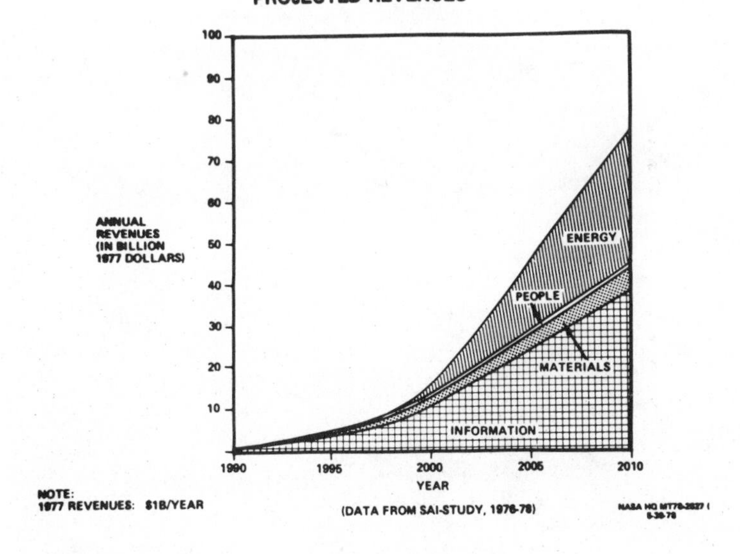
SPACE INDUSTRIALIZATION
PROJECTED REVENUES
ANNUAL REVENUES (IN BILLION 1977 DOLLARS)
100
90
80
70
60
50
40
30
20
10
ENERGY
PEOPLE
MATERIALS
INFORMATION
1990
1995
2000
2005
2010
YEAR
NOTE:
1977 REVENUES: $1B/YEAR
(DATA FROM SAI-STUDY, 1976-78)

Fig. 34

The class of activity designated "materials," being more speculative, depends on research carried out on early Shuttle/Spacelab missions. The activity designated "people" means "tourism," entertainment, and health; although not of large dollar volume, it will be a high-visibility area of space use.

Lastly "energy," the most speculative area assumes commercial feasibility of a satellite power system.

The primary significance of NASA study findings to date concerning this picture might be summarized as follows:

-Information services dominate the near-term commercial future of space.

-Even discounting "materials" and "energy" (Fig. 34) completely, revenues from commercial operations in space over the next 25 years will grow to tens of billions of dollars per year.

In short, anticipated commercial, scientific, and noncommercial uses of space spell a growing, challenging need for the wide variety of space transportation and operational systems described in this brief review.

In overall summary, the past twenty years have been exciting and rewarding, but the best in space is yet to come!

VII

TECHNOLOGICAL INNOVATION FOR SUCCESS: LIQUID HYDROGEN PROPULSION

John L. Sloop *

Liquid hydrogen has long been advocated for space flight, but its use was blocked by formidable disadvantages. A series of independent activities--low-temperature research, the atom bomb, rocket motor tests, the hydrogen bomb, ballistic missiles, and high-altitude aircraft--diminished the problems in the use of hydrogen. These, and a program's demise, led to Centaur; the same developments and the need for a large booster led to the bold decision to use liquid hydrogen in all the upper stages of Saturn vehicles for manned flight. Both were a great success.

Early in this century, Tsiolkovskiy, the Russian rocket pioneer, recognized the advantage of using liquid hydrogen and proposed its use in a space rocket but later rejected it. He was followed by many others who did the same thing, until 1958 and 1959 when liquid hydrogen was adopted for use in the Centaur and Saturn launch vehicles. Why did it take so long? To answer this, a study was undertaken for NASA and what was learned is summarized in this paper.[1]

IDEAL FUEL PROPERTIES

To understand better the attraction-rejection processes in considering liquid hydrogen, it is helpful to consider the properties of an ideal fuel--and these apply equally well to the oxidizer. The first and most important desirable fuel property is its ability, when burned with the oxidizer, to produce high rocket exhaust velocity. Tsiolkovskiy showed in 1903 that rocket flight velocity is directly proportional to the exhaust velocity. The higher the exhaust velocity the faster the vehicle velocity and when high enough, the vehicle can go into earth orbit or travel to the moon or planets. Hydrogen produces the highest exhaust velocity of all chemical fuels by a combination of high heat of reaction and low molecular mass of the gases.

The second most desirable property of an ideal fuel is high density. High density allows packing more energy into a given volume which means a lower structural mass for a given mass of fuel. Tsiolkovskiy also showed

*Author, formerly with NACA and NASA in aeronautics and space technology, is a principal of International Consultants on Energy Systems, Bethesda, Maryland.

that rocket flight velocity is proportional to the natural logarithm of ratio of full to empty mass, the latter including the payload and the structural mass. On density, liquid hydrogen is the worst of the liquid fuels, for its density is 1/14 that of water.

A fuel that has properties that make it a good coolant has an advantage. Liquid hydrogen scores well as a coolant although this was not recognized and verified until the late 1940's. A high reaction rate over a wide operating range is also desirable in converting chemical to thermal energy efficiently in a small volume. Hydrogen's high flame speed and wide flammability limits are advantages for efficient combustion.

Desirable handling and storage characteristics of a fuel require that it be non-toxic and non-corrosive and have low vapor pressure, low freezing point, high thermal stability, and high ignition temperature. In these, liquid hydrogen scores low for easy handling. Finally, an ideal fuel is available in quantity. Here liquid hydrogen scored low until there was sufficient incentive to build large liquefiers in the 1950's.

CONSIDERATION BY EARLY PIONEERS

Tsiolkovskiy proposed to use liquid hydrogen in a space rocket in 1903, five years after Dewar first liquefied it.[2] Tsiolkovskiy was aware of the high heat of reaction of hydrogen, first measured by Lavoisier and LaPlace in 1793-1794 and measured several times over by investigators during the nineteenth century. On using liquid hydrogen in 1903, he stated: "At the present time, the transfer of hydrogen and oxygen into their liquid states poses no special problem."[3] Goddard also recognized the energy potential of hydrogen. In 1909, he calculated that the energy of 45 kilograms was sufficient to carry a kilogram of payload into infinity.[4] In 1923, Oberth argued with remarkable foresight that liquid hydrogen-liquid oxygen should be used in the upper stages of space boosters.[5]

EARLY LIQUID HYDROGEN AVAILABILITY

With three rocket pioneers agreeing on the advantage of using hydrogen-oxygen, it would be reasonable to expect experimentation to follow soon but this was not the case. Tsiolkovskiy, a theoretician, soon became disenchanted with liquid hydrogen's other characteristcs. In 1911, he wrote: "The liquefaction of hydrogen is difficult (as of now) but it can be replaced with equal or even greater advantage by liquid or liquefied hydrocarbons such as ethylene, acetylene, etc."[6] Goddard, a theoretician and innovative experimentalist, wrote in 1910 about producing hydrogen and oxygen on the moon. He later took out a patent for producing hydrogen and oxygen where there was ice and snow at low temperatures. He proposed to generate the gases by electrolysis using solar energy.[7] In his experiments, however, Goddard did not use liquid hydrogen. First to use liquid oxygen in rocket engine testing and flight, he turned to gasoline as fuel rather than liquid hydrogen. The reason could have been nothing more than availability. According to his wife, Goddard had enough trouble in the early days getting liquid oxygen.[8] Oberth's contributions were theoretical but he may have operated a gaseous hydrogen burner at one time.

The early unavailability of liquid hydrogen appears to be simply a lack of sufficient demand. The basic technology of liquefaction and for storage vessels was developed by Dewar in the nineteenth century and subsequent developments have been refinements of his basic techniques. He used regenerative cooling (that is, hydrogen to cool hydrogen) and the Joule-Thomson expansion to liquefy hydrogen. His basic vacuum flask concept, the familiar dewar, for storing cryogenic fluids is in use today. Dewar was so confident about his flasks for storing and transporting cryogenic fluids that he predicted that liquid hydrogen could be stored and transported as easily as liquid air, which did not occur until the 1950's.[9] Aside from producers of liquid air, nitrogen, and oxygen, the principal users of gas liquefiers in the first part of this century were scientists engaged in low-temperature research, and their equipment was small. Until the 1920's, however, there was an unknown property of hydrogen which affects its storability, a property that would have frustrated any would-be rocket experimenter as we shall see next.

LOW-TEMPERATURE PHYSICISTS

In 1906, Nernst, a German chemist, postulated the third law of thermodynamics--that the total and free energies become equal at absolute zero. In 1908, Onnes reached 4.2°K when he first liquefied helium and by evaporating helium, scientists were soon within one degree Kelvin of absolute zero.

In the 1920's, a new era began in physics--quantum mechanics. Started by de Broglie, it was carried further by Davisson, Thomson, Schrödinger, and Heisenberg, all of whom won Nobel prizes for their contributions. Heisenberg postulated in 1926 that the hydrogen molecule existed in two forms, later called ortho and parahydrogen. In 1927, Dennison found that the two forms were not in equilibrium at low temperatures. In 1929, Bonhoeff and Harteck showed that the two forms of hydrogen had different specific heats and thermal conductivities as gases. These findings were significant for the practical use of liquid hydrogen. At room temperature, normal hydrogen is 75 percent ortho and 25 percent para. At the boiling point, 20.3°K, the equilibrium composition is 99.8 percent para. When gaseous hydrogen is first liquefied it is mostly orthohydrogen and begins to spontaneously convert into parahydrogen, releasing heat in the process. Unfortunately, the heat released is enough to vaporize all the liquid hydrogen. This means that even in a perfectly insulated vessel, the heat generated in the ortho-to-para conversion would boil away all the hydrogen. This problem can be overcome by using catalysts to convert ortho to para hydrogen during the liquefaction process, but this technology was not developed until the 1950's.

EARLY ROCKET EXPERIMENTS

The first rocket experiments with liquid hydrogen apparently occurred between 1937 and 1940 at Kummersdorf, Germany, and were not very satisfactory. Walter Thiel experimented with a number of propellant combinations, including liquid hydrogen-liquid oxygen with a small rocket. According to Wernher von Braun, who observed the tests:

"As to Thiel's liquid hydrogen tests with this set-up, I remember seeing liquefied (outside) air dripping from the super-cold liquid hydrogen line. In discussing liquid hydrogen's potential, Thiel fully endorsed Oberth's early optimism, but pointed out that tightness of plumbing connections was a critical problem and the ever-present explosion hazard caused by the accumulation of leaked-out hydrogen gas in an unvented structural pocket would require extreme care in the design of a liquid hydrogen-powered rocket or rocket stage."[10]

This work was dropped as the Germans concentrated on the development of the A-4 (V-2) rocket which used alcohol as fuel.

In 1945, both the U.S. Air Corps and the U.S. Navy initiated experiments on hydrogen for rockets. The Air Corps contract was with the Research Foundation of Ohio State University. Professor Herrick L. Johnston had built a cryogenics laboratory at Ohio State with university and government funds, the latter because of interest in deuterium for atom bomb development in the early 1940's.[11] A decade earlier, Johnston had worked with William Giauque of the University of California at Berkley when they discovered that atmospheric oxygen contains atoms of mass 17 and 18 as well as 16, a discovery that set in motion a chain of experiments leading to the discovery of heavy hydrogen by Urey in 1939.

Johnston produced his first liquid hydrogen at Ohio State in February 1943. His liquefier was modeled after the one developed by Giauque which was a refinement of the basic regenerative cooling process used by Dewar in 1898. Several heat exchangers were used to successively chill incoming gaseous hydrogen at high pressure to near the boiling point of liquid hydrogen, 20.3°K, and then expanding the very cold gas through a valve. The Joule-Thomson expansion through the valve provides the final cooling needed to liquefy part of the gas. Dewar, Giauque, and Johnston used boiling liquid air and its cold evolving gases for part of the cooling process and augmented this by cold hydrogen gas rising through heat exchangers and finally, liquid hydrogen in the last step of cooling the pre-expansion gas.

On June 13, 1947, Chief Engineer Marvin Stary and his rocket staff at Ohio State became the first in the United States to test a rocket using liquid hydrogen and liquid oxygen.[12] In the next two years, 156 runs were made, all on small rocket motors. Included, starting in August 1949, were a series of 37 runs using liquid hydrogen to regeneratively cool the rocket thrust chamber.[13]

Also part of the Ohio State investigation of liquid hydrogen were experiments on the feasibility of pumping it to high pressures using high-speed centrifugal pumps suitable for flight engines. The investigators found that it was feasible and also that precision ball bearings could be operated immersed in liquid hydrogen without lubrication. This innovation was independently rediscovered by Richard Mulready of Pratt & Whitney a decade later in developing the RL-10, the first flight model rocket engine using liquid hydrogen and liquid oxygen.

The Air Force work at Ohio State was not directed toward a specific application. The Navy, however, took a different tack; their interest in liquid hydrogen was to use its high energy for a single-stage-to-orbit booster for a satellite. The Glenn L. Martin Co. and North American Aviation were given design contracts. The Martin design went all out for a very lightweight structure with a concept to use integral liquid hydrogen tanks surrounding the thrust chamber in the aft section, much in the same way as did Tsiolkovskiy in 1903.[14]

Aerojet Engineering Corporation was given contracts for engine design and rocket engine feasibility tests on a small scale. David Young and Robert Gordon made the first recorded run of a hydrogen-oxygen rocket in the United States on October 15, 1945. The hydrogen was gaseous as liquid hydrogen was not then available on the west coast. In their gaseous hydrogen experiments, Aerojet found that the very high reaction rate of hydrogen-oxygen would allow a much smaller combustion chamber than with other fuels and they used a so-called tubular or throatless chamber. Their design concept for the flight engine of 1.3 meganewton thrust had a thrust chamber that was virtually all nozzle and looked like a giant ice cream cone 7 meters long. This concept was abandoned later and a more conventional shaped thrust chamber used. The rocket thrust chambers for the experiments varied in thrust from 45 newtons to 13 kilonewtons but the latter was still 1/100th the size of the design engine for the vehicle.

Aerojet, of course, wanted to experiment with liquid hydrogen and after some frustration in trying to get a commercial supply, convinced the Navy that it would be cost effective to build a liquefier at Aerojet. Johnston was a consultant and supplied the critical heat exchangers. The liquefier was capable of producing 30 liters per hour of liquid hydrogen and was the largest in the United States in the 1940's. It began producing in September 1948, but the first Aerojet rocket operation with liquid hydrogen did not occur until January 20, 1949. The results were disappointing. Unfortunately, less than 60 days after the first run, the Navy notified Aerojet to switch fuels from liquid hydrogen to anhydrous hydrazine but allowed experiments on liquid hydrogen to continue for three months until the end of the contract. In that period, three successful runs with liquid hydrogen were made with high performance. Two weeks before the end of the contract, an explosion occurred in the supply system--the second such mishap. Aerojet attributed the cause to the contamination of liquid hydrogen with solid oxygen. This ended Aerojet's tests with liquid hydrogen for the Navy.[15] No cooling experiments were conducted but the rapid progress made in the last three months of the contract indicated that Aerojet was on the right track to engine development. For example, one of their injector concepts was to prove successful in later work by others. Ironically, Aerojet's neighbor, the Jet Propulsion Laboratory, used Aerojet-furnished liquid hydrogen and made tests with it four months ahead of Aerojet. On September 21, Aerojet furnished JPL 75 liters of liquid hydrogen-oxygen.

Baker was appalled at how little liquid hydrogen he was actually able to use. Of the 75 liters, only 37 percent was used during the rocket operation. About 21 percent was lost cooling the transport dewar, 16 percent evaporated during the transport from Azusa to Pasadena, and 26 percent was lost cooling the propellant system. If Baker had not precooled his system with liquid nitrogen, the pre-run losses would have been greater. Baker was aware of the loss from ortho to para hydrogen conversion and suggested that liquid hydrogen be converted to para form by means of a catalyst, something that did not happen until the next decade. Baker was interested in regenerative cooling with liquid hydrogen and on April 15, 1949 became the first in the United States, if not the world, to operate a liquid hydrogen-liquid oxygen rocket chamber that was cooled entirely by liquid hydrogen.[16]

The successful runs of Ohio State University, Aerojet, and the Jet Propulsion Laboratory demonstrated high performance and regenerative cooling and that liquid hydrogen could be pumped. However, the experiments were on small-scale engines and there were no immediate applications. The Air Force and the Navy were considering more storable fuels for ballistic missiles and lost interest in liquid hydrogen. The technology was shelved insofar as propulsion was concerned, only to rise a short time later for another use.

THERMONUCLEAR RESEARCH

In January 1950, four days after Klaus Fuch told of giving U.S. atomic secrets to Russia, President Truman directed that development start on a thermonuclear weapon. This led to a significant advancement in liquid hydrogen technology in the 1950's. A key figure in this was Johnston at Ohio State. Hydrogen liquefiers were needed at Eniwetok. The Aerojet liquefier was disassembled and shipped there, and Johnston got a contract to assemble it and build a second liquefier.

The rise in interest in hydrogen led to the establishment of the National Cryogenic Engineering Laboratory of the National Bureau of Standards at Boulder, Colorado, and expansion of scientific and technological activities with regard to hydrogen.

As part of weapons development, the Air Force sponsored work by Johnston in developing mobile liquid hydrogen equipment. Both a liquefier and air transportable dewars were built.

MISSILE DEVELOPMENT

The 1950's were the golden years of long-range missile developments. Structures technology was developed that would make the use of low-density liquid hydrogen more feasible. In the United States, the Army developed the Redstone missile, patterned after the German V-2. It was built by a Redstone Arsenal team, the nucleus of which was 120 German missile experts headed by von Braun. A modified Redstone with two upper stages of solid propellants developed by JPL became Jupiter, which launched *Explorer I*, the first American satellite.

Although the Army got an early start, the Air Force became the dominant developer of the U.S. intercontinental ballistic missiles, the Atlas and Titan, using kerosene and oxygen as propellants. A technological breakthrough in weapons development made a comparatively small payload--about

700 kilograms--effective and the ICBMs were sized accordingly. The Atlas design was highly innovative with the major feature being the bold use of thin-wall, pressure-stabilized propellant tanks. This innovation was conceived, developed and proven by Karl Bossart, although others had also thought of the idea. Oberth proposed using such tanks in 1923 and both Martin and North American had proposed the same thing in their designs for the Navy using liquid hydrogen in the 1940's, as previously mentioned. No such tanks had been built and flown before Bossart, however. He overcame the skepticism about such seemingly fragile structures, but not before the Air Force initiated the Titan development with conventionally-designed propellant tanks as a back-up. An accident in 1957 proved the soundness of the concept. On the maiden flight of the Atlas, exhaust flames burned through a control wire, sending the missile tumbling while still in the atmosphere and imposing very severe loads on the tanks. They held. The use of thin-wall, pressurized tanks was a real breakthrough in achieving very lightweight structures, so essential in medium-sized vehicles for the effective use of low-density liquid hydrogen.

HYDROGEN-FUELED AIRCRAFT

During 1952-1955, four men of diverse backgrounds made independent proposals for achieving aircraft flight at very high altitudes. Two advocated the use of liquid hydrogen and by 1956, the other two also became advocates for its use.

In late 1952, an Air Force officer, Maj. John D. Seaberg, convinced his boss that the available new turbojet engines with high-altitude potential could be matched with light-wing loaded-aircraft for greater high altitude reconnaissance capability. This resulted in contract studies by the Fairchild, Bell, and Martin companies. A high-altitude B-57 was subsequently built by Martin. The Bell design, X-16, was initiated but cancelled in mid-1956.

In the midst of the studies of high-altitude aircraft, Clarence L. (Kelly) Johnson, the famed airplane designer of Lockheed Aircraft Corporation, proposed in 1954 to use the F-104 fuselage, larger wings, and a GE J-73 engine for a new high-altitude reconnaissance aircraft. Seaberg thought the J-73 was not the best engine for the purpose and seeing no advantage in developing a third airplane, recommended against it.[17] Kelly Johnson persisted and his proposal found acceptance as a result of recommendations of intelligence planners. It became the famous U-2 which has been in continuous service since 1956.

In March 1956, an inventor, Randolph Rae, presented an imaginative proposal for a new high altitude airplane to a group of Air Force experts at Wright Field. It was called "Rex I" and was a lightly-loaded, low-speed airplane for flight at 24000 meters which was well above the capability of other aircraft. Rex I was essentially a glider powered by a large propeller driven by a novel engine. The engine consisted of three turbines driven by hydrogen-rich combustion gases. The fuel was liquid hydrogen; liquid oxygen also was carried so combustion was independent of altitude. The turbines were geared down to drive the slow-speed propeller.[18]

The Rex engine was never built, but it stimulated great interest and activity in the Air Force for using liquid hydrogen for flight.

Sometime during 1954 or early 1955, Abe Silverstein, Associate Director of NACA's Lewis Flight Propulsion Laboratory in Cleveland, was struck with the idea of using liquid hydrogen for high-altitude flight. Experiments with liquid hydrogen for rockets were under way at his laboratory and he saw a way of using hydrogen's superior combustion characteristics and coping with its principal advantage, low density. At high altitudes and low speeds, large wings are needed and these allow a proportionally large fuselage. Under these flight conditions, drag is low. The large volume available in the wings and fuselage favored the use of low-density, liquid hydrogen, provided that light-weight hydrogen tanks were feasible. In March 1955, he co-authored a report with Eldon Hall on his ideas with analyses of their potential and set his laboratory humming in testing the feasibility of using liquid hydrogen.[19] This led to the first, and so far only, aircraft flight experiments with liquid hydrogen with the first flight on December 23, 1956. Three successful flights were made and in April 1956, Silverstein held a special conference to report on the findings. The 175 attendees heard 7 papers by 19 members of the project team covering hydrogen, consumption, fueling problems, airplane tankage, airplane fuel systems, and flight experiments. Subsequent flight experiments extended into 1959.

SUNTAN

The largest and most extraordinary propulsion effort in the 1950's involving liquid hydrogen was a project to use it in a special engine for high-altitude reconnaissance aircraft.[20]

Kelly Johnson had proven his ability to produce a high-altitude reconnaissance aircraft, the U-2. The Air Force had turned down the initial U-2 proposal and President Eisenhower had given the Central Intelligence Agency control of its development, with the Air Force providing operational support. The Air Force was unhappy with this arrangement, so when Kelly Johnson proposed an advanced reconnaissance aircraft to the Air Force in 1956, it was accepted fast. The U-2 had just begun flights and was, of course, a closely guarded secret. The Air Force went to great lengths to keep their new development super-secret; initially it was said that only about 25 people in the government were fully aware of what was going on. The Air Force team was headed by Col. Norman Appold and included Lt. Col. John Seaberg, Major Alfred Gardner, and Captain Jay Brill. Normal procurement regulations were waived and procurements were buried in other projects to preserve security. Kelly Johnson's "Skunk Works" at Lockheed was kept off limits to all except those working exclusively on the aircraft which was designated the CL-400.

Pratt & Whitney was selected to build the engine, with Perry Pratt, Richard Coar, and Richard Mulready as the principal engineers. They initially operated a standard J-57 turbojet engine with hydrogen, found it worked well, but did not utilize the full potential of hydrogen. A new design, designated 304, featured a large heat exchanger between the turbine and the afterburner. Beginning in September 1957, tests were conducted for a year at P&WA's new center west of West Palm Beach.

Suntan needed a large supply of liquid hydrogen and the responsibility for this was assigned to Captain Jay Brill. One plant was built by Air Products in Cleveland to serve Pesco Products, which was developing a hydrogen transfer pump, and to serve P&WA's initial needs in East Hartford. It was named "Baby Bear" and became operational in May 1957 with a capacity of 680 kilograms per day. The second plant was built by Air Products adjacent to P&WA's Florida test center. Named "Mama Bear", the 4500 kilogram-per-day plant went into operation in the fall of 1957. Even before it started operation, the Air Force contracted for a much larger plant, "Papa Bear", adjacent to "Mama Bear". The third plant had a capacity of 27,200 kilograms per day--the world's largest at the time. Kelly Johnson also needed liquid hydrogen for experiments at his Burbank plant and initially obtained his liquid hydrogen from the National Cryogenic Laboratory in Boulder. The Air Force contracted with Stearns-Roger for a liquefier at Bakersfield to support the Lockheed work and it was placed in operation in the fall of 1957.

Brill also was responsible for developing suitable liquid hydrogen transports. He initially scrounged excess Air Force equipment from the earlier thermonuclear tests. The Cambridge Corporation was given a contract for over-the-highway trailers and permission was obtained from the Interstate Commerce Commission to transport liquid hydrogen, code named SF-1 fuel, with the trucks labeled simply "flammable liquid". The first liquid hydrogen semi-trailers had a single axle because of the light weight of low-density hydrogen, but this produced so many problems from perplexed officials at truck weighing stations that it was replaced by a design with two axles.

The Suntan effort showed that liquid hydrogen could be handled safely, and although problems were encountered in engine development, they were not related to safety. By 1957, however, some of its proponents were having second thoughts and there were others who believed there were better ways to accomplish reconnaissance. Opposition to Suntan effectively doomed the project and like a summer tan, it just faded slowly away. It lingered through 1958 and was not cancelled until its management team, tired of waiting, requested cancellation in 1959. Surprisingly, one of the main opponents was the man who conceived and sold the project to the Air Force, Kelly Johnson. The main defendants were members of the management team, particularly Appold and Seaberg.

The Air Force had insisted on a minimum radius of 2800 kilometers and was convinced that this was feasible. On the other hand, Johnson was convinced that 2000 kilometers was about the best that could be achieved. He saw little growth potential in his aircraft design and Perry Pratt saw little for the 304 engine.

Suntan significantly added to the technology of liquid hydrogen but its most important contribution was four liquid hydrogen plants and a fleet of liquid-hydrogen transports. This ample supply of liquid hydrogen and the incentive of the Air Force management team to get maximum benefits from the money spent proved a great benefit for the space program.

ROCKET ENGINE EXPERIMENTS

Concurrent with the build-up of interest in liquid hydrogen for weapons and for aircraft in the 1950's, a renewal of interest in liquid hydrogen for rockets began at the NACA Lewis Flight Propulsion Laboratory in 1950. A group of rocket experts met there in May 1950 to discuss fuels for long-range missiles. The author, who headed the Lewis rocket research group, recommended liquid hydrogen as the primary fuel. If its density proved too great an obstacle, hydrazine, ammonia, and a mixture of the two were proposed as alternate fuels. Liquid fluorine was proposed as the oxidizer with liquid oxygen as an alternate.

Following the meeting, the Lewis rocket group sought a source of liquid hydrogen but encountered the same supply problem Aerojet faced in the second half of the 1940's. The only reasonable course was to build a small liquefier, and one was authorized for fiscal year 1952. Arthur D. Little contracted to supply the liquefier which became operational in 1954. On November 23, 1954, the first successful run was made and was followed by a series of runs with both oxygen and flucrine as oxidizers, using liquid hydrogen as a regenerative coolant. By then additional supplies of liquid hydrogen came from Johnston's mobile liquefier and the Air Force liquefier in Painesville. Two sizes of thrust chambers were used, 22 and 89 kilonewtons, and engine construction was of a lightweight flight type. The injectors were primarily a finely divided showerhead type and high performance was obtained. The Lewis research made two significent contributions. First, it strongly reinforced the views of Abe Silverstein who later played a key decision-making role in the space program and second, it was timely in influencing engineers from major propulsion manufacturers who visited the laboratory.

CENTAUR

The first rocket stage to fly using liquid hydrogen-oxygen was the Centaur stage on top of the Atlas intercontinental missile. Centaur was the brainchild of Krafft Ehricke. Within a month of Sputnik, he proposed a hydrogen-oxygen stage for the Atlas. He was able to move fast because of previous work on the Atlas and planning upper stages using hydrogen-oxygen.

Ehricke's proposal in December 1957 was for a four-engine, pressure-fed, hydrogen-oxygen stage with each engine developing 31-33 kilonewtons. The engine was proposed in cooperation with Rocketdyne. He was not aware of the work of P&WA in liquid hydrogen for Suntan. His proposal rested unanswered until the following August.

In July 1958, President Eisenhower signed the act creating NASA, which would begin official operation October 1, 1958, absorbing the NACA, its laboratories and personnel. T. Keith Glennan was named Administrator, Hugh Dryden, former Director of Research for NACA, was Deputy Administrator, and Silverstein became the third ranking official as Director of Space Flight Programs.

Prior to NASA operation, the Advanced Research Project Agency (ARPA) had taken the lead in space projects, both military and civilian, but the latter projects were to pass to NASA. In July 1958, Silverstein appointed an interagency committee, with himself as chairman, to coordinate plans for propulsion and launch vehicles. There was some competition between ARPA and the emerging NASA. In August 1958, the ARPA representative on the Silverstein committee, C. S. Cesaro, reportedly slipped out of a meeting of the committee to urge fast action on the Centaur proposal.[21] Whether for this reason or others, ARPA issued a directive on August 29, 1958 directing the Air Forces' ARDC to initiate a high-energy fuel stage for a modified Atlas. The propellants were to be liquid hydrogen and liquid oxygen. The propulsion system could be either pressure-or pump-fed, with a total thrust of 133 kilonewtons. The sole source for the engine development was Pratt & Whitney, and the sole source for the stage was the Convair-Astronautics Division of General Dynamics, where Ehricke was employed. The ARDC named its special projects office, then headed by Seaberg, to be the project manager. Gardner and Brill were still in the office along with another officer, Major Alfred J. Diehl. Thus, the former Suntan management team, with the exception of Appold, became the Centaur management team. Pratt & Whitney proceeded to use its fine liquid-hydrogen facilities built for Suntan, for the development of the first flight engine using liquid hydrogen-liquid oxygen, the RL-10.

In July 1959, Centaur was transferred from ARPA. Seaberg remained the project director and Milton Rosen became the NASA program manager. About the same time, the RL-10 engine ran for the first time. The Centaur stage was developed in the 1960's and has been a great success in the planetary exploratory missions and extending far beyond its predicted period of usefulness.

SATURN V

In the early days of the space era, it became obvious from comparing payload weights that the Russians were using a much more powerful launch vehicle. This convinced many people that the U.S. was far behind in vehicle development. What was overlooked, however, was the U.S. Intercontinental Ballistic Missiles (ICBMs) had been made smaller as a result of a breakthrough in weapons technology which resulted in a lighter payload. The public outcry to "catch up" and surpass the Russians in booster capability helped the NASA in its initial plans to build a large vehicle for manned flight. NASA was helped also by prior work by both the Air Force and the Army.

In 1955 the Air Force had contracted with Rocketdyne to build a single engine with two or more times the thrust of the largest engine in the ICBMs. The same year, Rocketdyne announced that a single engine of 4.5 meganewtons, over six times the thrust of the ICBM engines, was feasible. In 1958 the Air Force awarded a contract to Rocketdyne to design a 4.5 meganewton engine designated the F-1. This was to be transferred to NASA who elected to allow it to end in the fall of 1958 and opened competition on a 6.7 meganewton engine. Rocketdyne won and development of the F-1 engine at the higher thrust level began in January 1959.

The Army Ballistic Missile Agency (ABMA) had taken the lead in planning large vehicles, beginning in 1956. In August 1958, less than two months before NASA began official operation, ARPA issued a directive to ABMA to develop a large launch vehicle, originally called Juno V with the name changed to Saturn I in 1959. The NASA saw a need for a large launch vehicle, called Nova, which later evolved into Saturn V. The Air Force was planning a hypersonic boost glide airplane for suborbital flight, called Dynasoar, and proposed a large vehicle, Titan C, to launch it. Thus, in late 1958 and in 1959, there was competition to build three new large launch vehicles.

Top defense officials, especially Herbert York, Director of Defense Research and Engineering, could see little military value in large launch vehicles and favored the transfer of the Army's ABMA Saturn team to NASA.[22] This was favored also by President Eisenhower, but the Army opposition was so strong that a compromise was reached in December 1958 leaving the ABMA team with the Army. York, however, was not satisfied and kept pressing the arguments of little military justification and budgetary problems. In August 1959, he decided to cancel the Saturn program, and this forced the issue on large vehicle development. He formed a committee with NASA in September and all members agreed that it made sense to develop only one type of large vehicle. Saturn I emerged the winner, but not the Army. By October, agreement was reached to transfer the ABMA team and Saturn development from the Army to NASA, which President Eisenhower approved on October 19 to become effective the following March 15.

Remaining unsettled was upper stage configurations for Saturn I and successive, larger Saturns. Silverstein formed a committee to consider this, with von Braun, Appold, Abraham Hyatt, Thomas Muse and George Sutton as members and his vehicle analyst, Eldon Hall, as secretary.

Upper stage configurations for Saturn and Nova were not a new subject. Studies early in 1959 considered solids, storable propellants, kerosene-oxygen, and hydrogen-oxygen for second and third stages. The von Braun team initially selected Titan as the second stage of Saturn I, but considered Centaur, using hydrogen-oxygen, for the third stage. The much larger Nova was to use kerosene-oxygen in the second stage, hydrogen-oxygen in the third and fourth stages, and a storable fuel in its fifth stage. By the time the committee met, Hall and his small group at Headquarters had made a number of analyses of upper stages for a series of Saturns and so had the von Braun team at Huntsville. Silverstein and Hall favored the use of hydrogen-oxygen in all upper stages of Saturn, whereas von Braun's team favored kerosene-oxygen in the second stage, arguing that hydrogen-oxygen was unproven and that it would take several years to develop a reliable engine, whereas kerosene-oxygen engines were already developed. Silverstein, however, was able to convince von Braun and his team of the large performance advantage in using hydrogen-oxygen in all upper stages of Saturn and the committee so recommended.[23] It was a very bold and crucial decision to stake the success of the entire manned space program on a relatively new high-energy fuel, but subsequent developments proved it to be a sound decision and a key one in the success of the Saturn V and the Apollo missions.

SIGNIFICANCE

Recognition of liquid hydrogen's high energy potential as a propulsion fuel came early but in many considerations the disadvantages of low density, low temperature, low availability, and high hazards eventually led to its abandonment. This occurred over and over again. A series of independent activities, however, gradually increased the technology of liquefying hydrogen and storing, transporting, and handling it. This attraction-rejection process might have continued indefinitely except for a fortunate set of circumstances. The United States wanted a much larger launch vehicle for space missions than was available from missile developments. A proponent for liquid hydrogen who also had experimental experience with it played a key role in decisions about the upper stages of the large launch vehicle, Saturn. He was able to persuade his colleagues to accept the use of liquid hydrogen for these stages. It was a bold and crucial decision for the large vehicle was intended for manned flight. A failure of the hydrogen stages would have been disastrous for the U.S. space program. A more conservative approach would have been to use kerosene fuel as it had been tried and proven in the ballistic missile program. Using liquid hydrogen instead of kerosene, however, meant a larger payload capability for a given size vehicle. Balanced against this was the concern of many engineers that using liquid hydrogen would extend the development time; they favored an evolutionary process where high-energy fuels like hydrogen would be used for second generation launch vehicles. Making the bold step directly to hydrogen, however, turned out to be less difficult than many anticipated and the stages using liquid hydrogen have been very successful.

Today, many people are convinced that liquid hydrogen has an even greater role in earth applications than in space. Studies have been made on its feasibility for transport aircraft; probably first with cargo and after operational experience, with passengers. Opponents keep arguing about the great hazards of using hydrogen while its proponents argue that in many ways it is safer than gasoline.

Liquid hydrogen has been proposed also as an automobile fuel which would indeed be a bold step forward and require education of the public in its safe handling and use. In these applications of flight and ground transportation, hydrogen has a strong appeal for it can be made in a number of different ways including electrolysis of water and biological processes. It burns cleanly and produces water vapor as exhaust.

Storage of energy in the form of liquid hydrogen may be feasible in connection with large electric generating systems such as using ocean thermal energy, remotely located nuclear plants, and hydroelectric plants. This development of hydrogen technology is still in progress.

If liquid hydrogen turns out to be the fuel of the future, the technological developments of the space program, beginning with the crucial decision in 1959 to use it in vehicles for manned flight will have been a major contribution and milestone. As the technology progresses, which individual--like Silverstein in 1959--will be in the right position at the right time and have the courage of his convictions to make the decision for the next giant step forward in the use of liquid hydrogen? That will be a major story for some future historian to tell.

TECHNOLOGICAL INNOVATION FOR SUCCESS:
LIQUID HYDROGEN PROPULSION

REFERENCE NOTES

1. This paper is based mainly on material in: John L. Sloop, *Liquid Hydrogen as a Propulsion Fuel, 1945-1959,* SP-4404 (Washington: NASA, 1978).

2. K. E. Tsiolkovskiy, *Collective Works of K. E. Tsiolkovskiy,* ed., A. A. Blagonrovov, vol. 2, *Reactive Flying Machines,* NASA Technical Translation F-237 (Washington, 1965) pp. 78-79.

3. K. E. Tsiolkovskiy, *Works on Rocket Technology,* ed., M. K. Tikhonravov, NASA Technical Translation F-243 (Washington, 1965) p. 35.

4. Robert H. Goddard, *The Papers of Robert H. Goddard,* 3 vols., ed., Esther Goddard and G. Edward Pendray (New York: McGraw-Hill, 1970).

5. Hermann Oberth, *Rockets in Planetary Space,* NASA trans. TTF-9227, 1965; Oberth, *Ways to Spaceflight,* NASA trans. TTF-622, 1972.

6. K. E. Tsiolkovskiy, *Reactive Flying Machines,* NASA trans. F-237, p. 122.

7. Goddard, *Papers.*

8. Esther Goddard in conversation with the author, March 6, 1974, Chevy Chase, Maryland.

9. James Dewar, *Collected Papers of Sir James Dewar,* ed., Lady Dewar (Cambridge University Press, 1927).

10. Wernher von Braun in letter to the author, November 8, 1973.

11. Herrick L. Johnston, "The Cryogenic Laboratory," *Engineering Experiment News,* 18 (June 1946) pp. 3-21.

12. Herrick L. Johnston, "Rocket Motor Development--First Summary Report," Columbus: OSU Research Foundation, April 1949.

13. Johnston & William L. Doyle, "Final Report-Development of the Liquid Hydrogen-Liquid Oxygen Propellant Combination for Rocket Motors," TR-7, Columbus, OSU Research Foundation, December 1951.

14. Pedro C. Medina, "HATV--Summary Report," Engineering Report 2666, Glenn L. Martin Co., Baltimore, Maryland, June 1947.

15. David A. Young, "Research and Development of Hydrogen-Oxygen Rocket Engine Model XLR 16-AJ-2, Final Report," R-397, Aerojet Engineering Corp., Azusa, Calif., September 28, 1949.

16. Dwight L. Baker, "Regenerative Cooling Tests of Rocket Motors Using Liquid Hydrogen and Liquid Oxygen," Report 4-53, Jet Propulsion Laboratory, Pasadena, Calif., August 11, 1949.

17. Based on interviews and documents furnished to the author by John D. Seaberg, 1976.

18. R. S. Rae, "The REX-1, A New Aircraft System," Summers Gyroscope Co., Santa Monica, Calif., February 1954; interviews with Rae, 1974.

19. Abe Silverstein and Eldon W. Hall, "Liquid Hydrogen as a Jet Fuel for High Altitude Aircraft," RM E55C28a (Washington: NACA, April 15, 1955); interview with Silverstein, 1974.

20. From interviews with the principals, 1973-1974.

21. From interviews with Silverstein, A. O. Tischler, Richard Canright, and Cesaro, 1971-1974. Cesaro neither confirmed nor denied the account.

22. Herbert F. York, San Diego, Calif. to Eugene M. Emme, NASA, June 10, 1974.

23. "Report to the Administrator, NASA, on Saturn Development Plan by Saturn Vehicle Team," December 15, 1959. Reprinted without appendices as document IV-7 of "Documents in History of NASA," NASA History Office, Washington.

VIII

LESSONS OF APOLLO FOR LARGE-SCALE TECHNOLOGY

Robert C. Seamans, Jr. *
and
Frederick I. Ordway, III +

The Apollo project is generally considered as one of the greatest technological endeavors in the history of mankind. The managerial effort to achieve the lunar landings was no less prodigious than the technological one, and is here described. Many aspects to the successful management had to be learned during the project; and, more important still, they were to be changed as experience accumulated.

In this review, an evaluation is offered of the required political support, funding, manpower, and industrial team management; the visibility of the project; and the basic question of reliability of the millions of components. This article is condensed from one originally appearing in Interdisciplinary Science Reviews, London, with permission.

It was just three years ago that Dr. Anthony R. Michaelis, editor of Interdisciplinary Science Reviews in London, asked us to prepare an article on Apollo in terms of its being an object lesson for the management of large-scale technological endeavors. We reacted, and the resulting article appeared in late 1977 (Volume 2 No. 4).

We commenced our review by noting that the undertaking of the manned lunar exploration[1] involved one of the major technological endeavors in the history of civilization. In terms of financial, manpower and other resources applied and the time-frame allowed for its completion, Apollo had no parallel as far as single, goal-oriented efforts are concerned. It was unique.

---

* *Dr. Robert C. Seamans, Jr. is Dean, School of Engineering, Massachusetts Institute of Technology, Cambridge, Massachusetts.*

+ *Frederick I. Ordway is energy policy advisor, Office of Policy Coordination, Assistant Secretary for Policy and Evaluation, Department of Energy, Washington, D.C.*

At the time of preparation of the original, Dr. Seamans was Administrator of ERDA and Mr. Ordway a special assistant on his staff.

As the Apollo program was getting established back in the early 1960s, it was often said that the major breakthrough required to make the lunar mission a success was managerial rather than scientific and technological. Later, when six Apollo landing expeditions had been completed, this belief received widespread support. Indeed, the effort has been widely proclaimed as much a management triumph as an engineering and scientific one.

That a major legacy of Apollo was managerial seemed to us unquestioned. The sheer complexity of the program, the resources devoted to it, its pioneering nature, and the limited time available for its completion made optimum progress--and consequently optimum management--absolutely essential.

Apollo, like most large-scale enterprises, encompassed a broad spectrum of scientific, technological, social, and political forces and activities that had to be integrated and focused towards a carefully articulated and well supported goal. That it was reached within budgetary and time constraints in a manner satisfactory to planners and supporters alike attested to the program's success.

As impressive as the development and implementation of Apollo management philosophy and practices were, NASA nevertheless owed much to the experience of predecessor and contemporary organizations. Building upon the tried and the new alike, the space agency continually adjusted its managerial and organizational approaches as it grappled with an unprecedented technological challenge.

Large-scale technological endeavors[2] in general can profit from the NASA experience, just as NASA profited from that of the Department of Defense and other organizations. This is not to suggest, however, that the Apollo management system can be transferred wholesale from the environment of the 1960s to quite different settings today. The conditions are not the same, the players are not the same, the issues are not the same, and the goals and aspirations of the nation are not the same.

Despite differences inherent in each and every large-scale endeavor, all to a greater or lesser degree share some basic characteristics. To the extent that they do, management and other experience can be successfully transferred from completed enterprises to newer ones still in progress or being planned. The principal shared characteristics are grouped in Table 1.

Having examined some of the characteristics shared by large-scale endeavors, we narrowed our focus and looked at the NASA experience, paying as we did so, particular attention to the Apollo effort. We then selected areas that we felt were relevant to most, if not all, large-scale enterprises and then showed how they helped shape Apollo's success.

Table 1. Shared Characteristics of Large-scale Endeavors

Interdisciplinary Character

Array of scientific, technological, social, political and other personalities and resources brought into play, requiring regular crossing of disciplinary lines and mixing of individuals who would normally have no professional contact.

'State-of-the-art'

The basic science and technology involved are generally quite well developed, and thus do not have to be pushed excessively to justify program. But, appropriate elements of existing science and technology must be brought together, uncertainties impeding implementation removed, upgrading and focusing carried out, and results integrated into goal-oriented program that meets desired social, political, economic, military or other need.[3]

Selling

Promoters must demonstrate to potential sponsors how proposed endeavor can fulfill whatever need, ambition, or opportunity that may be addressed. Clear grasp of program content, time needed for accomplishment, and funding required is essential.

Funding

Adequate funding absolute requirement; moreover, it must be assured not only in crucial start-up phase, but over the lifetime of enterprise. Many aspiring programs commence life with adequate financial resources only to find them withdrawn when support vanishes.

Support

Having been 'bought' and provided with the needed resources, large-scale endeavor must be assured of *continuous* support. If troubled by constant exposure and repeated rejustifications, it may lose vital momentum and sense of urgency. No other shared characteristic of large-scale endeavors is so important not only at outset of activities, but all way through to successful completion.

Manpower

Large numbers of exceptionally well-qualified and highly motivated individuals are required to provide reserves in the event that unexpected problems or setbacks occur.

Planning and Analysis

Ordered, structured, precisely preplanned solutions to problems can rarely be counted upon in the large-scale enterprise, which may be very long in developing and which may involve many areas where knowledge is still hazy. Constraints and forces change continuously, new facts and discoveries must be accommodated, and the balance of organizational power may undergo significant shifts from time initial goal is articulated to time it is finally reached. Constant reassessment and redirection, therefore, are normal to planning and analysis process.

Communicating Information

Not only must vast quantities of data be made available upon demand to decision-makers at all levels of management, but widely-scattered field and contractor networks must be knit together through communications if they are to function effectively as an integrated force.

Visibility

To gain and maintain support, large-scale enterprise must be kept 'visible'[4] and hence subject to almost endless public scrutiny and legislative review. Visibility has advantages in that it tends to make organizational elements and individuals react to problems they might otherwise be tempted to ignore. This is particularly true when sheer organizational size inhibits flow of information and where problems may become magnified as their effects ripple throughout the endeavor. By maintaining visibility, strong motivations are created to solve problems rather than avoid them.

Decentralization

Because of complexity of large-scale enterprise, management must contemplate a significant degree of decentralization in decision making and in project execution. No single individual or even small group of individuals can possibly cope with all relevant facts and issues.

Flexibility

Management must continually adapt and reorganize in the face of changing pressures, priorities, restraints, policies, developments, budgets, manpower and general support. This helps insure ability of enterprise to recover if things go wrong, if support suddenly evaporates in one sector or another, or if major contractor fails to deliver on time or within established specifications.

Control and Integration

Despite need for flexibility and some degree of organizational autonomy, diverse elements that make up large-scale endeavor must be carefully controlled and integrated.[a]

Risk and Uncertainty

Some failures are to be expected along the complex research, development, and demonstration route towards a workable, reliable, on-line technological system.

[a] As Sayles and Chandler pointed out in their study of large-scale enterprises, '. . . a wide array of intellectual and economic commitments must be simultaneously focused on a very explicit task without destroying the motivations that release energy and commitment.' They also stress that such enterprises '. . . come more and more to resemble large-scale business systems. They require not less but more human ingenuity, improvization and negotiation than old style business and government organizations . . . human intervention, confrontation and compromise are indispensible to their governance.'[5]

When we prepared the article for Interdisciplinary Science Reviews several years ago, we thought it would be interesting to compare the legislation that gave rise to NASA with that of the then two-year old Energy Research and Development Administratio We pointed out that NASA and ERDA were the largest non-military government-run research and development organizations ever esta lished in the United States and hence invited comparisons. Sin the publication of the article, ERDA and its R&D missions were absorbed into the new Department of Energy.

Despite the demise of ERDA, the comparison between its legis lation (many facets of which, incidentally, are still in force within the Department of Energy) and that of NASA is quite instructive.

With these introductory thoughts, we now proceed with the ma points of the 1977 article, describing in turn the legislative comparison, the creation of the goal, funding, manpower, planni and analysis, the management process, the industrial team, managing change, visibility of the program, reliability, and lesso from Apollo.

LARGE-SCALE ENDEAVORS

Both agencies were established as a result of foreign events: the October 1957 orbiting of Sputnik 1 by the Russians in the case of NASA and the Arab-imposed winter 1973–1974 oil embargo in the case of ERDA. Partly because of perceived threats from abroad, both commenced operations within an environment of strong presidential, congressional and public support. As a direct corollary to these threats, whose seriousness has never been doubted, both space and energy legislation drew attention to the security of the Nation as well as to the general welfare of the American people. In the light of this background, it is not surprising that the framers of the National Aeronautics and Space Act of 1958 would declare that:

> It is the policy of the United States that activities in space should be devoted to peaceful purposes for the benefit of all mankind [and that] the general welfare and security of the United States require that adequate provision be made for aeronautical and space activities.[6]

In the same vein, 16 years later in 1974 it w observed that 'the general welfare and the comm defense and security' all require that action be tak across a broad energy front:

> To develop, and increase the efficiency and reliab ity of use of all energy sources to meet the needs present and future generations, to increase t productivity of the national economy a strengthen its position in regard to internation trade, to make the Nation self-sufficient in energ to advance the goals of restoring, protecting a enhancing environmental quality and to assu public health and safety.[7]

A couple of months later in related legislation, th Congress referred to 'the urgency of the Nation energy challenge', one that 'will require commi ments similar to those undertaken in the Manhatta and Apollo projects ....'[8] Interestingly, it was est mated that the 'total Federal investment ... ma reach or exceed $20 thousand million over the ne decade', which was about what Apollo cost during th 1960 decade.[9]

Although NASA and ERDA were new organiz

In April 1957, the Army Ballistic Missile Agency in Huntsville, Alabama commenced studies of a clustered engine rocket booster developing 1.5 million pounds of thrust. Based on Redstone and Jupiter military missile technology, the giant rocket would be capable of orbiting large military payloads into space. A large booster program first designated Juno and later Saturn was formally approved by the Department of Defense in August 1958. In October of the following year, President Dwight D. Eisenhower ordered the transfer of the Saturn development team under Wernher von Braun to the newly-created NASA. On September 8, 1960, a couple of months after the transfer was carried out, Eisenhower formally dedicated the George C. Marshall Space Flight Center, new home of the von Braun team. Here, von Braun briefs the President and NASA's first administrator, T. Keith Glennan, on Saturn's potential for civilian space activities.

Although he somewhat reluctantly got America started on the road to space, as Glennan would later recall, Eisenhower was no "space cadet." The Kennedy Administration contrasted sharply with that of its predecessor. Two years after the Eisenhower visit to Huntsville, President John F. Kennedy and Vice President Lyndon B. Johnson travelled there (September 11, 1962) to review progress on the Saturn program. With them were key personalities in the evolving effort to land men on the Moon before the end of the 1960 decade. Left to right, President Kennedy; Wernher von Braun; James E. Webb; Vice President Johnson; Secretary of Defense Robert S. McNamara; Jerome B. Wiesner, Kennedy's special assistant for science and technology; and Harold Brown, DOD director of research and engineering.

Immediately after Kennedy's whirlwind September 1962 visits to NASA installations at Cape Canaveral and Huntsville, the President toured the Manned Spacecraft Center in Houston. During the course of his Texas stopover, he talked to 35,000 people at the Rice University stadium. "In the last 24 hours," he reported, "we have seen facilities now being created for the greatest and most complex exploration in man's history. We have felt the ground shake and the air shattered by the testing of a Saturn C-1 [later called Saturn 1] booster rocket...We have seen the site where five F-1 rocket engines...will be clustered to make the advanced Saturn missile [later designated Saturn 5], assembled in a new building to be built at Cape Canaveral as tall as a 48-story structure, as wide as a city block, and as long as two lengths of this field." (NASA Photo No. 62-ADM-14)

tions with wide-ranging mandates, each was created from a major core module to which other elements were added. The space agency started operations on 1 October 1958, by receiving *en masse* the personnel and facilities of the former National Advisory Committee for Aeronautics. But this was only the beginning. In accordance with Section 302 of the cited Space Act of 1958, the President enjoyed the authority 'for a period of four years after the date of enactment of this Act, [to] transfer to [NASA] any functions . . . of any other department or agency of the United States . . . which relate primarily to the functions, powers, and duties of [NASA].' This is striking evidence of the accommodating mood of Congress in the Sputnik-dominated environment of 1958.

Exercising the freedom given him, President Eisenhower's Executive Order 10783, dated 1 October 1958, brought the Navy's Vanguard rocket/satellite team and some personnel from other parts of the Department of Defense to NASA. Shortly afterwards, Executive Order 10793, of 3 December, transferred to NASA the Army's Jet Propulsion Laboratory in Pasadena, California. Then, on 14 January 1960, a so-called 'transfer plan' was delivered to the Congress which called for NASA's absorbing the Development Operations Division of the Army Ballistic Missile Agency at Redstone Arsenal in Alabama. When it became effective two months later, organizationally NASA's growth was complete.[10] No other new agency of the Executive Branch of the U.S. Government had been created by the transfer of as many units and programs of other departments or agencies as had NASA.

Legislation establishing ERDA was somewhat more rigid with respect to transfers in the sense that the Reorganization Act of 1974 spelled out precisely what would be turned over to the new energy agency – and when. Perhaps the Arab oil embargo did not seem quite as frightening to Congress as the orbiting of the early Sputniks had a decade and half before.

Except for regulatory functions,[11] ERDA received at the time of its creation all the personnel and facilities of the former Atomic Energy Commission. At the same time, the Department of Interior's Office of Coal Research transferred to ERDA along with fossil fuel energy research and development programs handled by the Bureau of Mines; the solar heating and cooling and geothermal power activities undertaken by the National Science Foundation; and alternative automotive power system efforts under way at the Environmental Protection Agency.

Just as NASA was given greater discretionary authority with respect to the elements that it could – and subsequently did – absorb, so the space agency enjoyed a freer hand than ERDA in organizing them into a new administrative entity. Whereas both the space and energy reorganization acts provided for presidentially-nominated, Senate-confirmed administrators and deputy administrators, NASA was permitted to set up an internal organization along the lines it deemed most appropriate to its missions. Specifically, the Space Act stated that 'Under the supervision and direction of the President, the Administrator shall be responsible for the exercise of all powers and the discharge of all duties of [NASA] and shall have authority and control over all personnel and activities thereof.'[12]

The Congress was in a less liberal mood when it wrote the legislation that established ERDA, stating that 'there shall be in [ERDA] six Assistant Administrators'[13] representing the major energy disciplines or areas. Moreover, a Director of Military Application was also specified as well as a person responsible for international cooperation.

How much importance should be attached to flexibility and openness in the large-scale research and development organizations charged with novel responsibilities is difficult to assess, but, in view of the success of Apollo and other NASA programs, these were likely to have been contributory factors.

CREATING THE GOAL

The first couple of years of NASA passed in a somewhat uncertain manner, for no clear-cut space goal had been established for the United States other than somehow to 'catch up with the Russians' and secure American competency in a brand new field. Then, during the opening months of 1961, an analysis and evaluation of a possible national space goal was carried out under the supervision of the White House. As the study was nearing completion in the spring, the American people were stunned to learn that the Russians had scored another important milestone in space exploration: on 12 April 1961, Yuri Gagarin became the first man to orbit the Earth. Vostok 1's success reverberated around the world as quickly as had Sputnik's three and a half years earlier.

Just as a foreign event, Sputnik, had shaped America's initial response to the advent of the Space Age, so another, Vostok, helped set the stage for President John F. Kennedy's historic address to Con-

In 1961, man on the Moon represented an ambitious national objective. Eleven years later, it had been fulfilled and, surprisingly to many, came to a complete halt. In this next-to last Apollo flight *(Apollo 16)* we see in the foreground in the shadow of the lunar module the ultraviolet camera-spectrograph, part of the array of instrumentation effected on the surface. Space-suited Astronaut John W. Young stands in the center and to the rear left is the lunar roving vehicle. (NASA Photo No. 72-H-685)

gress on 25 May 1961. An exciting new national goal was to be announced. 'Now is the time to take longer strides,' the President said,

> 'and the time for a great new American enterprise – time for this Nation to take a clearly leading role in space achievement which in many ways may hold the key to our future on Earth . . . Space is open to us now; and our eagerness to share its meaning is not governed by the efforts of others. We go into space because whatever mankind must undertake, free men must fully share . . .
>
> I believe that this Nation should commit itself to achieving the goal before this decade is out, of landing a man on the Moon and returning him safely to Earth.'

With these ringing words, the lunar exploration enterprise – a large-scale endeavor *par excellence* – got under way.

SUPPORT

Certainly a major factor in Apollo's success was the generally strong support enjoyed from the beginning of the program right through to completion. More than any other effort in US peacetime history, the lunar landing illustrated the value of such support – which could accurately be described as a mandate – and the degree it could be generated when public interest was aroused.

In the wake of the early Sputniks, but before NASA actually came into existence – and a full three years prior to President Kennedy's announcement of the national goal to land a man on the Moon – space flight was being presented as an imperative to which our society ought to respond with a strong sense of urgency. Let us give an example of how this feeling was expressed at the time by an element of the US Senate.[14]

> **The imperative**: Space is presented at this junction, as a frontier. It is a dimension, not a force, a dimension enlarging the sweep and scope of all our established activities to the measurement of infinity – as the frontier of the American West enlarged the potential of the colonies to the limits of a continent.
>
> For any frontier, as Americans know from their national history, the imperative is exploration, not control. Only by exploration – by pioneering the unknown, by venturing the uncertain – can the promise of any frontier be realized. This must now be our imperative for the space frontier.
>
> **And the sense of urgency**: While the goals remain unchanged and the imperative is familiar to Americans, the nature of space is quite apart from the nature of land frontiers. We were permitted, in the development of the West, to roll back the unknown at our leisure, confident as we did so that such gains as were claimed would be permanently ours. With space, neither such leisure nor such confidence is permitted . . . .
>
> We shall, perpetually, ascertain where we stand in space not by lines on a map, but rather by intangibles of imagination, skill and techniques. The abundance of our talents, not the abundance of our resources will be the index of our national strength, and freedom will be secure in relation to our ability to out-think freedom's adversaries rather than out-produce them.
>
> This is the special urgency of our space adventure . . . . Space has given priority to a new denominator of national strength: the human mind. Unless this resource receives continuous exercise, our presumptions of security are meaningless . . . .
>
> To 'marshall our resources and order our course' is a task of the greatest delicacy which must be accomplished under the most unrelenting urgency.

Table 2. Interest in Space Program During 1969

General inquiries received by NASA	485 300[a]
Mail directed to astronauts	483 530
Publications distributed gratis by NASA	5 000 000
Publications sold by the Government Printing Office	500 000
Sales of NASA picture sets	500 000
Number of persons viewing NASA exhibits	37 600 000
Visitors to NASA facilities	2 600 000
Audience for NASA films (excluding TV)	9 800 000
Television audience for NASA films	248 500 000
Speeches delivered by NASA personnel to non-technical groups	2 049
whose audience was	265 000

[a] Including 205 100 from educational institutions and 68 000 from abroad.

Twelve years after this Senate committee report had appeared and nine months after the first successful Apollo landing on the moon, Dr Thomas O. Paine, then serving as NASA's third administrator, said that 'The space program has engendered almost incalculable interest on the part of the general public . . . as great perhaps as there has been in any single effort in this Nation's history.'[15]

It is estimated that more people on this planet Earth were aware of and were actively following the

As soon as the inhabitants of Earth had seen their own planet from space, as was first achieved by *Apollo 8* in December 1968, support for Apollo solidified. Here, half Earth is recorded from the Moon four years later by the *Apollo 17* crew. (NASA Photo No. 72-H-1609)

At the same time, hundreds of millions of people thrilled as they watched over television American astronauts explore our neighboring world. Scientist-Astronaut Harrison H. Schmitt is seen anchoring a geophone at the Taurus-Littrow landing site during the first *Apollo 17* extravehicular activity. (NASA Photo No. 72-H-1646)

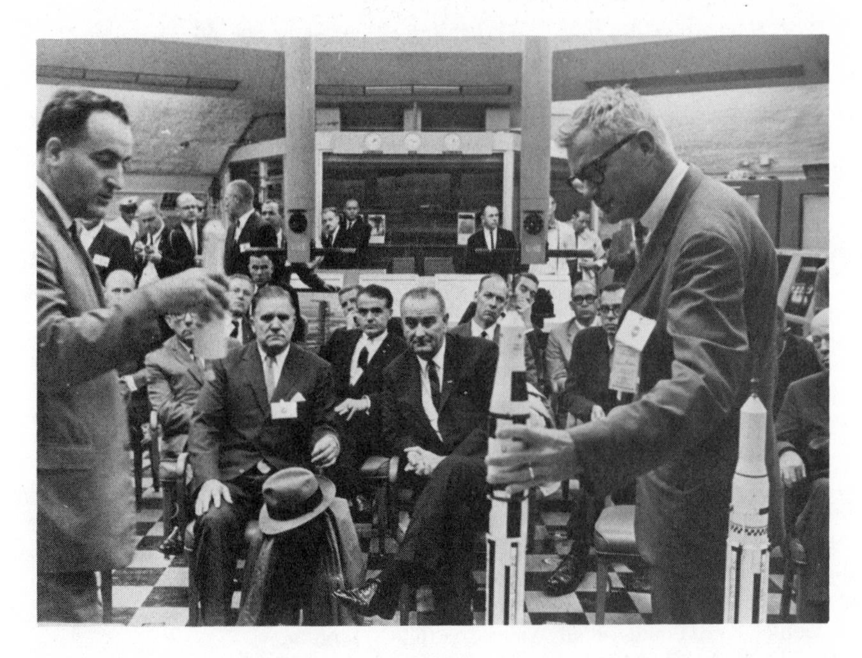

President Lyndon B. Johnson and NASA administrator James E. Webb attend briefing at Cape Canaveral, Florida by Rocco Petrone (left foreground), assistant director for program management, Kennedy Space Center, and Robert C. Seamans, Jr., associate administrator of NASA, September 15, 1964. (NASA Photo No. 64-H-2362)

And all the while, in innumerable speeches, briefings, and testimony, NASA management and scientists kept the American public and the Congress thoroughly informed on the trials and triumphs of Apollo. Seamans, Webb and Mueller (left to right) testify in 1967 before the Senate Committee on Aeronautical and Space Sciences. (NASA Photo No. 67-H-778)

events associated with the July 1969 landing of Apollo 11 astronauts on the Moon than any other occurrence in the history of mankind. At least one thousand million persons heard over radio or saw on television the spectacle through the worldwide satellite communications network, itself a marvel of space science and technology.

Interest generated by Apollo and other space activities during the year 1969 can be illustrated in a number of ways (see Table 2).

As interest in the space program peaked in 1969, mail or telephone inquiries for story information went up dramatically to a total of 112 643, not including queries made during launch activities. News accreditation at launches from Apollo 7 through Apollo 11 also increased substantially. (See Table 3.)

Also during 1969, NASA's monthly TV newsreel, *Aeronautics and Space Report*, was used by 734 of the 840 stations then on the air in the United States; and, in addition, more than 7700 TV showings of 28-minute-long NASA films reached an estimated 347 million people. About half of all American radio stations aired NASA features, notes and spots further augmenting coverage.

Table 3. News Accreditation for Apollo Missions 7–11

Apollo flight	Mission	Total news accreditation
7	Earth orbit	646
8	Lunar orbit	1500
9	Earth orbit	1403
10	Lunar orbit	1519
11	Lunar landing	3497

We do not want to imply that Apollo did not have its critics, that everyone in Congress gave it full support all of the time, that the Bureau of the Budget was not carefully monitoring our activities and seeking ways of reducing or stretching out our expenditures, that Mr James E. Webb, NASA's second Administrator and responsible during most of the Apollo years was not constantly protecting the program from the opponents of manned space flight. But, whenever NASA reached the situation where it had to tell a detractor that if its resources and support were cut beyond such and such a level there would be no way to put a man on the Moon by the end of the 1960 decade, he would usually back off. The lunar mandate was deeply embedded in the national psyche, and would not be satisfied until the objective was achieved.

Naturally, all large-scale endeavors cannot b up the intrinsic interest that NASA's programs co especially during the days of Apollo. After the s cess of Apollo 11, Paine was careful to point that the space agency did not 'translate interest ir support for,' though the former is clearly essentia secure the latter. However, without adequate di sion of information, particularly concerning b immediate and long-term benefits, interest in endeavor will sag and when that happens suppor bound to taper off.

FUNDING

Funding is one of the three principal variables in a large-scale endeavor, the other two being ti required for completion and program content. E is understandably sensitive to changes in the othe but the one that typically receives the most scrutin funding.

NASA managers were well aware of the cruc importance of adequate funding and never ceas emphasizing that along with the establishment of new national space goal must exist the long-te financial obligations to make it possible. They oft referred to Kennedy's prudent words:

> Let it be clear that I am asking the Congress and t Country to accept a firm commitment to a n course of action, a course which will last for ma years and carry very heavy costs . . . If we are to only halfway or reduce our sights in the face difficulty, in my judgment, it would be better not go at all.

This note of caution can well be re-read by t planners and supporters of any large-scale enterpri for its message is universal.

During the late 1950s planners estimated that t total funding necessary to land the first two astrona on the Moon might range anywhere from $20 thou and million to as much as $40 thousand million. Th after NASA conducted more detailed analyses duri the winter of 1961, Webb advised President Kenne and the Congress that the costs through the fi manned lunar landing were estimated to be $ thousand million. In actual fact, NASA assessed t accrued costs of the Apollo program through 31 Ju 1969 – that is, through the Apollo 11 mission th met Kennedy's goal – at $21 349 million. (See Tab 4.) After the completion of the first lunar landing, considerable quantity of Saturn 1B, Saturn 5 ar

The most spectacular of the many costly preparations for the manned lunar landing effort were the Vertical Assembly Building and the Saturn-Apollo launch vehicle-payload combination. At the time, the VAB was the largest building in the world, and was used to bring together and mate all the space vehicle components. Here, it is receiving the giant *Apollo 16* Saturn launch vehicle aboard its "crawler." (NASA Photo No. 72-H-116)

While funding had to be assured for Saturn, Apollo spacecraft, and the VAB, myriad other crucial items had to be developed and paid for, including space suits and cameras. Spacecraft Commander Neil A. Armstrong (front) and lunar module pilot Edwin E. Aldrin (rear) practice a lunar surface mobility exercise in Houston for the forthcoming *Apollo 11* mission. Armstrong has his camera attached to the chest area of his space suit. (NASA Photo No. 69-H-666)

Table 4. Cost Breakdown of Apollo Through First Landing on the Moon

Element	Accrued costs ($)
Apollo spacecraft	6 939 000 000
Saturn series of launch vehicles	7 940 000 000
Rocket engine development	854 000 000
Operations support[a]	1 137 000 000
Tracking and data acquisition	541 000 000
Facilities	1 810 000 000
Manned space flight center operations	2 128 000 000
Total	21 349 000 000
Less leftover hardware	−2 000 000 000
Apollo costs through first landing on Moon	19 349 000 000

The last figure above represented about half the accrued money appropriated to NASA since its formation in the autumn of 1958 up to 31 July 1969.

[a] Mission control systems, launch operations, flight and crew operations and technical support.

Apollo spacecraft hardware remained for later use. This, NASA felt, was worth approximately $2 thousand million, so the adjusted accrued costs through Apollo 11 came to approximately $19 350 million.

The program, therefore, was carried out at slightly less than the lowest original cost estimate. This was possible because of cost-consciousness on the part of NASA management and because the 1960s were not as beset as the 1970s by inflation.

MANPOWER

The existence of abundant, highly-trained human resources was integral to NASA's philosophy of doing business and was indispensible to carrying out the Apollo and other space missions. These resources were spread throughout the entire NASA civil service and contractor network. And how large was this network?

NASA – Contractor Employment

When NASA was created from the old NACA back in October 1958, the space agency employed slightly less than 8000 persons plus an estimated 37 000 working as contractors. By 1967, NASA employment had risen to a high of 36 169 individuals, a 450% increase in a decade. As for total overall employment on NASA programs, the peak of 411 000 persons was reached in 1965; at that time, the agency had 34 3(on its in-house payrolls and the rest were contracto in industry and academic institutions. The fact th only 8.3% of the work force was made up of govern ment employees illustrates the extent to which NAS relied on contracted services.[16]

At peak employment, combined NASA and cor tractor strength in the US space program, about 70% or 300 000 individuals were involved in the manne space flight effort that led directly to the Apoll landings. In all, over 20 000 industrial prime, secon and third tier contractors were involved in th endeavor along with some 200 universities and 8 foreign nations. After the personnel peak had bee reached in 1967 – a full two years before the Apoll 11 landing on the Moon – employment began a stead decline for the simple reasons that no large, post Apollo mission was approved for the space agency.

Far more important than raw numbers of individu als was their quality and the fact that a strong *esprit d corps* evolved among them during the Apollo period As the vast interdisciplinary government–industry university team was built up, it tended to draw in dynamic-thinking individuals who were attracted t the very spirit that created it in the first place. Thus the system was self-reinforcing, somewhat easin recruitment problems.[17]

The University Connection

In our opinion, one of the best and most far-reaching decisions made by NASA was to work closely with and rely heavily upon, the academic community to supply a critical portion of Apollo's manpower needs As Dr Hugh L. Dryden, NASA's deputy adminis trator from 1958 to 1965, pointed out, 'Our educa tional institutions bear a major responsibility for the success of our national effort to explore space. Our universities and colleges are called upon to produce a body of scientists and engineers of unexcelled compe tence.'[18] So outstanding was the response to NASA's call that during the peak Apollo years, about 10 000 individuals in several hundred universities became involved.

While NASA wanted to tap university brains, it was important that professors should not abandon their academic environment and transfer to work at NASA installations. It was essential that most of them stayed where they were and not become divorced from their teaching and close working rela tionships with their students.

NASA could not, of course, ask universities to

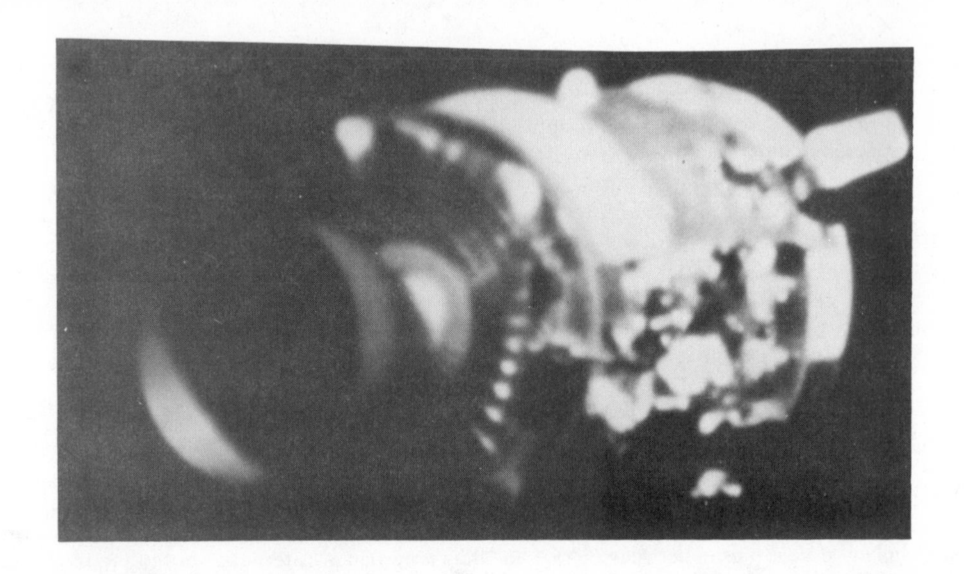

The only serious accident in space during the manned lunar exploration program was the explosion of an oxygen tank in *Apollo 13's* service module, here photographed in its damaged state by the crew in their lunar module/command module following SM jettisoning. The availability of reserves of highly skilled manpower on the ground helped to avert what could have become a tragedy. (NASA Photo No. 70-H-696)

Close-up view of the *Apollo 13* lunar module following its separation from the command module. (NASA Photo No. 70-H-661) Above is an interior photograph showing the "mail box" or jerry-rigged arrangement built by the astronauts to use the command module lithium hydroxide canisters to purge carbon dioxide from the lunar module. It was designed and tested on the ground at the Manned Spacecraft Center before it was suggested to the problem-plagued astronauts in space. (NASA Photo No. 70-H-702)

embark on critical basic and applied research programs unless they were provided with appropriate laboratories and other facilities. Accordingly, the practice of awarding facility grants grew. It was also apparent that, as NASA requirements for academically-trained manpower increased, universities would become strained in their abilities to meet them. Thus, in the spring of 1962, a program of predoctoral training grants was initiated building up to a level of over 3000 doctoral candidates in the mid-1960s.

As the Apollo build-up got under way, a broader 'Sustaining University Program' was brought into being. It included, under a single umbrella, a mechanism for a given academic institution to receive training grants, facilities grants and some special purpose research grants. The program also facilitated the establishment of close liaison between NASA and the research community. The inspiration to inaugurate the Sustaining University Program was based on estimates in the early 1960s that by 1970, as much as a quarter of US scientific and engineering manpower would be engaged in space activities.

Focusing Manpower in Emergencies

NASA's emphasis on seeking and obtaining outstanding government, industry, research institute, and university personnel paid off time and again during the Apollo program. We will give two examples, one quite typical and the other not so typical, of how reserves were concentrated on areas of concern or even danger.

The Saturn 5 Second Stage. Back in mid 1966, a structural design weakness in the launch vehicle, the Saturn 5's second (or S-2) stage, became known, indicating that development might fall behind by some six months. An S-2 Project Task Team made up of technical and program management specialists was quickly formed and sent to the resident management office at the contractor's site. With the full backing of laboratory and program office capabilities at the Marshall Space Flight Center, on-the-spot corrective decisions and actions based on them were taken; and, after six months of uninterrupted, seven-days-a-week effort, three of the six months internal loss were recouped. In all, some 120 actions were taken, including structural design assessments, improvements in welding procedures, upgrading tooling practices and scheduling reviews. By making a number of schedule adjustments, the S-2 stage was made ready to meet target launch dates.

The Apollo 13 Saga. The second example involved the Apollo spacecraft itself during the course of the third attempted manned lunar landing. The Apol 13 flight began routinely enough at 2.13 p.m. Easte Standard Time on 11 April 1970. However, 55 h a 55 min later, it had become anything but routin First, telemetry contact with the spacecraft w almost completely lost for just under two secon Immediately afterwards, the crew was alerted to a lo voltage condition on main bus B; and, at about t same time, a loud 'bang' sounded. It was then not that oxygen tank number 2 in the service module h lost pressure integrity. As oxygen was dissipated, t fuel cells became inoperative, leaving the comma module with batteries as the sole power source a only a small amount of reserve oxygen contained i surge tank and in repressurization packages.

Faced with a clearly catastrophic situation, t crew had no alternative but to abandon the comma module and crawl into the lunar module. The feat w especially remarkable because the two-man lun module was called upon in the emergency to provi electric power, water and oxygen consumables for crew of three.

The crew took immediate steps to activate t lunar module, the MIT Instrumentation Laborator inertial guidance reference was aligned with the co mand module guidance system, and the comma module was shut down until a little before reentry. return trajectory leading to splashdown in the Paci was selected. This required two lunar module desce engine burns, followed by a third to correct t normal maneuver execution variations in the fi two. In addition, one small velocity adjustment w carried out with the reaction control system thruste In order to program the guidance system for the maneuvers, a computer at the MIT Instrumentati Laboratory in Massachusetts was connected into t control system at Mission Control in Texas and t signals were in turn transmitted to the Apollo spac craft. All systems operated flawlessly as did t guidance in the command module when it w turned on prior to reentry. After 142 h and 40 m ground elapsed time, astronauts Lovell, Haise a Mattingly found themselves safely floating in t Pacific Ocean.

Behind these events in space were thousands up thousands of highly trained individuals ready a able to back up the ground control team. Many these same people later became involved with t Apollo 13 Review Board, whose final report w released within only two months after the accident. The accompanying organizational chart of the Boa illustrates the kind of competence applied to t awesome task of saving a manned space mission fro disaster. (See Table 5.)

After the Apollo 13 saga was over, NASA convened a special review board to determine the full nature of the accident and to assure that it would not be repeated on later flights. Part of the space agency's success was attributed to its ability to focus manpower in such emergency situations and to take follow-up corrective actions. Vincent L. Johnson stresses a point during deliberations. Seated in front of the blackboard is Edgar M. Cortright, chairman of the review board. To his right is Marshall's Karl Heimburg, while Astronaut Armstrong sits to Johnson's left. To Armstrong's left in short sleeves is Goddard Space Flight Center Director John F. Clark. (NASA Photo No. 70-H-745)

Table 5. Organization of the Apollo 13 Accident Review Board. The Four Panels, Mission Events, Manufacing and Test, Design, and Project Management, are Shown Along with the Subpanels and Supporting Office Structure

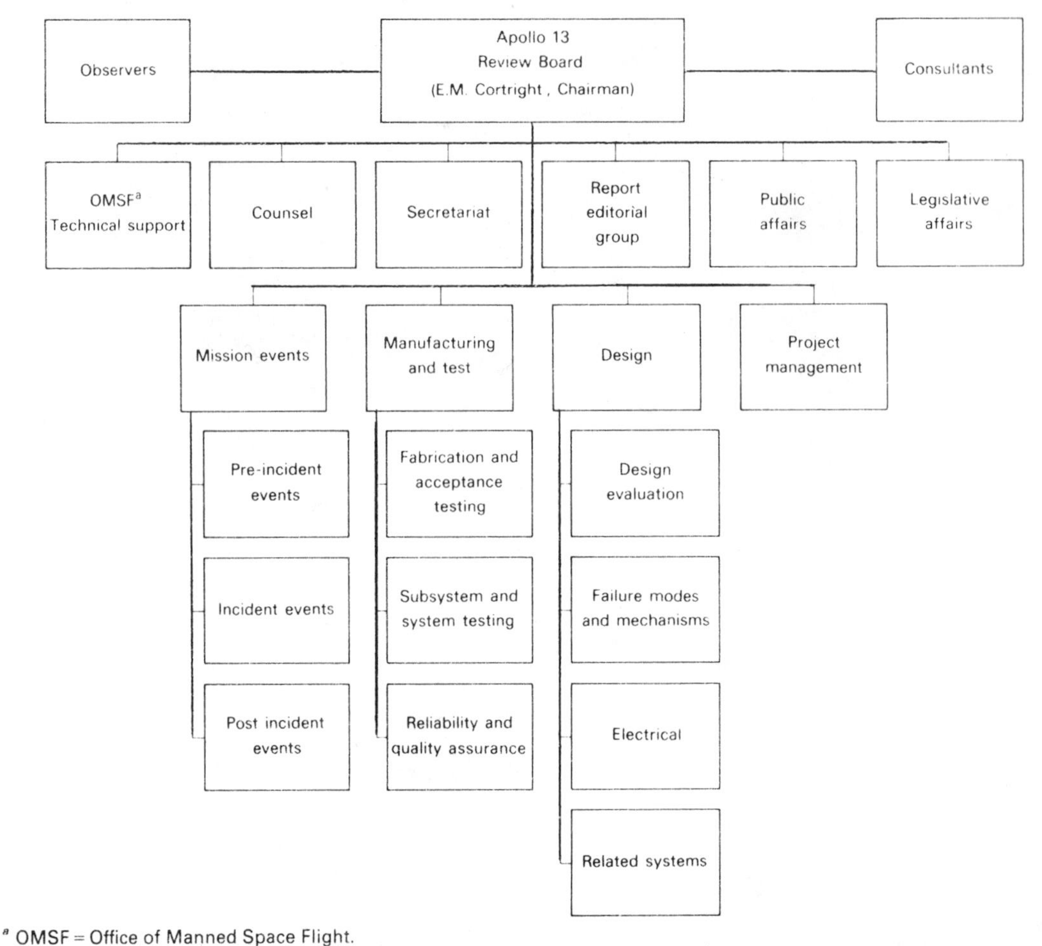

[a] OMSF = Office of Manned Space Flight.

Praise for the rescue of Apollo was all but universal. The British scientific journal, *Nature*, was particularly impressed by NASA manpower resources.

> One of the striking features of the past days has been the sheer competence of the people who have been concerned on the ground as well as in the sky . . . A part of the secret [of the success in saving the spacecraft] has been that there have been enough men on hand to think of everything or nearly everything. Another has been the logic of the planning.[20]

PLANNING AND ANALYSIS

Planning and analysis which, for brevity's sake we ca group under the single word, planning, were centr to the ordered evolution of our space programs and the exercise of our management responsibilities.

The Early NASA Years

During the initial years of operation under th administration of Dr T. Keith Glennan, NAS

George E. Mueller, associate administrator for manned space flight (head of table) flanked by his three manned space flight center directors in the autumn of 1968. Left to right: Wernher von Braun (Marshall), Robert R. Gilruth (Manned Spacecraft), and Kurt Debus (Kennedy).

Top management, NASA headquarters, mid-1965. Left to right: Hugh L. Dryden, deputy administrator, James E. Webb, administrator; and Robert C. Seamans, Jr., associate administrator. Following a lingering illness, Dr. Dryden died on December 2 of that year at the National Institute of Health at the age of 67. He was replaced by Dr. Seamans. (NASA Photo No. 66-H-93)

Meeting of NASA's Office of Manned Space Flight Management Council in mid-1963, a couple of years after the Apollo program officially got under way. Clockwise around table: Eberhard Rees, Marshall Center deputy director: James Sloan, deputy to Shea for integration; William Lilly, manned space flight program control director; Robert Freitag, OMSF director of launch vehicles; Walter Williams, deputy to Gilruth; Joseph F. Shea, director of integration, OMSF; Robert R. Gilruth, Manned Spacecraft Center director; Brainard Holmes, chairman and deputy associate administrator for manned space flight centers and also director of manned space flight; Wernher von Braun, director of the Marshall Center; Kurt Debus, director, Launch Operations Center at Cape Canaveral; Clyde Bothmer, OMSF director of administration; Albert Siepert, deputy director, Launch Operations Center; Charles Roadman, OMSF director of space medicine; James Elms, deputy director, Manned Spacecraft Center. Although OMSF director of spacecraft George Low signed the photo at lower left, he had left his chair when the picture was made. The Management Council was established in December 1961 to coordinate the manned space flight program. According to Holmes, the mission of the Council was to "spot and identify problems as early as possible and to resolve them quickly."

Apollo management team, 1969, left to right: George E. Mueller, NASA associate administrator for manned space flight; U.S. Air Force Lieutenant General Samuel C. Phillips, Apollo program director; Kurt Debus, director, Kennedy Space Center; Robert R. Gilruth, director, Manned Spacecraft Center; and Wernher von Braun, director, Marshall Space Flight Center. (NASA)

directed its attention to scientific and applications satellites and the unmanned spacecraft exploration of the Moon. In addition, a modest but courageous beginning in manned flight was made on 7 October 1958 when the single-occupant Mercury spacecraft program was officially started. Its aim was to serve as the focus for the later and more ambitious manned efforts, though how ambitious they were to become was far from apparent at the time.

Planning the Lunar Mandate

When NASA was given the mandate to land a man on the Moon by the end of the 1960 decade, planning took on new dimensions. But even so, the agency knew that one could never establish a rigid master plan and then sit back and assume that preconceived events would unfold serenely. There were too many ifs, too many unknowns, too many scientific and technological uncertainties, too many happenings beyond one's power to govern or to predict. As a single example, US space planners could never be certain what the Russians were going to do next, yet they were only too aware that Soviet continuing space achievements would inevitably affect the US public and its leaders. These, in turn, could – and usually did – color US planning and decision-making.

What NASA did was to establish broad program objectives to serve as a foundation within which detailed planning could take place. It was recognized that this latter facet of planning had to be flexible, for it was a continuing exercise influenced by myriad political, technological, financial, and other inputs. NASA planning had to reflect Webb's repeated observation that 'At NASA, the name of the game is uncertainty.'

Not only did planning have to account for – and to an extent accommodate – inputs from within the space community and the White House, but had to be responsive to the Congress as well. A program with Apollo's approximately $3 thousand million a year budget occupying some 300 000 American scientists, engineers, technicians, managers and support personnel clearly had to involve the legislators in the planning process. It was therefore fundamental to management philosophy that all facets of the Apollo program be thoroughly understood by the Congress and that its judgments be carefully considered as NASA exercised its history-making mandate. NASA maintained close liaison with Congress through program planning and budget hearings, by special briefings within NASA facilities and at contractor installations and by a variety of other mechanisms.

The Apollo program graphically demonstrated the need for phased development planning so that managers could define, at an early date, the large number of emerging work packages or tasks in terms of objectives, performance, priorities, manpower requirements, time schedules, and relationships of each to the others. To get a feeling for the sheer numbers involved, when President Kennedy proposed the lunar landing goal back in May 1961 already more than 10 000 such work packages had been identified! By then, each had been programmed in terms of probable schedules and estimated resources on a digital computer to enable their individual criticality to be assessed in terms of achieving the President's goal.

Phased Project Planning

Gradually, the concept of phased project planning (PPP) gained currency since it offered a convenient way of basing management decisions on the extent to which project activities should be undertaken and contractual commitments made. In brief, PPP was designed to provide a steady accretion of knowledge and understanding of all aspects of the particular development project under investigation, taking into account such diverse areas as technical, management, funding, scheduling, manpower, operational support, facilities, procurement, and relationship to the larger enterprise of which it constituted a vital part.

Essential to PPP were four major decision points which occurred when it is desired to:

(A) Initiate the analytical work required to establish the overall feasibility of undertaking the specific project (several previously identified approaches could be compared at this stage);
(B) Engage in study, preliminary 'breadboarding' and engineering analysis to sort out from the several feasible approaches the single approach to be followed;
(C) Start detailed definitions of the project including preliminary design; develop the necessary supporting data; and prepare a firm plan for project development and operation; and
(D) Initiate final design and hardware development, and carry out operations for the achievement of project objectives.

These decision points divided the planning and definition processes into four discrete phases: Phase A, preliminary analysis; Phase B, definition; Phase C

The Saturn launch vehicle-Apollo spacecraft combination achieved universal admiration for their performance during the course of the lunar exploration program. We see *Apollo 11* systems being monitored from the Operations Management Room, Launch Control Center at Cape Canaveral in Florida. (NASA Photo No. 69-H-1153)

The *Apollo 15* lunar roving vehicle undergoes compatability fit checks with the lunar module that carried astronauts David R. Scott and James B. Irwin to the Hadley Apennine region of the Moon in the midsummer of 1971. (NASA Photo No. 71-H-113)

design; and Phase D, development/operations.

Phase A – Preliminary Analysis. *Phase A* was primarily an effort undertaken in one or another NASA facility to analyze alternate overall project approaches or concepts for accomplishing a proposed technical objective or mission. In short, it attempted to determine if a given mission objective was achievable. It identified, from the more promising concepts that had been examined in an earlier advanced mission study effort or elsewhere, those project approaches felt to be worthy of further refinement.

Phase A also identified such project elements as major facilities, operational and logistical support, and advanced research and technology effort required to support the proposed project. Furthermore, it assisted in determining whether the proposed technical objective or mission were feasible and worthy of further definition. Contracted effort was limited to auxiliary studies in support of in-house NASA analyses.

Phase B – Definition. *Phase B* involved detailed study, comparative analysis and preliminary system design directed toward facilitating the choice of a single project approach from among the alternate approaches selected from the *Phase A* activity. It was, in effect, a 'weeding-out' process. Also included in *Phase B* was the identification of facilities, logistics, operations, and additional advanced development tasks required to support the specific approach selected. As a rule, the major effort was accomplished through contracted studies, which provided the necessary data needed for analyses. These, in turn, determined the nature of the next recommendation to management.

Phase C – Design. The detailed definition of the objectives and final project concept occupied *Phase C*. System design with mockups and test articles of critical systems and subsystems was undertaken, as necessary, to assure that the hardware was within state-of-the-art limits, that technical milestone schedules and resource estimates for *Phase D* were realistic, and that definitive contracts could be negotiated. Also included was the identification of alternate or back-up system/subsystem development requirements, facilities, logistics and operational support requirements.

A unique feature of *Phase C* was the provision for additional work by contractors during the period between submittal of their reports and action on *Phase D*. Thus, *Phase C* contractors competed for *Phase D* contracts, performing all the work except for NASA supporting studies and efforts required to monitor the contractors and analyze the results of their efforts.

Phase D – Development Operations. *Phase* covered final hardware design and developme (including alternate and back-up system subsystems), fabrication, test, and project operation This phase possessed the normal balance and con position of contractor/NASA work associated with development/operations activity.

It is important to remember that it was very hard predict the detailed nature of systems to fulf requirements that had never existed in the past. Ma times in space vehicle development one had to inve the approach, not just the techniques and equipme needed to carry out the mission. Even after this w accomplished, it was often necessary to develop ne methods or new materials, or both, to accommoda the manufacturing process. Finally, innovative wa had to be devised of moving some of the very lar manufactured items, such as Saturn 5 stages, fro where they were fabricated and tested to where th were ultimately used.

As each program activity passed through the var ous steps from definition to operation, the ability managers to understand its many ramifications wou increase. This, in turn, meant that resource estimat could be refined as well as the time needed f completion. In other words, forward planning was continually iterative process.

Checkpoints

To aid in assessing the principal phased elemen applicable to a particular effort, five checkpoints we established. In order, these were:

(1) *Preliminary design review*: The basic desig approach was approved at this point for a major program elements.

(2) *Critical review*: Held following detaile engineering design, it involved the approval manufacturing specifications and drawings.

(3) *First article configuration inspection*: Her flight articles were reviewed to assure th manufacturing was according to the approve design and could hold up under simulate testing.[21]

(4) *Design certification review*: The above testin permitted this checkpoint to be assessed t insure that all flight and ground systems were in fact, qualified for flight.

(5) *Flight readiness review*: This review was con ducted on the basis of the results of actua

countdown procedures, including the fueling and arming of launch vehicle pyrotechnics up to the time of simulated launch. This review established whether all ground and flight systems were fully operable and ready for the actual mission.

These are but a few examples of some of NASA's planning methods and tools. There is no question that they improved as time progressed (see Table 6), and ultimately they were effective in allowing NASA to meet the national objective established by the President in May 1961. Leonard Sayles, a student of NASA planning and management had this to say about them: 'It is very hard to go halfway to the Moon. NASA did not allow its objectives to be plundered, or distorted by all of the agencies and vested interest groups that are normally associated with a large system.'[22] In other words, NASA lived up to the saying that planning and its management represent the art of doing the work that is projected.

THE MANAGEMENT PROCESS

In a mid-1969 issue of *Fortune*, Tom Alexander wrote that:

> The really significant fallout from the strains, traumas, and endless experimentation of Project Apollo has been of sociological rather than a technological nature; techniques for directing the massed scores of thousands of minds in a close-knit, mutually enhancive combination of government, university, and private industry.[23]

Maturing of Management

The techniques that led to the effective concentration of vast numbers of minds onto a single objective did not mature overnight. But mature they did, forced in part by the extremely rapid expansion of NASA. Throughout the agency's first five years of existence,

Table 6. Apollo Technical Review Process

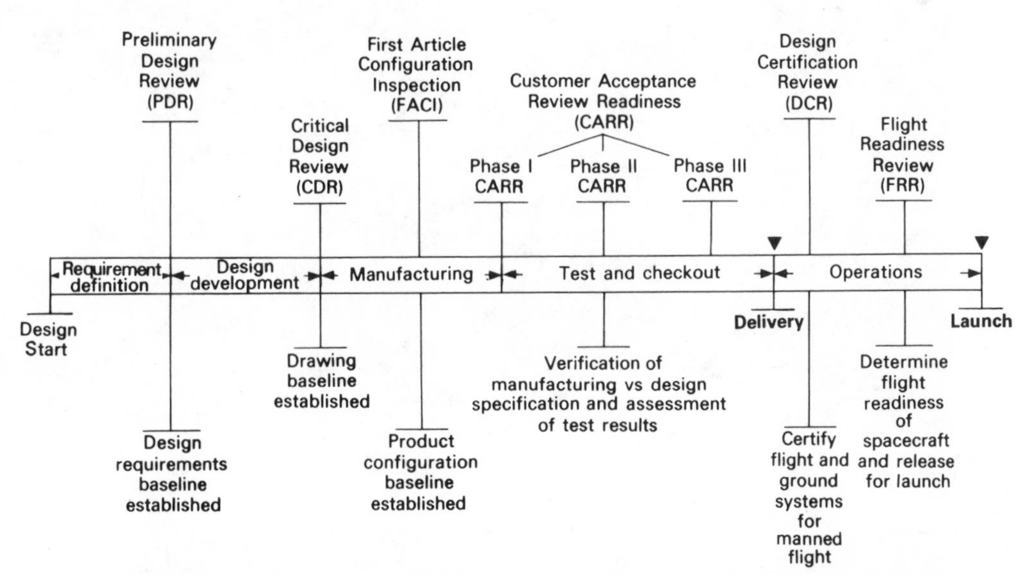

The first step is *program definition* when decision must be made as to what technical objectives are and what kind of schedule is desired. Then comes the *preliminary design review*, which is scheduled as program begins to take shape. Therein, general agreement is reached between government and industry that culminates in the approval of basic design approach. *Critical design reviews* are scheduled as point of detailed drawing release is reached. *First article configuration inspection* is held at a point at which configuration is beginning to take shape and it is desired to limit configuration changes. *Customer acceptance readiness review* may involve up to three phases: (1) upon completion of manufacturing; (2) following individual subsystem testing and before integrated testing; and (3) after integrated testing some two weeks before delivery of space vehicle to customer. Sometimes *CARR* can be in two phases or even one phase, depending on nature of hardware, configuration changes involved, and so on. Following *CARR*, *design certification review* takes place, though not necessarily for all vehicular elements. Its purpose: to review all aspects of the particular mission about to be flown, including hardware, software, support, and operational elements needed for success. Last, *flight readiness review* occurs so that an over-all look at hardware, software, operations, and so on can provide confidence that the vehicle is ready to fly. After *FRR* but before launch, a meeting of key management takes place to review all open items, and to make absolutely certain that the flight can be initiated.

Close-up of another element of the overall lunar effort: the central station of the ALSEP array of instruments left on the Moon by *Apollo 14* astronauts Alan B. Shepard and Edgar D. Mitchell. The craft landing in February 1971 in the hilly upland region north of the Fra Mauro crater. (NASA Photo No. 71-H-388)

its programs doubled in size each year.

During much of this early growth period, lines and areas of responsibility were often ill-defined, job descriptions had not and could not yet be written, ground rules were only beginning to be established and much improvization and reorganization proved necessary. Webb would later recall that 'The process of management became that of fusing at many levels a large number of forces, some countervailing, into a cohesive but essentially unstable whole and keeping it in a desired direction.'[24] There were no structured and tradition-proven areas of authority and no well-oiled plans to rely on. Almost everything was new.

Realigning Management for Apollo

By the beginning of the 1960 decade, it was becoming apparent that NASA might be called upon to handle a very large-scale program. In anticipation of this possibility, in January 1961 – four months before the Kennedy lunar landing speech – NASA realigned and streamlined its management procedures in three key areas or classes of activity:

(1) Project planning and implementation which were designed to control the content of its programs;
(2) Financial operating plans which were designed to control expenditures; and
(3) Program management which involved the systematic reporting of progress and problems.

Since the early days of Apollo, the names of these procedures have varied, refinements and additions have been made and the formats used have been altered. But, the need for such management systems has endured, for in every large-scale enterprise one must plan and control resource budgets, apply resources wisely against specific objectives, systematically review progress toward these objectives and make schedule adjustments in the light of resources available, changing emphasis and rates of progress. In synthesis, then, Apollo management's goal boiled down to maintaining a logical, realistic balance between performance, time, and cost.

For ultimate success to be realized each and every segment of the launch vehicles and their payloads had to be continuously monitored and timed so that all program elements could fall precisely into place. To assure that this, in fact, occurred, a whole system of reporting methods, controls and crosschecks had to be implemented. Without these and other management tools, the complex space missions could never have been carried out.

The management of Apollo relied heavily on effective communications and the rapid generation of technical and management information.[25] Both, as has already been seen, are key factors in the success of any large-scale enterprise. Government and industrial experience had long ago demonstrated that a large percentage of management problems is due to breakdowns in the rapid flow of information.[26]

In the light of this, one of the first activities undertaken in the Apollo program was to structure NASA's manned space flight development organization so as to give each major element a single clear line of reporting and authority.

Systems Management

Many names have been given the kinds of management that brought Apollo from concept to reality. Of them, systems management has been given considerable publicity. Described as primarily a way of thinking about the job of managing, the approach provides

> a framework for visualizing internal and external factors as one integrated whole. It allows recognition of the functions of subsystems, as well as supersystems . . . and fosters a way of thinking which, on the one hand, helps to dissolve some of the complexity and, on the other hand, helps the manager to recognize the nature of complex problems and thereby to operate within the perceived environment. It is important to recognize the integrated nature of specific systems, including the fact that each system has both inputs and outputs and can be viewed as a self-contained unit.[27]

Apollo Program Organization

Operating within this broad philosophy, NASA opted for 'projectized' organizations to guide Apollo to fruition (see Table 7). In essence, this involved the establishment of Apollo program offices at the manned space flight centers in Huntsville, Houston and Cape Canaveral, all directed by the Apollo Program Office, and in turn overseen by the Associate Administrator for Manned Space Flight at NASA Headquarters in Washington. The acceptance of projectization meant that individuals working within the various Apollo program offices were really serving two bosses, the local program managers and the host center directors. Occasionally, this dual-boss situation could lead to disputes that would have to be resolved at the Washington level.

Within NASA as a whole, authority moved from

the Administrator and Deputy Administrator to the Associate Administrator, who was the agency's general manager with responsibility for overall program objectives. The Associate Administrator was required to approve program plans and to allocate resources not only for manned space flight, but also for automated scientific and applications spacecraft, ground tracking and data acquisition, aeronautics and advanced research programs.

In the key manned space flight area, authority passed to the Associate Administrator for Manned Space Flight who had cognizance over NASA's three manned space flight centers and their vast contractor network.[28]

The Manned Space Flight Centers

Actually, that was the situation from November 1963, onward. Earlier, all the centers had reported directly to the Associate Administrator. The placin of the field centers under the line command of th headquarters program directors increased the authority and responsibility which was precisely wh was needed as the Apollo program moved into hig gear. According to a staff paper prepared at the tim this fundamental change would make 'clearer an more direct [the] lines of authority and responsibilit between Headquarters and field installations . . .'[2 and at the same time would permit the Associat Administrator to have more time to discharge hi general responsibilities and to consult on polic matters with the Administrator and Deput Administrator.

As things stood from November 1963 onward launch operations were placed under the control o the Launch Operations Center at Cape Canaveral i Florida (renamed the John F. Kennedy Space Cente a month later), while the Manned Spacecraft Cente in Houston (later designated the Lyndon B. Johnso

Table 7. Management Organization for the Apollo Program[a]

Apollo
program office
Washington, D.C

Program control
Systems engineering
Test
Reliability and quality
Flight operations

Director
MSC[b]
Houston, Texas

Director
MSFC[c]
Huntsville, Alabama

Director
KSC[d]
Cocoa Beach, Florida

Apollo spacecraft
project office
MSC

Saturn I/IB
project office
MSFC

Saturn 5
project office
MSFC

Launch operations
project office
KSC

Contractors

[a]Functional offices at centers are shown coded to correspond with functional offices in Washington program office.
[b]MSC = Manned Spacecraft Center, later Johnson Space Center.
[c]MSFC = Marshall Space Flight Center.
[d]KSC = Kennedy Space Center.

Space Center) was called upon to handle the development of the Apollo spacecraft payload as well as flight control once the Saturn 1B and Saturn 5 launch vehicles cleared the launch tower at the Cape.

For its part, the George C. Marshall Space Flight Center in Huntsville, Alabama, was in charge of launch vehicle and related development, including the post-Apollo Skylab program. Working closely with these centers in a key supporting role was the Goddard Space Flight Center in Beltsville, Maryland, responsible for ground, sea and air tracking and data acquisition. All other NASA centers were involved in specialized Apollo activities.

During its first decade of operations, almost $32.5 thousand million were appropriated to NASA by Congress and somewhat more than $32 thousand million were actually spent. During the same decade, slightly under 80% of these funds were for research, development, test and operations, with the rest covering administrative operations and construction of facilities. The manned space flight program element consumed a little more than 67% of the first decade's funding. Most of the monies expended by NASA on Apollo were in the form of contracts let by the three manned space flight centers.[30]

As the Apollo program evolved, each center developed a strong competitive spirit. At the same time, each learned to think in terms of broad system objectives as contracts were placed with industry and universities. Moreover, the centers took advantage of their unique talents and facilities to do important in-house research and development.

NASA Headquarters in Washington had to execute all-important functions in the areas of policy and implementation, overall program definition, configuration control, major change approval, budget control, flight mission planning and the provision of central supporting services and review.

Cooperation at the Top

Needless to say, the coordination and direction of these diverse elements, as they grew up during the Apollo program, called for a closely-knit team at the top. During much of Webb's nearly eight years as Administrator of NASA, he, Dryden as Deputy Administrator (to late 1965) and Seamans as Associate Administrator (to late 1965) worked closely together as a troika. (Following Dryden's death in December 1965, Seamans served as both Deputy Administrator and Associate Administrator through to 1968.)

Webb was fond of underscoring the closeness of top management, and often spoke of an informal partnership within which all of NASA's principal policies and programs became a joint responsibility. However, the execution of each policy and program was expected to be undertaken by only one of the three. He also stressed that

> In every major matter, we worked intimately together to establish a sound foundation for our policies and actions. Each of us helped to bring capable and valued associates into positions of responsibility. When one of us found the burden of his work too heavy, the others stepped forward to share it.[31]

To illustrate how group decision-making was carried out, in May 1965, Seamans reminded Webb[32] that extra vehicular activity (EVA) was, according to the Gemini Project Approval Document,[33] a primary objective of that two-man spacecraft program. At the time, the fourth mission was being planned and astronaut and spacecraft time in orbit was to be extended to four days. Pointing out that EVA 'reduces by a small but finite amount the chance of success,' Seamans nevertheless believed that EVA could be carried out without unduly reducing the probability that all primary goals could be achieved. Seamans therefore recommended that EVA should be undertaken during the Gemini 4 flight.

Characteristically, Webb wrote on the memo: 'Approved after discussing w. Dryden. JEWebb 5-25-65.'

THE INDUSTRIAL TEAM

More than 90 cents out of every NASA dollar earmarked for space flight was spent under contract. At NASA's peak in the mid-1960s, some 20 000 industrial units were at work, the majority on the Saturn launch vehicles and their Apollo payloads. They included approximately 2000 prime contractors and an additional 18 000 subcontractors.

In the following subsections, we review first NASA's procurement plan; then we will focus on the use of an innovative form of contracting with built-in incentives; next we will discuss briefly performance, cost and time parameters as they were affected by contracting; finally we will summarize the functions of the source evaluation board that assessed proposals for the Apollo spacecraft.

It should, however, be emphasized that NASA's objective during Apollo was to place principal reli-

Built by the Grumman Aircraft Engineering Company, Bethpage, Long Island, New York, the lunar module performed admirably throughout the Apollo program. *Apollo 15* mission commander David R. Scott walks to a training model of the Boeing-developed lunar roving vehicle to practice getting in and out of a full-scale LM mockup, rear, at the Kennedy Space Center Flight Crew Training Building. His space suit was developed by ILF Industries. (NASA Photo No. 71-H-1144)

ance on the American industrial and university system. NASA sought to work side by side with its contractors on a peer basis, to be involved with them technically as well as in an overseer sense from the beginning of a procurement through to the delivery of a flight-ready article. In other words, NASA desired to develop a far more intimate role with its contractors than simply receiving reports on their progress at arm's length. Webb summarized it well when he said that

> . . . we have found that we must be able to speak and understand the language of those on whom we rely, to know as much about the problems they are dealing with as they do, to check and supplement their work in our own laboratories and, in some cases, help untangle snarled situations.[34]

The Procurement Plan

This plan involved the preparation and utilization of a detailed document describing the objectives of a given procurement in terms of technical requirements, hardware deliveries, launch schedules, general methods of contracting, the type of contract envisaged and the administrative procedures to be followed. On the basis of the plan, a request for proposal (RFP) would be prepared by whatever center had been assigned cognizance over the particular procurement.

After approval at the headquarters level, RFPs would be released to industry, bidders' conferences would be held to assure that all concerned received ample and equal information and subsequently proposals would be received. All proposals were then evaluated by a source evaluation board which (in the case of procurements in excess of some $5 million – the figure varied over the years) would present its findings to the Administrator. He, in consultation with key members of top management, would later make the final selection, thereby permitting the procurement of hardware and services to begin.

Before describing the workings of the source evaluation board, we review the innovative type of contracting that was used throughout Apollo.

The Incentive Contract

NASA was always concerned with meeting its mission objectives on time and at the least possible cost to the taxpayer. To do this, advantage was taken of the contracting experience of other agencies, particularly the Department of Defense. Indeed, General Samuel C. Phillips of the US Air Force was assigned to NASA as the Apollo Program Director together with many experienced program managers and contracting officers.

At the same time, new and often innovative contracting ideas were tested. One was the incentive contract which provided an increased fee if costs were reduced, schedules were improved and/or performance specifications were exceeded. But there was the stick as well as the carrot. If the contractor did not perform as well as contemplated under the terms of the contract, penalties were applied.

There was an underlying reason why the incentive approach to contracting was developed. Traditionally, defense contractors do not receive a large fee for research and development. But they do expect an appropriate return from extended production contracts occurring as a result of their research and development.

During the Apollo program, NASA could not offer contractors large production orders: Saturns and Apollos were few in number and could never compare in quantity with military aircraft or missile system orders. Consequently, the agency could not base its incentives to industry in terms of production runs but had to resort to special awards.

As contractors moved into risky and challenging areas of rapidly advancing space technology, it was not unusual for industry to be overly optimistic. This kept NASA technical management on its toes at all times, and meant that the agency had to be intimately aware of what was going on.

Occasionally, when it was found that a contractor was headed for trouble, in-house expertise would be called upon to help resolve the difficulties. NASA was much more interested in encouraging contractors to admit their errors and quickly correct them than in fixing blame for deficiencies. This philosophy of mutual understanding helped immeasurably to weld the NASA-industry spirit of teamwork.

Juggling Performance, Cost and Time

As we have already noted, three variable factors had to be continuously considered in managing Apollo: performance, cost, and time. Any change made in one factor most likely would have a perturbing effect on another. Performance, for example, was reflected by figures such as propulsion efficiency or dry weight, while cost and time were closely related to manpower levels.

Because of this, engineering managers were forced to make liberal use of a 'trade-off' or sub-optimization system in dealing with contractors,

Apollo 16's command (top) and service (below) modules developed by North American Aviation are being lowered onto the Saturn 5 McDonnell-Douglas Astronautics Company third stage at the Vertical Assembly Building at the Cape in Florida.
(NASA Photo No. 72-H-118)

wherein something was traded off or given up in one area for a gain in another. It also meant that one had to analyze very carefully to ascertain that the trade-off did not result in a net loss.

A number of techniques evolved during the Apollo program that enabled management to appraise what kind of trade-off might be necessary, and acceptable, for given situations. One of them, known as *FAME*,[35] featured a large number of charts that were continuously updated to display such information as performance, schedules, weights and costs relative to Saturn 5 stages and Apollo spacecraft modules. Carefully defined limits were noted on these charts, beyond which each parameter could not extend without endangering other elements of the launch vehicle or spacecraft or perhaps their ground support or other equipment.

Since computers were continuously fed data relevant to the various parameters, trend lines could be rapidly produced thereby alerting managers to possible trouble ahead. This management tool thus enabled them to take anticipatory corrective action with the appropriate contractors before such trouble actually developed.

In general, NASA program directors enjoyed more freedom to trade-off than managers working on more tightly configured military systems. The space agency, as we have already noted, took maximum advantage of military management experience; but, in addition, improvised, revised and adapted for the special needs that resulted from the unprecedented nature of its missions.

The agency also realized that its contractors often had to solve problems on the frontiers of human experience. Consequently, it was important to work jointly, as partners, and not expect the industrial teams to take unfair risks. All members of the team were striving for the same national goal and all should have an opportunity to contribute and to benefit accordingly.

A great deal of management time was spent during the days of the Apollo build-up to develop effective ways of dealing with industry. As a logical first step, NASA had to devise a means of evaluating proposals that were reaching management in ever increasing numbers.

Source Evaluation Board Approach to Contractor Selection

Webb felt that one of the real secrets of NASA's success was that its source evaluation boards could bring into play every facet of judgment.[36] Such boards themselves were quite small, but were supported by technical management and other review sub-committees whose total personnel could range up to nearly 200 individuals. The board and its sub-committees were free to select their own methods of judging the contractor proposals assigned to them. Once their evaluation had been completed in accordance with a ranked numerical system, the entire evaluation process was reviewed by top management, from the specific method chosen to the results of the evaluation itself.

The SEB's function can best be illustrated by reviewing the procedures that led to the selection of the prime contractor for the Apollo command and service modules.[37]

On 28 July 1961, following NASA headquarter's approval of both the statement of work and the procurement plan, requests for proposals were sent out to 12 companies. Later, upon request, four additional companies were provided with the same material. A little over a fortnight afterwards, a pre-proposal conference was held for the companies at which time technical and business aspects were reviewed by NASA personnel. Some 400 questions were answered orally and later documented and confirmed by mail.

Of the 16 companies, only five ultimately submitted proposals by the 9 October deadline. Three of the five proposed to take on team members; only North American Aviation and The Martin Company preferred the conventional prime-subcontractor approach. Two days later, representatives of all five companies made their oral presentations.

The source evaluation board established for the Apollo spacecraft was advised by technical and business subcommittees that represented all major NASA elements plus some of the Department of Defense. Each of the subcommittees was in turn aided by panels of specialists. In all, some 190 persons were involved.

The *technical evaluation* involved two assessment areas: (1) technical qualifications and (2) technical approach. The former dealt with experience, facilities, personnel and the technical ramifications of the proposed project organization, while the latter comprised 11 areas covering mission and system design; systems integration; development, reliability and manufacturing plans; and operational concepts. Then there was the *business management and cost evaluation* assessment area which was concerned with organization and management, logistics, subcontract administration and cost.

The weighing factors assigned the technical and business-cost evaluations were as follows:

Technical evaluation 60
(Technical qualifications 30)
(Technical approach 30)
Business management and cost evaluation 40
Total 100

Ratings were made on a zero to ten basis, as follows:

8–10	Excellent
5–7	Good
2–4	Fair
1	Poor
0	Unsatisfactory

On the basis of this system, between the 9 and 21 October 1961, a detailed assessment of the five proposals was made by panels, following which the subcommittee reviews took place from 23 to 28 October. Finally, during the three-week period between 1 and 22 November, the source evaluation board made its assessment with the results noted in Table 8.[38]

Since North American Aviation and General Dynamics/Astronautics received the same second-place rating, considerable effort was made to study the relative merits of the respective proposals of these companies. The relevant NASA report had this to say:

> In assessing the ratings, the board recognized that all the proposals had received high ratings in the Business area, the lowest rating (7.59) being higher than the highest rating (6.66) received in either the Technical Approach or Technical Qualifications areas. Since those ratings established that all the companies could more than adequately handle the business aspects of the program, the board turned its considerations to the Technical Evaluation for further analysis of the ratings.[39]

Two other areas aroused particular interest in the minds of the evaluators: cost and possible conflicts of resources (for example personnel and facilities). North American's low cost was noted, of course, but it was not felt that its cost estimating was as thorough as either General Dynamics or Martin's. Clearly offering the lowest cost was not, *per se*, a decisive feature.

As for possible conflicts of resources, NASA queried both Martin and North American as to whether or not other on-going programs would require personnel and facilities then being utilized by their respective Titan 3 and Saturn 5 second stage contracts. Both companies assured NASA that there would be no such conflict.

On the basis of this thorough evaluation, the source evaluation board concluded that 'The Martin Company is considered the outstanding source for the Apollo prime contractor. Martin not only rated first in Technical Approach, a very close second in Technical Qualifications, and second in Business Management but also stood up well under the further scrutiny of the board.' Moreover, the source evaluation board liked Martin's management approach.

> Martin's proposed management arrangement of a prime contractor with subcontractors [the report concluded] appears technically to be the most sound both as far as reaching technical decisions quickly and properly and also for implementing these decisions. Short lines of communications

Table 8. Source Evaluation Board Rating of Apollo Spacecraft Proposals

	(1) Ratings by area		
	Technical approach (30%)	Technical qualifications (30%)	Business management (40%)
The Martin Company	5.58	6.63	8.09
General Dynamics/Astronautics	5.27	5.35	8.52
North American Aviation, Inc.	5.09	6.66	7.59
General Electric Company	5.16	5.60	7.99
McDonnell Aircraft Corporation	5.53	5.67	7.62

	(2) Summary ratings
The Martin Company	6.9
General Dynamics/Astronautics	6.6
North American Aviation, Inc.	6.6
General Electric Company	6.4
McDonnell Aircraft Corporation	6.4

involved in their proposed arrangement will minimize interface problems and required documentation and thereby result in few opportunities for errors.[40]

The board also looked with favor upon the strong project organization proposed by the Martin Company.[41]

North American was considered 'the desirable alternate source' for Apollo. Not only did it come out No. 1 in technical qualifications but had a wealth of manned aircraft and missile experience. However, in the source evaluation board's view, 'Their project organization . . . did not enjoy quite as strong a position within the corporate structure as Martin's did.' Despite North American's shortcomings, the board nevertheless was 'convinced that North American is well qualified to carry out the assignment of Apollo prime contractor and that the shortcomings in its proposal could be rectified through further design effort on their part.'[42]

Despite the board's conclusion that the Martin Company was a slightly better contender for the Apollo command and service module development contract, North American Aviation ultimately won out. After receiving the board's evaluation and listening to oral presentations, Webb, Dryden and Seamans, with the assistance of Dr Robert R. Gilruth, Director of the Manned Spacecraft Center, made a thorough review of the situation and determined that North American had lost points because it had not participated in government-sponsored studies which included spacecraft design work. Also, it was felt that the board had not given sufficient weight to North American's experience with the high altitude X-15 rocket-powered aircraft. Webb summarized the factors that influenced the decision to choose North American over Martin during the course of Senate space committee hearings.

> They [North American] were second in the numerical evaluation which the Board reported, but when we questioned members of the Board with respect to, first, the method they had chosen to evaluate the various proposals; second, the actions they had taken to apply their own methods; and third, what they had found, we discovered that they had inadequately, in the judgment of all four of us, applied their numerical equations or numerical ratios to the element of experience. The experience element we felt particularly pertinent was in the area of high-performance systems involving man, such as the X-15. We therefore made a correction of what we believed to be an error in the numerical ratings of the Board; or shall I say evaluations of the Board.[43]

Other factors influencing NASA's decision in favor of North American were its lower proposed cost and the company's reputation for handling low-cost, high-quality engineering production.

MANAGING CHANGE

Configuration Management

Since large-scale technological systems generally press against, and hence tend to advance, the state-of-the-art, changes in designs and specifications must be contemplated – and often accommodated – by management.

Configuration management was instituted in order to deal with a flow of change-order requests that continued right into the hardware-cutting stage. This approach created for managers a carefully controlled environment in which change recommendations could be assessed and approvals made or denied.

First, each element in the Apollo system would be defined in terms of its own intrinsic specifications as well as in terms of the schedules and budgets that its development dictated. Then, as change requests were generated during the development cycle, they would be logged and studied. In undertaking their reviews, managers were naturally concerned as to how the requested changes would affect other elements of the program as well as schedules and appropriated funds.

Dealing with change requests in a system as complex as the Apollo spacecraft was no simple matter. To enable technical management to visualize what the completed vehicle would be like, a breadboard was developed – a design layout used to determine the eventual configuration.[44] This enabled restraint to be exercised in approving changes, for unless managers knew exactly what was in the spacecraft they could easily lose management control over it. For example, they had to know that the latest version of a given replacement part had been flight tested – and that information concerning it had been circulated to all affected organizations and individuals.

There was a saying at NASA that 'better is sometimes the enemy of good.' This meant, in simple terms, that NASA tried to settle as early as possible in the development cycle for a good component and did not always keep striving to make it better. Once an Apollo-related component had been verified and thoroughly tested, it was usually unnecessary and undesirable to make changes.

Of course, this idealized situation could not always exist in such large and complicated systems as Saturn

The "pogo" phenomenon appeared on this unmanned Saturn 5 *Apollo 5* test flight of April 4, 1968. Within a few months, the problem was identified, changes made, and a potentially serious delay brought under control. (NASA Photo No. 68-H-320)

5 and its Apollo payload. Design, component testing and flight operations often overlapped, and changes might be required as a result of a failure somewhere in the system. This meant that continual re-examinations of designs were necessary so that situations could rapidly be identified that might lead to failure at a later date. Then, too, the possibility had to be faced that some managers might resist making changes when they were really necessary. To maintain a sense of balance between approving changes for change's sake and not making them when they were required, configuration management panels were convened where views would be aired, trade-offs proposed, conflicts resolved, and actions recommended.

'Pogo'

An example of a problem calling for configuration management response occurred late in the development cycle of the Saturn 5. Referred to as 'pogo,' it involved longitudinal vibrations that built up during flight. Such vibrations, termed regenerative system oscillations, are generally caused by the coupling of the vehicle structure and the engine thrust through the propellant feed system. The problem had been studied for a long time, particularly during the course of the Titan 2 launch vehicle flights with the two-man Gemini spacecraft payload.

Pogo first appeared in the Saturn 5 program during the unmanned AS-502 (Apollo 6) mission of 4 April 1968, 100 seconds after launch. Post-flight analysis of telemetry revealed that oscillations began when the resonant dynamic frequency of the vehicle's structure tuned with the resonant frequency of the oxidizer feed system. These frequencies change in flight and pogo is created only if they coincide – as they did so dramatically in AS-502.[45]

Pogo provided an excellent example of how NASA could rapidly concentrate expert personnel to resolve a serious problem.[46] It also showed how a variety of technical and management tools could help managers keep track of many different events and responses in a situation where later actions depended heavily upon the timely completion of earlier efforts.

Everyone involved with pogo recognized that a solution was absolutely mandatory. It was entirely possible, they knew, that Saturn 5's oscillations could become so serious that they might break up the launch vehicle structure. Even if they did not, severe vibrations could cause damage to the astronauts' capabilities and even to their long-term health.

Events moved rapidly. Pogo was identified on the 15 April, task teams were organized a day later, the first meeting of the NASA/contractor pogo working group took place on 8 May, and later in the month engine testing got under way. The pogo solution decision was made on 15 July, verification testing was completed on 31 July and operational checkout occurred on 13 August. The effort involved not only investigating shifts in resonant frequencies that modifications would provide for the Saturn 5 but assuring that changes did not affect engine performance and reliability.

Many approaches were tried. Although test data from earlier flights were available, additional data could not be secured until the new firings were undertaken – the pogo phenomenon occurred only during the course of flight and could not be simulated, for example, in a static test stand. Mathematical models were employed to augment the information obtained from dynamic tests, various theories were tried out and eventually the pogo situation was brought within safe and acceptable limits.[47]

Change Control Boards

To supervise changes such as those dictated by pogo, change control boards were established at NASA headquarters, the manned space flight centers and key contractor plants. Within each level of command, certain changes could be made without higher approval providing they did not affect multiple elements of the Apollo system.

Thus, if a change recommended for a given stage of Saturn 5 affected nothing but that stage, headquarters approval would not be sought. If, on the other hand, such a change would affect another stage, or the Apollo payload, or perhaps elements of the ground support equipment, change order permission would have to be directed to the appropriate higher echelon.

Lunar Orbit Rendezvous

Pogo was not, of course, the only major problem that came up during the Apollo program to test out NASA's ability to adapt to desirable or mandatory changes. Of the many others, the best known are the decision to go to the Moon via the lunar orbital rendezvous technique and the partial redesign of the Apollo command module as a result of the tragic fire at Cape Canaveral in 1967.

As US plans for going to the Moon began to gel, four basic means of putting astronauts on the surface were considered: (1) direct flight from the Earth's surface onto that of the Moon; (2) ascent from Earth's surface to Earth orbit where rendezvous and refueling of vehicular components would take place, and then flight to the lunar surface; (3) flight from Earth's surface to lunar orbit, descent onto the Moon and then rendezvous between the lander's ascent stage and the Earth-return module that had been kept in lunar obit; and (4) lunar surface rendezvous. The latter concept, which was advanced by the Jet Propulsion Laboratory, was quickly abandoned,[48] but the other three continued to be studied for some years. The basic idea of a lunar orbital rendezvous, as finally used by Apollo, had already been suggested in 1949 by a British rocket pioneer, H. E. Ross.[48a]

Each of the three major modes – direct flight, Earth orbital rendezvous and lunar orbital rendezvous – had its champions in the United States. The Manned Spacecraft Center under Gilruth preferred the first, the Marshall Space Flight Center under Dr Wernher von Braun the second, and the Langley research Center's Lunar Mission Steering Group established by Clinton Brown of the Theoretical Mechanics Division the last.

The Langley group, which was formed back in 1959, championed the lunar orbit rendezvous cause through 1960 and 1961 in the face of skepticism and opposition voiced at NASA headquarters, the established manned space flight centers and special task groups convened to study the lunar mission.[49] Ultimately, however, the lunar rendezvous advocates were successful in proving their case.

In a paper published back in the spring of 1960, a full year before Kennedy's pronouncement of the lunar landing goal, the advantages in mission weight reduction that resulted from lunar rendezvous were revealed.[50] Seven months later, on 14 December a full-scale briefing of the benefits of lunar orbit rendezvous was given to NASA headquarters personnel. From that point on, recalls Langley's John D. Bird,

> We began . . . to study lunar orbit rendezvous rather substantially. More or less continuously, I guess. At this time, we felt that something good was shaping up. [John C.] Houbolt became very interested about this time period. From that time to the successful conclusion of the matter it was a crusade with him.[51]

After an estimated million man-hours of study, the decision was made in July 1962 to follow the lunar orbit rendezvous route to the Moon. This meant a number of major changes in NASA planning had to be accommodated:

First, there would be no need to develop the Nova rocket, which would not only of itself have been a gigantic task, but would have required mamm static testing facilities, handling equipment a launch installations.

Second, lunar orbit rendezvous required a re tively small lunar landing module, one that needed accommodate only two men instead of three as v ualized by both direct flight and Earth orbit rend vous approaches.

And third, the requirements for Saturn 5 (th known as C-5 or Advanced Saturn) launch vehic would be lowered by choosing lunar orbit rendezvo over Earth orbit rendezvous. This was because t latter required dual launches, one to loft Apollo and partially filled C-5 third stage into orbit and anoth to carry up a liquid oxygen tanker to refuel that sta so that it could propel Apollo on to the Moon. Lun orbit rendezvous, on the other hand, required but single Saturn C-5 to send Apollo to lunar orbit fro where the landing module would be detached descend onto the surface below.

The decision to go lunar orbit rendezvous re resented a momentous change from earlier plannir and illustrates not only NASA's inherent flexibili but its ability to create an environment in whic competing concepts and ideas could flourish. If th agency had rigidly adhered to direct flight or Ear orbit rendezvous, Apollo's cost would have unden ably been higher than it was and the Kennedy goal reaching the Moon by 1969 might not have bee achieved.

Apollo 204 Fire

The most traumatic event in the program was th Apollo 204[52] flash fire that swept a command modul undergoing routine tests at the Cape on 27 Januar 1967. As is well known, the disaster not only result in the death of astronauts Virgil I. Grissom, Edwar H. White II and Roger B. Chaffee – who were to hav been boosted into orbit by a Saturn 1B rocket on 2 February – but caused a major review of th management team and management philosophy res ponsible for America's march towards the Moon.

Immediately following the tragedy, an Apoll Review Board was convened under the chairmanshi of Floyd L. Thompson, director of Langley. As result of exhaustive investigations, a series of change in the spacecraft design were ordered involving cabi and spacesuit materials; emergency egress; groun and spacecraft communications; the environmenta control, thermal control, and electrical systems launch complex emergency equipment; and a numbe of additional revisions.[53] Beyond reworking the cap-

NASA was forced to make sweeping changes in the Apollo command module as a result of the traumatic 204 fire that killed astronauts Virgil I. Grissom (left), Roger B. Chaffee (center), and Edward H. White (right), in January 1967. (NASA Photo No. 66-H-1277)

The charred interior of the 204 command module, near the floor in the lower forward section of the left hand equipment bay below the environmental control unit. (NASA Photo No. 67-H-380) (Right), lowering the damaged AS-204 command module at Complex 34, Kennedy Space Center. (NASA Photo No. 67-H-216)

sule itself, the accident forced NASA to make a searching review of its safety and quality control practices.

Terrible as it was, the fire did have a beneficial side.

> Suddenly everything changed. Not only was the entire design of the spacecraft re-analyzed, but the entire manned spaceflight organization and its attitudes as well. From this analysis emerged virtually a new organization, new states of mind, new approaches.[54]

The result of these changes, in the author's words, was

> an intimate new sociology of space, a new kind of government-industrial complex in which each interpenetrates the other so much that sometimes it is hard to tell which is which. Frequently now the government and corporate participants in Apollo display an emotional – at times almost mystical – comradeship that seems unique in industrial life.

VISIBILITY

The Apollo 204 fire illustrated the extent to which Apollo was open to the scrutiny of the entire Nation – and to the entire World. It also showed that NASA relationships with the centers and contractors were open to question and that serious problems had remained hidden from view. Somewhere, communications had failed and visibility had not been achieved.

In dealing with such an immense program as Apollo, management knew that difficulties were going to crop up, that important considerations were going to be overlooked and that errors were going to be made. This made everyone the more insistent that methods be developed to get all the relevant information out into the open, to identify little problems before they became big ones and to work out a strategy to deal with them no matter what size they might be.

To accomplish the objective of minimizing problems and catching errors before they became serious, NASA developed a kind of goldfish-bowl approach to its entire operations. NASA emphasized and re-emphasized that all participants were members of a single national team and that our mandate could not be achieved if our problems and failures were kept hidden. It was felt that if difficulties were out in the open there would be a strong incentive to overcome them.

All possible steps were taken to ascertain that no individual, no office, no committee, no organizational element of any kind ever got isolated from the main stream of events. NASA wanted to insure that problems or difficulties cropping up anywhere in the NASA-contractor system were not ignored or glossed over. Management sought to draw people out into working groups, *ad hoc* committees, task forces, panels or whatever, so that a continual cross-feeding of ideas could take place at countless interfaces and checkpoints. By working within such an open atmosphere, it was reasoned, the chances of anything going wrong, of errors remaining below the surface without becoming generally known, would be considerably reduced.

To create the highly visible environment which NASA sought, project managers established networks and bridges across many organizational and disciplinary boundaries. At the same time, they were constantly on the alert not to get trapped into possibly debilitating traditional moulds of thought and action. The practice of interacting laterally as well as up and down the organizational ladder became a way of life in the space community, not only in NASA headquarters but throughout center and contractor operations.

There is one other point that we would like to make concerning this matter of visibility. NASA was dealing on a daily basis with all sorts of people from all kinds of disciplines. Many of them were not used to working together, much less in the exposed environment characteristic of the Apollo program. In addition to hard-boiled engineers, technicians and construction workers, NASA was involved with theoretical scientists representing a broad spectrum of fields, legislators from Congress, lawyers, businessmen – a whole gamut of motivated individuals. NASA's goals, problems, failings, and ways of doing things had to be explained to one and all in a language each could understand. Everyone had to be encouraged to work in an open, time-constrained, team-oriented and often stressful interdisciplinary environment that was new and puzzling to many.

RELIABILITY

The importance of reliability was clearly understood although views differed on the best method for achievement. It was recognized that the lives of astronauts would depend on the near flawless functioning of all flight systems and that to insure such functioning broad scale ground testing, failure

During the Apollo program, people from many disciplines and walks of life came together to work towards a common goal. Their work was kept highly visible throughout the entire lunar exploration effort, partly through lectures and other public activities, partly through testimony to Congress, partly through the press and television, and to some extent by people actually being on hand at the launch site. Here are some of the 6,000 persons who viewed *Apollo 15* from a site adjacent to the Vehicle Assembly Building. Another 20,000 witnessed the event from the Kennedy Parkway, while an estimated million lined up along highways adjacent to the spaceport. (NASA Photo No. 71-H-1200)

In the depths of space and on the Moon, astronauts counted on built-in reliability to make their safe return to Earth virtually guaranteed. The *Apollo 14* command and service modules piloted by Astronaut Stuart A. Roosa circle the Moon while his comrades Alan B. Shepard and Edgar D. Mitchell approach in their lunar module climbing to lunar orbit from the surface below. (NASA Photo No. 71-H-382)

detection devices, redundancy and tight management control would have to be developed and put into practice. This meant, in turn, that an outstanding system of integration would have to be developed wherein the major elements of the launch vehicles and their space payloads all over the Nation would have to be checked out by the same testing equipment and computer networks. By so doing, a complete record of checkout would be available for recall at any point from the factory right down to the launch pad at the Cape.

To determine the reliability of a given vehicular system, subsystem, part or component, a list was prepared of the environmental conditions to which each would be exposed during its in-flight lifetime. For example, on the launch pad the Saturn 5 launch vehicle would be exposed to salt spray, winds and lift-off acceleration. Then, during flight, its upper staging and the Apollo payload would be exposed to hard vacuum, temperature extremes, ionization, and the possibility of micrometeoroid penetration. After the environmental conditions during all portions of the flight had been assessed, each potentially affected system, subsystem, part or component would be exposed under test conditions to the estimated level of severity plus an additional 50%.

In estimating Apollo reliability, some arbitrary estimates had to be established. For example, it was felt that not more than one lunar landing out of ten attempts would fail in the sense that the scientific and technical objectives of the mission might not be carried out. A case in point was Apollo 13 which was unable to descend onto the Moon, though it did return crippled but safely to Earth with its crew unharmed.

In order to achieve this degree of mission safety, reliability figures had to be assessed for various systems, subsystems, and components. For the Saturn 5 launch vehicle, for example, it was determined that 90% of potential failures were likely to be of a non-catastrophic type. Of the remaining 10% judged to be catastrophic, 1% was allotted to the risk of a disastrous explosion in the propulsion system of each stage. That 1% permissible-failure figure shrank to a tiny fraction of a per cent as it was further distributed over the pumps, turbines and valves of each engine of that stage and so on.

A difficulty sometimes arose in establishing a component or system reliability that was sufficiently high. In these cases redundancy was absolutely essential – that is, it was necessary to have several ways of performing the same vital function to protect against failure in one or another segment of the vehicular system.

An interesting and unusual case where redundancy was used to help solve a difficult reliability problem was the 'voting computer' located in the Saturn 5's control system. Critical operations were computed in parallel in three identical circuits. If two results agreed but differed from the third, the computer accepted the 2 to 1 majority vote as correct.

Evidence of NASA's improving reliability can be seen from its launch record. In 1959 – the first full year of operations of the space agency – the agency had eight successes against six failures. By 1972, success had reached 100% – 18 launches without a hitch. By the end of the Apollo program that same year, 12 astronauts had safely spent almost 400 h on the Moon's surface, of which more than 90 h were devoted to walking and riding on the surface. They brought back to Earth in excess of 400 kg of lunar material for scientific study, an overall achievement not anticipated in 1961.

Lessons from Apollo

It seems clear that once a given enterprise has been found technologically sound and long-term support for it seems assured, much of what was learned from Apollo is germane. Certainly, the interdisciplinary team approach; management and contracting techniques; emphasis on good communications and smooth information flow; the need for visibility and reliability; and much more can contribute to the success of any large-scale technically-oriented endeavor. But, without the support we have repeatedly mentioned and a solid use of relevant technology, no managerial wizardry can save a program. In the final analysis, the presence or absence of support is the single, most crucial element that spells success or failure.

Of course, every large-scale endeavor must offer something useful, something that benefits large numbers of people. The benefits may be essentially intangible, such as national security or national pride; or they may be tangible, such as a major urban renewal effort or a pipeline connecting Alaska with the lower 48 states. They may also involve new knowledge and understanding, such as resulted from Apollo.

Apollo, and its immediate – and closely related – successor Skylab, taught NASA the valuable lesson that today's triumphs do not automatically signal support for tomorrow's follow-on ventures. Almost as amazing, to many observers, as the success of these two programs was the failure of a large-scale, post-

Apollo manned space flight effort to gel in their aftermath. The attempt to build major new missions failed not because of lack of imagination and drive on the part of the space community but rather because public support was diminishing. And this was reflected by cooled-off attitudes in Congress.

A couple of months after the stunning success of Apollo 11, Associate Administrator for Manned Space Flight, George E. Mueller, reviewed[55] some post-Apollo goals that had just been recommended by the Space Task Group[56] to President Nixon. Noting that the Apollo program 'had served our Nation well in providing a clear focus for the initial development and demonstration of manned space flight capabilities and technology,' he went on to express the need for 'a balanced program' that would provide for 'the sustained development and use of manned space flight over a period of years focused on an eventual manned planetary landing in the 1980s.' By 'planetary landing,' he was thinking of Mars.

The two major directions identified for attention in the decade immediately following Apollo 11 were first the further exploration of the Moon, including perhaps the establishment of a lunar base; and second the development of a permanent manned space station supported by an Earth-to-orbit shuttle system.

Recognizing that costs are 'of paramount importance,' Mueller said that 'unless we can substantially change our current way of doing business we will not be given the opportunity to demonstrate the unique capabilities that space provides.' This meant, among other things, minimizing the number of space vehicles in use, introducing reusable launch vehicles – the Shuttle – and developing manned spacecraft for longer duration flights.

The Space Task Group carefully examined five possible programs for NASA, of which three were consistent with the group's recommendations. One, which was not recommended, would have led to a $10 thousand million peak in funding in 1976 and a manned launch towards Mars in 1981. This was considered the maximum pace at which NASA could possibly operate. The minimum pace, which was also not favored by the group, would have suspended manned space flight altogether after Apollo and Skylab[57] and subsequently maintained NASA funding at a $2 to $3 thousand million a year level.

Discarding these maximum and minimum approaches to post-Apollo manned space flight, the Space Task Group focused on three options that it considered feasible. These are summarized in Table 9. It illustrates comparative program accom-

Table 9. Comparative Post-Apollo Program Options[a]

Option I. Mid 1980's Mars launch with an $8 to $10 thousand million peak funding by 1980; Fiscal Year 1971 decision required on space station, space shuttle, and space tug, followed by a FY 74 decision on the manned Mars mission.

Option II, III. Stretched out accomplishment of above missions with a 1986 Mars launch (II) or one deferred (III) to some later date; maintain funding levels of late 1960s for several years, then gradually increase them; FY 72 decision needed on space station and space shuttle programs.

Milestone	Maximum pace (not recommended)	Option I (consistent with S.T.G. recommendations)	Options II, III (consistent with S.T.G. recommendations)	Low level (not recommended)
	Manned space vehicle systems			
Earth orbital elements				
Initial space station	1975	1976	1977	n.a.
50-man space base	1980	1980	1984	n.a.
100-man space base	1985	1985	1989	n.a.
Lunar elements				
Orbiting station	1976	1978	1981	n.a.
Surface base	1978	1980	1983	n.a.
Mars				
Initial expedition; Mars orbital rendezvous and landing	1981	1983	1986 Open (III)	n.a.
	Manned space transportation elements			
Space transportation elements				
Earth-to-orbit shuttle	1975	1976	1977	n.a.
Space tug	1976	1978	1981	n.a.
Nuclear Earth orbit – lunar orbit transfer stage	1978	1978	1981	n.a.

[a] Manned only; in addition, unmanned scientific and applications missions were recommended, most of which have been, are being, or soon will be carried out.

plishments if either the maximum pace; or Option I or Options II and III; or the low funding approach had been followed.

As it turned out, none of the designated manned systems recommended by the Space Task Group was approved. Moreover, of the space transportation systems identified, only the Earth-to-orbit shuttle was ultimately approved for development at a much slower pace than originally anticipated.

The reasons why the United States failed to undertake aggressive space missions based on the splendid Apollo–Saturn–Skylab foundation established in the 1960s and early 1970s are varied and complex, including substantial scientific, biomedical and technical issues. But one factor was dominant: the post-Apollo climate was not propitious for another great surge into space. America's priorities were shifting elsewhere.

The failure of a large-scale post-Apollo manned space program to start illuminates dilemma of how funds are to be allocated among many scientific and technological claimants. And once they have been allocated, what process ought to be followed to permit sponsors to decide if support should continue? How do signals appear that suggest a program is going to cost more than it is worth? And who, in our pragmatic society, can determine the worth of any large-scale scientific technological endeavor, no matter how well thought out and efficiently managed it may be?

The answer is that no one individual or group can make the determination. Support for large-scale scientific and technical endeavors must involve a preponderance of the public. US citizens must view the effort as worthy of their tax dollars if support is to last for extended periods.

NOTES AND LITERATURE CITED

1. Often referred to as the Apollo–Saturn or Saturn–Apollo program. The spacecraft that went to the Moon was known as Apollo, while the huge three-stage rocket that launched it was the Saturn 5. In this article, the word Apollo is often used to cover the overall Saturn and Apollo system. By extension, the Skylab embrionic space station program is included as it consisted largely of Saturn 5 and Apollo-developed technology, hardware and experience, and followed directly on the heels of the lunar landings.
2. Among the most impressive endeavors undertaken by the US in World War II were the Manhattan Engineer District Project (atomic bomb), the marshalling of vast petroleum resources for delivery to far-flung battle areas and the development and manufacture of tens of thousands of aircraft. For the 1950s and 1960s, one recalls the advent of intercontinental, intermediate range and fleet ballistic missiles and the protective Ballistic Missile Early Warning System. Today, the Air Force is proposing a $21.4 thousand million program for a fleet of 244 B-1 bombers, while the Army is budgeting $4.5 thousand million for 3300 of its 58 ton XM-1 Abrams tank. Such costs, which are expressed in current inflated dollars, are comparable to those applied to Apollo in the 1960s and early 1970s. A modern civilian comparison to these government-sponsored programs is the Alaskan pipeline which is expected to cost $7.7 thousand million to the user companies. Since the mid-1960s through to 1976, oil companies have spent an estimated $3.4 thousand million on exploration and development of the Alaskan North Slope, including $900 million for lease-sale rights and $2.3 thousand million for capital expenditures.
3. The US Tennessee Valley Authority would represent a real or perceived social need; the Marshall Plan, political; the Alaskan pipeline, economic; and the ballistic weapons program, military. The space program was in great part inspired by international political events, Sputnik, and certainly had national prestige and security overtones. In the broadest sense, however, it responded to an opportunity made possible by decades of research and development in rocketry, materials, guidance and control, computers and other areas of technology and science.
4. Security considerations often make exceptions of wartime enterprises.
5. L. R. Sayles and M. K. Chandler, *Managing Large Systems: Organizations for the Future*, p. 6, Harper and Row, New York (1971).
6. National Aeronautics and Space Act of 1958, Public Law 85–568, 85th Congress, 29 July 1958, amended 20 June 1960, Section 102.
7. Energy Reorganization Act of 1974, Public Law 93–483, 93rd Congress, 11 October 1974, Section 2(a).
8. Federal Nonnuclear Energy Research and Development Act of 1974, Public Law 93–577, 93rd Congress, 31 December 1974, Section 2(c).
9. The Second World War Manhattan Engineer District Project was roughly a $2 thousand million effort. Broken down, through the end of December 1945, a total of $1.4 thousand million was spent on plant build-up and $500 million on operations. If the project had been delayed several decades, it would of course have cost several times that amount due to the effects of inflation.
10. The actual transfer of personnel occurred on 1 July 1960.
11. These were taken over by the new Nuclear Regulatory Commission.
12. Space Act, Section 202(a).
13. Energy Reorganization Act, Section 102(d).
14. National Aeronautics and Space Act of 1958, US Senate Special Committee on Space and Aeronautics Report No. 1701, 85th Congress, 2nd Session, Washington, 11 June 1958: US Government Printing Office, p. 2.
15. Public Interest in the Space Program, Appendix 5 to a statement presented by Thomas O. Paine in *Hearings* before the Committee on Aeronautical and Space Sciences, US Senate, 91st Congress, 2nd Session. Washington, 6 April 1970: US Government Printing Office, p. 354.
16. NACA's 1958 budget of about $100 million was spent largely in-house, whereas most of NASA's 1967 budget of over $5 thousand million, a 5000% increase in a decade, ended up in contractor labs and plants.
17. In November 1961, NASA announced a drive to recruit 3000 engineers and scientists. By 1 July of the following year, the objective was almost met. It was learned that the most difficult recruitment problem involved finding experienced technical managers. In late 1962, an 'executive search service' was initiated, but soon aborted.

18. H. L. Dryden, The Role of the University in Meeting National Goals in Space, *NASA and the Universities: Proceedings of the NASA-University Conference on the Science and Technology of Space Exploration*, Chicago, 1 November 1962. NASA Report EP-5, Washington (1962).
19. *Report of Apollo Review Board.* Washington, 15 June 1970: National Aeronautics and Space Administration.
20. Long Shadow from Apollo, *Nature* **226**, 197 (1970).
21. Under conditions that duplicated to the extent possible those met during actual launch, orbital or lunar flight, and/or reentry into the Earth's atmosphere at the close of a mission. Included were static firing, vibration, and thermal vacuum tests.
22. L. Sayles, The NASA Management System: A Critique, in *The Proceedings of a Symposium on the Application of NASA Management Technology to the Management of Urban Systems*, Edited by S. R. Siegel, p. 48. Rutgers University, Camden, New Jersey (1972).
23. T. Alexander, The Unexpected Payoff of Project Apollo, *Fortune*, **LXXIX**, No. 7, 114 (1969).
24. J. E. Webb, *Space Age Management*, pp. 135–139. McGraw-Hill, New York (1969).
25. It is estimated that during the course of the Apollo development, some 300 000 tons of documentation were generated. In a single year, the Marshall Space Flight Center alone put out some 22 railway boxcars of data.
26. An outstanding documentation system was introduced early in the Apollo program to help assure that engineering specifications and the decisions of technical management were implemented in a timely and correct manner. In addition, an open-loop communications system was employed to make certain that all affected persons and echelons within the space agency and throughout the contractor network were informed of important changes and other factors that could influence or affect their operations or responsibilities.
27. R. A. Johnson, F. E. Kast and J. E. Rosenzweig, *The Theory of Management of Systems*, McGraw-Hill, New York (1973).
28. Within a couple of years of its founding, NASA made the significant decision to strengthen the role of its various field centers. One of the results of this policy was the development of increased expertise within the centers which allowed them even greater latitude in establishing scientific and technical specifications, letting and monitoring contracts with industry and universities, and otherwise supervising large-scale projects and programs from the moment of conception to the completion of whatever mission they might involve. These centers all had to learn to manage large-scale research and development contracts, something the predecessor NACA had little experience in doing.
29. Adapting NASA's Organization and Management to Future Challenges, staff paper, Office of Administration, p. 1-4. NASA Headquarters, Washington (October 1963).
30. The Marshall Space Flight Center in Huntsville, Alabama had the largest 10-year expenditures, totalling $8.359 thousand million. It was closely followed by the Manned Spacecraft Center (later, the Johnson Space Center) in Houston with obligations of $7.901 thousand million. Thus, these two manned space flight installations together handled more than half of NASA's funding. As would be expected, a majority of the monies were spent on the Saturn 5 launch vehicle and the Apollo spacecraft.
31. In his foreword to R. L. Rosholt, *An Administrative History of NASA, 1958–1963*, p. iv, NASA Report SP-4101, Washington (1966).
32. In a Memorandum to the Administrator from the Associate Administrator entitled *Extra Vehicular Activity for Gemini IV.* NASA, Washington (24 May 1965).
33. The Project Approval Document was prepared and signed off by general management. Typically, a PAD would include the name of the project under consideration, its objectives, the recommended technical plan, principal procurement items, management responsibilities, schedules, resource requirements, reporting procedures, and the like. In effect, a PAD would be a contract between general management and the director of the specific program.
34. J. E. Webb, address at Harvard University, 30 September 1968.
35. Forecasts and Appraisals for Management Evaluation.
36. J. E. Webb in Siegel, Ref. 22, *op. cit.*, p. 37.
37. *Project Apollo: Source Evaluation Board Report – Apollo Spacecraft*, Report RFP 9–150, NASA-Manned Spacecraft Center, Langley Air Force Base, Virginia (24 November 1961).
38. *Ibid.*, p. 10.
39. *Ibid.*, pp. 10–11.
40. *Ibid.*, p. 13.
41. A Project Apollo Division was proposed, managed by a vice-president, who would report directly to the corporate president.
42. Apollo SEB Report, Ref. 37. *op. cit.*, p. 14.
43. *Hearings* before the Committee on Aeronautical and Space Sciences, Part 6, US Senate, 90th Congress, 1st Session, US Government Printing Office, Washington (9 May 1967).
44. The Apollo command module alone had more than 2 million functional parts compared to fewer than 3000 for a typical motor car. To cite one element of the craft, the command sector panel display incorporated 24 instruments, 566 switches, 40 event indicators, and 70 on–off lights. When the three-stage Saturn 5 was assembled in the Vertical Assembly Building with its three-module Apollo payload at Cape Canaveral, one became aware of how complications were multiplied by bringing the launch vehicle into consideration as well. Saturn and Apollo together stood 111 m high.
45. Pogo was not noticed in Saturn 5's maiden flight, AS-501, of 9 November 1967. Following AS-502, the problem was corrected with the result that AS-503, the first manned flight, Apollo 8 into lunar orbit, December 1968, was pogo-free.
46. All possible technical competence within the responsible centers and industry was brought to bear so that an acceptable definition of the problem could be prepared. Because of time criticality, it was necessary to know as soon as possible if a redesign would be required. Moreover, individuals had to be identified as responsible for planning, analyzing, and conducting such test activities as proved necessary. The pogo situation impacted on PERT, the Program Evaluation and Review Technique or progress reporting system adapted from the Navy Polaris fleet ballistic missile program. PERT provided a means of synthesizing a large number of judgments into a comprehensive plan while taking into account an even larger number of variables. Since PERT enabled management to determine constraints and critical points in the total Saturn–Apollo program and to evaluate corrective action, the system had to interact with efforts to control pogo.
47. The solution involved shutting down the center engine ahead of the four outside engines, and using gas-filled accumulators in four of the five liquid-oxygen feed-lines. The technique selected called for employing the stage prevalve as an accumulator by bleeding helium into the prevalve annulus cavity.
48. This scheme involved the landing of unmanned crawling vehicles that would group together on the lunar surface to form a vehicle capable of returning to Earth. Later, a manned one-way craft would descend onto the Moon, the astronaut would accomplish his scientific duties, and then

would enter the previously assembled return vehicle and fly back to Earth.

48a. H. E. Ross, The Genesis of Orbital Rendezvous, *Spaceflight* **18**, No. 5, 185–186, (1976). See also: Orbital Bases, *J. Brit. Interplan. Soc.* (January 1949).

49. On 16 June 1961, an *ad hoc* task group under William Fleming of NASA Headquarters recommended the direct ascent approach based on a huge launch vehicle concept known as *Nova*. Seven months later, however, a Large Launch Vehicle Planning Group under the chairmanship of Nicholas Golovin of NASA and Laurence Kavanau of the Department of Defense concluded that the development of a *Nova* would take too long to meet President Kennedy's goal of a manned lunar landing by the end of the 1960 decade. Their recommendation: Earth orbital rendezvous using what were then termed 'Advanced Saturn' launch vehicles. (These were later designated Saturn 5.)

50. W. Michael, 'Weight Advantages of Use of Parking Orbits for a Lunar Soft Landing Mission,' Theoretical Mechanics Division, NASA-Langley Research Center, Langley Air Force Base, Virginia (26 May 1960).

51. John D. Bird, remarks, NASA-Langley Research Center, 20 June 1966. These are reproduced in *Documents in the History of NASA*. An Anthology, issued by NASA History Office, Washington (August 1975). See also: *Spaceflight* **18**, No. 5. 183–184 (1976).

52. Named because of its designation AS-204 (Apollo–Saturn–204).

53. Manned Space Flight Report: Actions Taken as a Result of the AS-204 Accident, Section IX, NASA Headquarters, Washington (1967).

54. Alexander, Ref. 23. *op. cit.*, p. 117.

55. In a letter dated 11 September 1969, to the three manned space flight center directors.

56. Formed by President Richard M. Nixon on the 13 February 1969, the group was given the task of developing definitive recommendations for the direction of the American space program in the post-Apollo period. Vice-President Spiro T. Agnew served as chairman; Dr Thomas O. Paine, NASA Administrator, Dr Lee A. DuBridge, presidential science advisor and Dr Seamans, the then Secretary of the Air Force, served as members. Mr U. Alexis Johnson, Under Secretary of State for Political Affairs, Dr Glenn T. Seaborg, Chairman of the Atomic Energy Commission and Mr Robert P. Mayo, director of the Bureau for the Budget were observers. The STG's report, *The Post-Apollo Space Program: Directions for the Future*, was issued in September 1969 and was 25 pages long. It responded to many of the recommendations contained in a NASA report released to the public the same month entitled: *America's Next Decades in Space: A Report of the Space Task Group* (84 pages). Six months later, the Space Science and Technology Panel of the President's Science Advisory Committee (Executive Office of the President, Office of Science and Technology) published *The Next Decade in Space* (March 1970, 63 pages) which covered much the same ground.

57. At the time, Skylab was an unnamed element in the so-called Apollo Applications Program.

GENERAL BIBLIOGRAPHY

H. L. Dryden, *The U.S. Space Program – What Is It? Where Is It Going? Why Is It Important?*, Franklin Society, New York, 1963.

D. Ertel and M. L. Morse, *The Apollo Spacecraft, A Chronology, Volume I* (through 7 Nov. 1962). The NASA Historical Series, Washington, D.C. (1969): Scientific and Technical Information Division, Office of Technology Utilization, National Aeronautics and Space Administration. Report SP-4009. (See Part III: Lunar Orbit Rendezvous: Mode and Module, pp. 131–202.)

H. B. Finger and A. F. Siepert, NASA's Management of the Civilian Space Program, presented at the 16th International Conference of the Institute for Management Sciences, New York, 26–28 March 1969.

F. E. Kast and J. E. Rosenzweig, Organization and Management of Space Programs, in *Advances in Space Science and Technology Vol. 7,* edited by F. I. Ordway, p. 273, Academic Press, New York (1965).

R. C. Seamans Jr, Action and Reaction, the 1969 Minta Martin Lecture, Massachusetts Institute of Technology, Cambridge, Massachusetts (1969).

R. C. Seamans Jr, The Management of a National Space Program, presented at the United Nations Conference on the Exploration and Peaceful Uses of Outer Space, Vienna, August 1968.

J. Van Nimmen, L. C. Bruno and R. L. Rosholt, *NASA Historical Data Book, 1958–1968,* Vol. 1: *NASA Resources*, Scientific and Technical Information Office, National Aeronautics and Space Administration Report SP-4012, Washington (1976).

J. E. Webb, An Overview of the NASA Management System, in *Management Technology Applied to Urban Systems*, edited by S. R. Siegel, Drexel University, Philadelphia (1972).

J. E. Webb, NASA As An Adaptive Organization, John Diebold Lecture on Technological Change and Management, Harvard University Graduate School of Business Administration, Boston, 30 September 1968.

J. E. Webb, commentary in *Harmonizing Technological Developments and Social Policy in America*, edited by James C. Charlesworth, p. 113, American Academy of Political and Social Science, Philadelphia (1970).

J. E. Webb, Foreword, in *Managing Large Systems: Organizations for the Future*, Leonard R. Sayles and Margaret K. Chandler, Harper, New York (1971).

Apollo Program Management, Subcommittee on NASA Oversight of the Committee on Science and Astronautics, US House of Representatives, Ninety-first Congress, First Session, Serial C, US Government Printing Office, Washington (1969).

The Apollo 13 Accident, Hearings before the Committee on Science and Astronautics, US House of Representatives, Ninety-first Congress, Second Session, 16 June 1970, No. 19, US Government Printing Office, Washington (1970).

Pacing Systems of the Apollo Program, Subcommittee on NASA Oversight of the Committee on Science and Astronautics, US House of Representatives, Eighty-ninth Congress, First Session, Serial K, US Government Printing Office, Washington (1965).

IX

ASTRONAUTICS AND ART: A SURVEY

Frederick C. Durant, III *

First off, I should reveal that I am neither an artist nor an art critic. I have, however, been involved in rockets, missiles and space flight for more than thirty years. During this time I have become increasingly interested in art as a form of communication.

Like most of you, I react to works of art subjectively. Each of us has a background differing in age, environment, professional activities and experience. Art relates to each individual in many ways; to both the conscious and subconscious. A work of art stimulates or does not. I tend to view art with regard to its interpretation of a subject. What is the artist trying to tell me? Do I like it? Why? Perhaps the subject matter is ingeniously expressed, emotionally exciting. Perhaps it stretches my imagination. Or, perhaps the colors or balance of design are pleasing. Or it may be satisfying because of an elegance of presentation. Thus, we are treating a very personal subject.

Some fifteen years ago, when I joined the National Air and Space Museum to organize a Department of Astronautics, one of the subject areas approved by the Director, S. Paul Johnston, was space art. Although there were four Bureaus of the Smithsonian Institution which collected and displayed art, none was interested, more than casually, in the subject of space flight. So we initiated, in a small way, a collection of space art. We were aided greatly by the Curator of Painting of the National Gallery of Art, the late H. Lester Cooke, and James Dean, head of the space art program at the National Aeronautics and Space Administration. Jim Dean later joined the Museum to become Curator of Art. By that time, I had had the fun of educating myself on the subject and had acquired about 150 works for the Museum. All were donations, generally by the artist.

This presentation is not only a personal view, but an incomplete one. I intend to show you some examples of space art of well-known and great artists and to introduce you to the works of an exciting group of young artists. Several of the latter are still in their twenties. The scope of this presentation is incomplete since many competent artists will not be mentioned because of time limitation. Nor will a number of specific kinds of space art be explored. Let us consider it an introduction to the subject seen through the eyes of one individual.

* *Special Assistant to the Director, National Air and Space Museum, Smithsonian Institution, Washington, D.C. 20560*

Well, space art; what is it? An omnibus description might be: "Artistic renderings of the environment of the solar system and beyond; illustrations of contemporary space flight projects; rockets and spacecraft; or future program concepts, and/or visions of the imagination. Such renderings may be intensely factual and realistic or variously impressionistic, surrealistic, abstract and subjective emotional statements by the artist; or combinations of the above." Media used may be charcoal, pastel, pencil, oil, tempera, acrylic or water color paints; wood block, lithograph or other modes of printing; woven tapestry, needlework, stained glass, or other.

Art is a form of communication. A piece of space art tells the viewer (to a greater or lesser degree) what the artist perceives. It may be factual, but usually not photographic. The artist, through line and color, may lead the eye, strengthen some elements and weaken others.

Honoré Daumier observed more than a hundred years ago, "The camera sees all, and understands nothing." Wonderfully, each artist sees and interprets differently the same scene. His choice of media, color, mood, mode of rendering, result in an interpretive creation different from that of the artist next to him. Each artist's interpretation, quite apart from his (or her) skill in drawing, is a product of an individual background--training, personal experience, technical understanding, imagination and emotional response.

The artist discerns not only with his eyes but, in varying degree, with all of his physical senses, lending emphases. Emotions affect his mental image. What he sees is colored by memory, technical understanding and imagination at a particular time.

ART, HISTORICAL TO WORLD WAR II

Today I plan to share with you some art of several varieties that treat aspects of astronautics. The first category is what illustrator Ron Miller in his splendid book *Space Art* (Starlog, New York, 1978) calls: "The Hardware Artists."

Here are some examples ranging from about a hundred years ago to 1950: There are illustrations from imaginary space voyages of Jules Verne, H. G. Wells and John Jacob Astor, or art in popular scientific magazines. Past concepts of space vehicles, space suits, etc., elicit smiles and amuse us now because of the superior knowledge of hindsight. Yet, at the time they were made, it was the idea, the concept, which was stimulating and mind-stretching. It is my view that there is a strong romantic streak in those of us who dream and fantasize about the future. As members of the American Astronautical Society, I suggest that you all have romantic--in the broad sense of the word--and adventurous impulses or you would not be members.[1]

ART, POST WORLD WAR II, PRE-SPUTNIK

Let us look now at some more recent concepts of hardware for future voyages into space. This art was produced in the post-World War II era when knowledge of German V-2 technology caused many of us to dream of its extrapolation to achieve orbital and escape velocities.

The technical design, structure and propulsion elements are not fantasy but based upon a logical extension of technical knowledge of the early 1950's.[2]

POST-SPUTNIK ART

When *Sputnik I* ushered in the space age, manned space flight, Apollo lunar landings and close-up views of the planets stimulated many artists. Aerospace industry illustrators and popular magazines gave opportunities to translate concepts and design studies into exciting works of art.

Here are some examples of visions of space projects of the future and the distant future.[3]

ASTRONOMICAL ART

The next category is "Astronomical Art." Two masters of this genre are Chesley Bonestell and Ludek Pesek. Their views of the planets and stars have inspired many thousands of the young, and what I term the "young-minded." Such art gives us close-up views of planets and their moons, or stars, based upon our best understanding at the time they are made.[4]

IMPRESSIONISTIC ART

Another style of space art is the impressionistic, bordering on the abstract. Such works stress that which is important to the artist in his view of space flight and the mental vision he perceives. Elements of illustrative detail are often lacking or down-played as less important to the artist's mental picture while other elements are accentuated. In some cases, the artist shuns the treatment of hardware actuality and presents a mood or sense of motion or awe at cosmic events.[5]

SPECULATIVE ART

Finally, we have art which is totally speculative, such as concepts of extraterrestrials, hypothetical planets of distant suns and spacecraft with design elements that do not fit today's technology.[6]

CONCLUSION

I am certain that each of you has responded differently to these works of art. Quite probably some of them you have found beautiful, some you would enjoy living with and some which irritate (for various reasons), and some you wouldn't give wall space to. This is my feeling also.

During the next several years, I intend to broaden my knowledge of space artists and their works. Contracts are being sought in Japan, Germany, Yugoslavia, The Netherlands and elsewhere. There seems to be a flowering of interest in artistic expression of the great adventure of space exploration and utilization which has been realized.

As yet, little has been published on the subject in our lifetime. Further, I know of but one small art gallery (in Boston, MA) devoted solely to exhibition and sale of space art. There are devotees but their number is limited. If this presentation has stimulated and interested you, I have accomplished my purpose.

EXAMPLES OF SPACE ART

Contemporary Art. *Gemini VI* Astronauts Frank Borman and James A. Lovell, Jr. walking up ramp at Complex 19, Cape Kennedy, by Artist Paul Calle, 1966. (NASA Photo No. 66-H-420)

Contemporary Art. *Apollo 12* Astronauts Charles Conrad, Jr. and Alan L. Bean on the Moon, by Artist Pierre Mion, 1970. (NASA Photo No. 70-H-1090)

Portraiture Art. *Apollo 11* Astronaut Neil A. Armstrong, by Artist Paul Calle, 1969. (NASA Photo No. 69-H-1233)

Portraiture Art. Commander of *Apollo 14* mission, Astronaut Alan B. Shepard, by Artist Captain Ted Wilbur (USNR), 1970. (NASA Photo No. 71-H-527)

Astronomical Art. Leonid Meteor Shower, by unknown artist, 1833. (NASA Photo No. 65-H-1594)

Astronomical Art. Surface of Mercury, by Artist Chesley Bonestell, Ca., 1947.
(Smithsonian Institution Negative No. A 4598)

Astronomical Art. Saturn from Rhea, by Artist Chesley Bonestell, Ca., 1948.

Imaginative Astronomical Art. Our Galaxy from a Hypothetical Planet 400,000 Light Years Distant, by Artist Chesley Bonestell, Ca., 1960.

Impressionistic Art. Lift-off of *Faith 7* (Project Mercury) with Astronaut Gordon Cooper aboard, by Artist Paul Calle, 1963. (NASA Photo No. 63-MA9-303)

Impressionistic Art. Project Mercury Astronaut Gordon Cooper steps onto the deck of the recovery ship after his conquest of space, by Artist Mitchell Jamieson, Ca., 1965.
(NASA Photo No. 67-H-599)

<u>Impressionistic Art</u>. "The Space Mural--A Cosmic View" by Artist Robert T. McCall (July 1975-Jan. 23, 1976) covers 2100 square feet of canvas and can be seen in the Independence Avenue Lobby of the National Air and Space Museum. (NASA Photo No. 76-H-85)

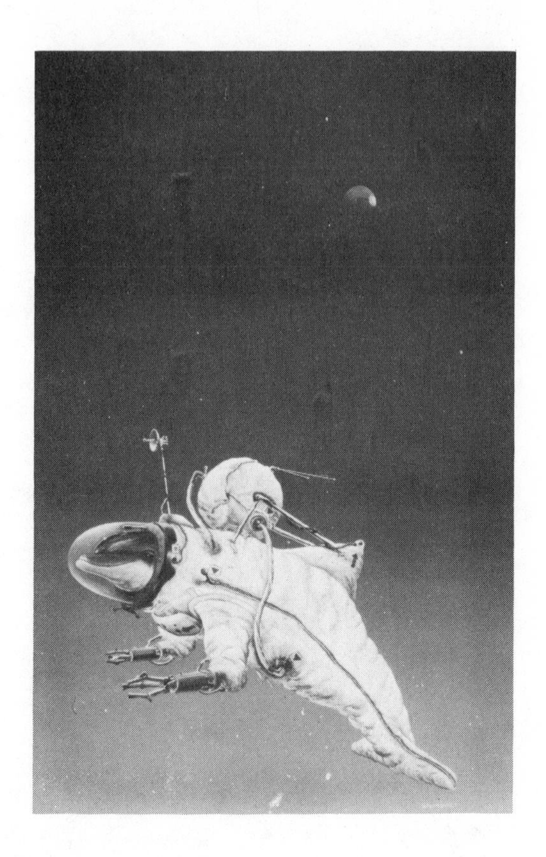

Fantasy Art. Porpoise in Extra-Vehicular Activity (EVA), by Artist Rick Sternbach, 1973.

ASTRONAUTICS AND ART

REFERENCE NOTES

1. Historical, to World War II. Works by: Emile Bayard and A. de Neuvill (from Jules Verne's "From the Earth to the Moon"); P. Phillipoteaux (from Jules Verne's "Off on a Comet"); Dan Beard (from John Jacob Astor's "A Journey to Other Worlds"); Lucien Rudaux; Frank R. Paul.

2. Post World War II, Pre-Sputnik. Works by: Chesley Bonestell; Rolf Klep; Fred Freeman; R. A. Smith.

3. Post Sputnik. Works by: Keith Ferris, Andrei Sokolov; Pierre Mion; Jay Mullens; Jack Olson; Chesley Bonestell.

4. Astronomical Art. Works by: Chesley Bonestell; Ludek Pesek; John W. Clark; Ernest and Anne Norcia; Don Davis; William K. Hartman; David Egge; Rick Sternbach.

5. Impressionistic. Works by: James Cunningham; John W. Clark; David Egge; Inge Keks; M. Mategot; Hans Paul Cremers; Paul Calle; Mitchell Jamieson; Robert Rauschenberg; Robert Shore; Robert T. McCall.

6. Speculative. Works by: Chesley Bonestell; Robert T. McCall; Rick Sternbach; Don Davis; David Egge; Andrei Sokolov; Jon Lomberg.

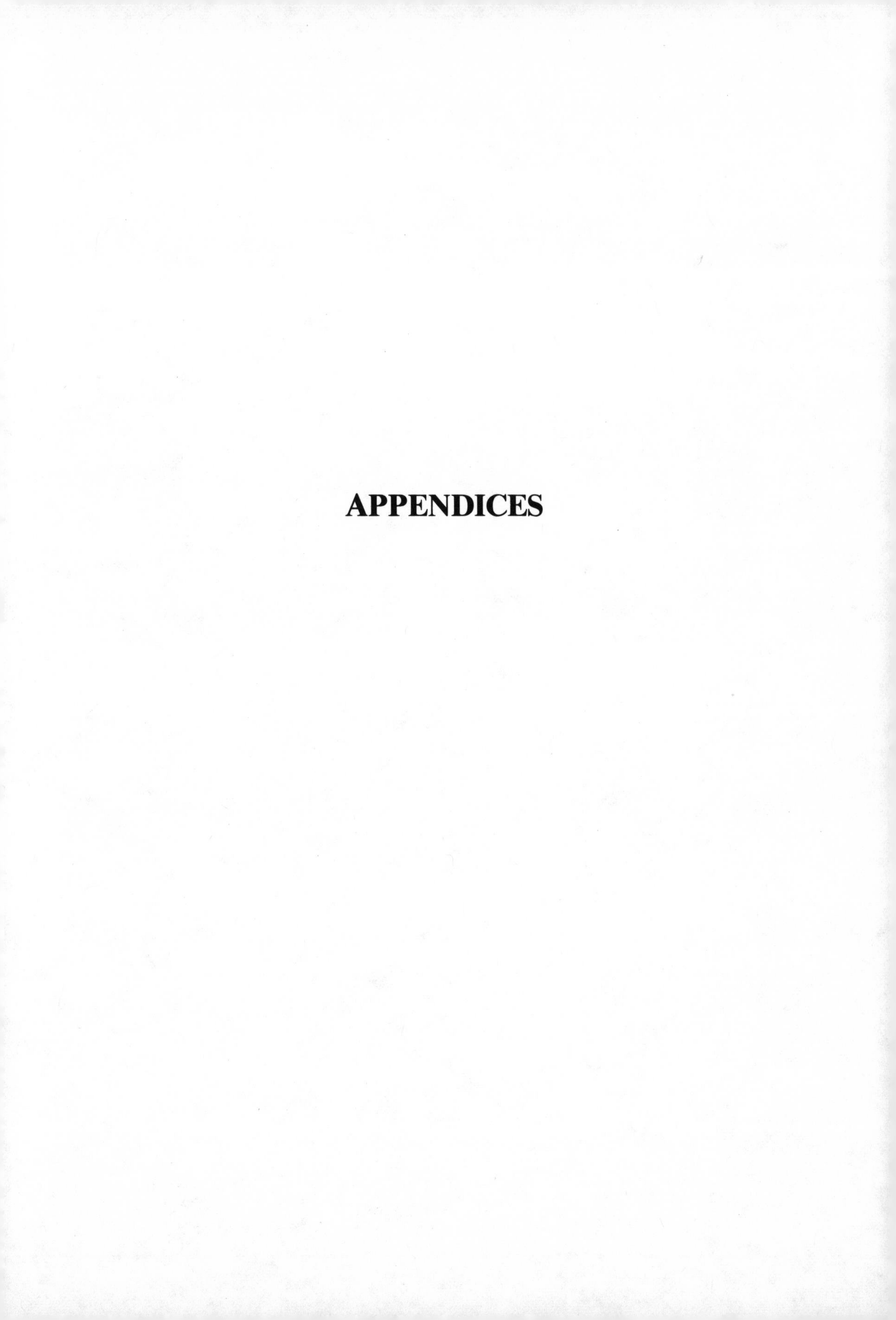

APPENDICES

TABLE OF ILLUSTRATIONS

Page

Page

PROFILES OF CONTRIBUTORS

Disher, John H. Director, Advanced Programs, NASA Office of Space Transportation. Involved with space program from the beginning as assistant director for Apollo spacecraft development, director of the Apollo test program, and deputy director of the Skylab program. Among his many awards are NASA's Exceptional Service Medal for his work on Apollo and NASA's highest honor, the Distinguished Service Medal, for his work on Skylab. He is the author of numerous technical papers.

Doyle, Stephen E. General Counsel of the AAS and Vice President, Aerojet Corporation, Sacramento, California. He was Group Manager of Telecommunications and Information Systems of the U.S. Congress Office of Technological Assessment (1978-80), Deputy Director of International Affairs Office, NASA Headquarters (1974-78), and previously held telecommunications positions in the White House, Department of State, and the FCC. He served as alternate representative to committees serving the UN Committee on Peaceful Uses of Outer Space (1969-74, 1978-80). He is a graduate of the University of Massachusetts (B.A.), and holds LL.B. and J.D. degrees from Duke University. Author of numerous papers and articles, he is corresponding member of the International Academy of Astronautics, and a member of the IAF Institute of Space Law. He was author of the Goddard Prize Historical Essay for 1980, for his "Original Contributions to Concepts of Space Law," to be published in the *Journal of the British Interplanetary Society*.

Durant, Frederick C., III. Assistant Director for Astronautics, National Air and Space Museum, Smithsonian Institution. Fellow of AAS, 1971 Chairman of AAS Awards Committee, and now member of History Committee. Graduate Lehigh University, Chemical Engineering, attended Philadelphia Museum School of Industrial Arts. Naval Aviator and Test Pilot in WWII, and 1948-1952 in rocket engineering at NARTS. Associated with DuPont Nemours, Bell Aircraft, Arthur D. Little, AVCO Everett Research Laboratory, Bell Aerosystems before joining the Smithsonian in 1964. President of American Rocket Society (1953), President of the International Astronautical Federation (1953-1956). Fellow of ARS (AIAA), and AAS, and member of numerous other societies and history committees of the International Academy of Astronautics and AIAA. Author of numerous articles, including "Rockets" and "Space Exploration" in the *Encyclopedia Britannica*, co-editor of *First Steps in Outer Space* (1975), and a recent series on space art in *Omni*.

Emme, Eugene M. President, CLIO Research Associates, and a Director of the American Astronautical Society. Founder of the NASA History

Program, he is a fellow of the AAAS, AAS, BIS, and AIAA (Associate). Dr. Emme is co-chairman of the history committee of the International Academy of Astronautics. U.S. Naval Aviator, and historical officer WWII. Author and editor of *A History of Space Flight, The History of Rocket Technology*, and many other works. Colonel USAF Reserve (Ret.).

Galloway, Eilene. Vice President, International Institute of Space Law, International Academy of Astronautics. She is President, Theodore von Karman Memorial Foundation, Inc., and for many years a special consultant to Congressional committees on international space activities as well as author of many professional papers.

Holman, Mary A. Professor Economics, The George Washington University; and professorial lecturer at the Industrial College of the Armed Forces, National War College, U.S. Naval School of Health Sciences. She is a consultant to the National Aeronautics and Space Administration, National Science Foundation, Food and Drug Administration, Cost of Living Council, and the National Bureau of Standards. Dr. Holman is author of *The Political Economy of the Space Program*.

Ordway, Frederick I., III. Policy Planning Staff, Department of Energy. He is the author or editor of numerous historical works on the history of rocketry and astronautics. He was editor of *Astronautics*, Journal of the American Astronautical Society, soon after the AAS was founded. For many years he was on the technical staff of the NASA Marshall Space Flight Center before joining ERDA. One of his books, *The Rocket Team*, coauthored with Mitchell R. Sharpe, with a foreword by von Braun, has been highly praised. He was also technical adviser for the production *2001 Space Odyssey*.

Seamans, Robert C., Jr. Dean, School of Engineering, Massachusetts Institute of Technology. He is the former President of the National Academy of Engineering, Administrator of ERDA, the Secretary of the Air Force, and Associate Administrator of the National Aeronautics and Space Administration. Before his governmental service he was a professor at M.I.T. and then with R.C.A. He is a fellow of the American Astronautical Society.

Sloop, John L. Manager of International Consultants on Energy Systems (ICES), is the author of 45 publications including the distinguished history, *Liquid Hydrogen as a Propulsion Fuel, 1945-59* (NASA SP-4404, 1978) which explains the development leading to the success of the Apollo-Saturn and the Centaur outer planetary missions. He is a graduate of the University of Michigan in electrical engineering, joined NACA Langley Laboratory in 1941, moving to the NACA Lewis Engine Laboratory in 1941, moving to the NACA Lewis Engine Laboratory in 1942. He served 31 years in NACA/NASA as a leader in rocketry research on high-energy propellants, moving to NASA headquarters in 1960. In 1964, Mr. Sloop became associate assistant administrator for advanced research and propulsion, retiring in 1972. He held offices in the American Rocket Society, became a Fellow in AIAA, and sharing in its Goddard Award in 1974. He is also a member of the Society for the History of Technology.

Suranyi-Unger, Theodore, Jr. Research Professor of Economics, The George Washington University. He received his doctorate at the University of Vienna, Austria, and served as Economist of Stanford Research Institute, Federal Price Commission, National Science Foundation, and the National Planning Association. He is currently a consultant and lecturer to various governmental agencies.

AAS HISTORY COMMITTEE

Dr. Eugene M. Emme, Chairman
NASA Historian Emeritus
11308 Cloverhill Road
Wheaton, MD 20902
(301/593-2938)

President Philip H. Bolger
Cranston Research, Inc.
6060 Duke Street
Alexandria, VA 22304

Dr. Tom D. Crouch
Historian of Flight
National Air and Space Museum
Washington, DC 20560

Dr. Albert E. Eastman
Hq. AOPA
4616 Winston Place
Alexandria, VA 23310

Mr. Frederick C. Durant, III
Historian of Astronautics
109 Grafton
Chevy Chase, MD 20015

Mr. R. Cargill Hall
Historian, Hq. Military Airlift Command
230 Lakeland Hills Drive
Belleville, IL 62221

Professor Richard P. Hallion
University College
University of Maryland
College Park, MD 20742

Mr. George S. James
Communications Programs (ISPT)
National Science Foundation
Washington, DC 20550

Mr. Frederick I. Ordway, III
Policy Planning, Dept. of Energy
2350 N. Taylor Street
Arlington, VA 22207

Mr. Mitchell R. Sharpe
Historian of Rocketry and Astronautics, Alabama Space and Rocket Center
2215B Jonathan Drive
Huntsville, AL 35810

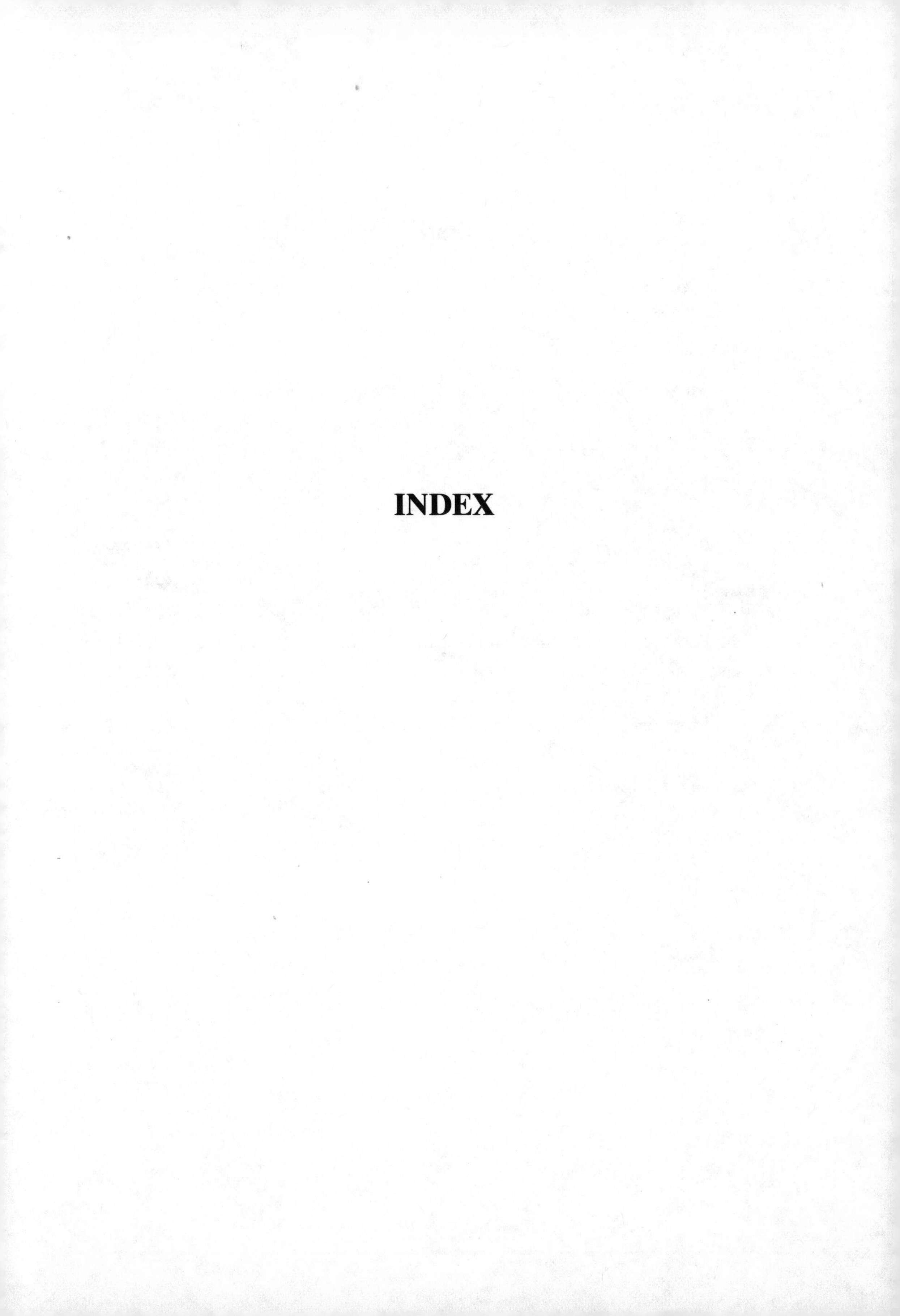

INDEX

REFERENCE NOTES

Users of this index should not overlook the extensive reference notes which are appended to most chapters as follows:

INDEX